SOOT IN COMBUSTION SYSTEMS

AND ITS TOXIC PROPERTIES

NATO CONFERENCE SERIES

VI MATERIALS SCIENCE

SOOT IN COMBUSTION SYSTEMS
AND ITS TOXIC
PROPERTIES

Edited by

J. Lahaye

and

G. Prado

Center for Research on the
Physico-Chemistry of Solid Surfaces, C.N.R.S., and
University of Haute-Alsace
Mulhouse, France

Published in cooperation with NATO Scientific Affairs Division

PLENUM PRESS · NEW YORK AND LONDON

Library of Congress Cataloging in Publication Data

NATO Workshop on Soot in Combustion Systems (1981: Le Bischenberg, Obernai, France)
    Soot in combustion systems and its toxic properties.

    (NATO conference series. VI, Materials science; v. 7)
    "Proceedings of a NATO Workshop on Soot in Combustion Systems, held August 31—September 3, 1981, in Le Bischenberg, Obernai, France"—T.p. verso.
    Includes bibliographical references and index.
    1. Combustion—Congresses. 2. Soot—Congresses. 3. Soot—Toxicology—Congresses. I. Lahaye, J., 1937–   . II. Prado, G. III. North Atlantic Treaty Organization. Scientific Affairs Division. IV. Title. V. Series.
QD516.N32 1981   621.402′3   82-22552
ISBN 978-1-4684-4465-0   ISBN 978-1-4684-4463-6 (eBook)
DOI 10.1007/978-1-4684-4463-6

Proceedings of a NATO Workshop on Soot in Combustion Systems, held August 31—September 3, 1981: Le Bischenberg, Obernai, France

© 1983 Plenum Press, New York
Softcover reprint of the hardcover 1st edition 1983

A Division of Plenum Publishing Corporation
233 Spring Street, New York, N.Y. 10013

PREFACE

        Our interest in Mulhouse for carbon black and soot began
some 30 years ago when J.B. Donnet developed the concept of
surface chemistry of carbon and its involvement in interactions
with gas, liquid and solid phases. In the late sixties, we
began to study soot formation in pyrolytic systems and later on
in flames.

        The idea of organizing a meeting on soot formation originated
some four or five years ago, through discussions among Professor
J.B. Howard, Dr. A. D'Alessio and ourselves. At that time the
scientific community was becoming aware of the necessity to
strictly control soot formation and emission.

        Being involved in the study of surface properties of carbon
black as well as of formation of soot, we realized that the
combustion community was not always fully aware of the progress
made by the physical-chemists on carbon black. Reciprocally,
the carbon specialists were often ignoring the research carried
out on soot in flames. One objective of this workshop was to
stimulate discussions between these two scientific communities.

        During the preparation of the meeting, and especially during
the review process by the Material Science Committee of the
Scientific Affairs Division of N.A.T.O. the toxicological aspect
emerged as being an important component to be addressed during
the workshop.

        To reflect these preoccupations we invited biologists, physical-
chemists and engineers, all leaders in their field. The final
programme is a compromise of the different aspects of the subject
and was divided in five sessions.

        The first session presents the reasons which make it necessary
to control soot and PAH (polyaromatic hydrocarbons) formation and
destruction: the toxicological aspect and the formation of
these materials in practical systems are examined.

    The second one is concerned with the mechanisms of PAH and
soot formation and burnout. In this session three presentations
out of nine are referring to ionic effects in flames: it does
not mean that the scientific committee considers that ions are of
major importance in PAH and soot formation but it appears that the
understanding on that particular point is ambiguous and needs to
be clarified.

    Aerodynamics of sooty flames is very important with regards
to emissions from practical systems. The third session is devoted
to this aspect and special thanks are due to the speakers who have
accepted to introduce this difficult subject to an audience for
the major part not familiar with it.

    The fourth session describes optical diagnostics. In the
programme it appears as an appendix to the former parts of the
meeting, but we must realize that without the development of
these techniques, progress in the understanding of soot formation
would have been very limited.

    The last part of the meeting is devoted to a synthesis of
the workshop, first in sub-committees, and finally in a plenary
session.

    The questions, comments, and answers submitted in a written
form have been edited in the present volume.

    Special thanks are addressed to the Scientific Committee, Drs.
A. D'Alessio, W. S. Blazowski, Profs. N. Chigier, R. Truhaut,
H. Gg. Wagner, and F. Weinberg, for their efficient collaboration
to the preparation of the meeting.

    We are grateful to N. Bomo, A. Garro, B. Prado, S. Wagner
and E. Wozniak, for their technical assistance before and during
the workshop.

                                        J. Lahaye
                                        G. Prado

                                        November, 1981

CONTENTS

AERODYNAMICS OF SOOTY FLAMES

OPTICAL DIAGNOSTICS

SOOT COMPONENTS AS GENETIC HAZARDS

William G. Thilly

Department of Nutrition and Food Science
Massachusetts Institute of Technology
Cambridge, Massachusetts 02139

The study of combustion effluents potential health effects
involves identifying the substances present and estimating the
probable health hazards of each. Unfortunately, there are no
methods which permit such estimates of health hazards. We are
presently using techniques for which we have no means of assessing
accuracy.

In our program at MIT, we have chosen genetic changes induced
in bacteria and human cells as an indication of the inherent
hazard associated with a chemical or complex mixtures. I will
not attempt to defend this choice here beyond pointing out that
humans do indeed suffer from a myriad of diseases of demonstrable
genetic origin; this is accompanied by an assumption that environmental chemicals cause an appreciable amount of these diseases.
If the assumption is correct, it thus follows that human exposure
to chemicals which do, in fact, cause significant genetic damage
should be reduced.[1]

Our approach to complex mixtures depends on the interaction
and cooperation of a number of scientists and engineers. In this
paper I will draw illustrative data from an ongoing study of
emissions from an ordinary home oil burner, which is the principle
means of home heating in New England. Engineers studying combustion
(Dr. Gilles Prado, Prof. Jack Longwell) in the Department of
Chemical Engineering were responsible for setting up, operating
the burner under reproducible conditions and devising means to
assure collection of all the effluent material. Extraction and
preliminary chemical identifications were followed by capillary
column gas chromatography separations and identifications by mass
spectrometry in the Department of Chemistry (Ms. Julie Leary,

Prof. Klause Biemann). Bacterial mutation assays were performed in the toxicology testing laboratory (Ms. Barbara Andon) and, when indicated, human cell studies in my laboratory.

Basically, our approach is very simple. We start with the hypothesis that, at low concentrations, such as those to which people are expected to be exposed, there will not be significant interactions among the chemicals effecting the amount of genetic damage induced by each. We have tested this hypothesis fairly vigorously in the case of a turbulent diffusion flame with kerosene as a fuel[2] and found it to be useful in describing our observations. Thus, our expectation for home oil burner emissions was stated as

$$M = \sum_i C_i A_i$$

in which M was the fraction of bacterial cells mutated by a known amount of the complex extracts under study, C_i was the concentration (molar) of each chemical present in the test mixture and A_i was the specific mutagenic activity of each chemical (mutants x liter/survivor x mole), at the concentration tested.

MATERIALS AND METHODS

The burner used in my example of work in progress was a modern flame retention burner, a Beckett, Model A. The burner was fitted with a Monarch nozzle delivering 1.0 gallon per hour.

The fuel used was a No. 2 heating oil containing about 30% aromatics. (A complete description of the burners and fuel characteristics will be offered in a future publication by Profs. Prado and Longwell).

The burner was operated under conditions of varying excess air which effected the amount of particulate produced. Both continuous burning and cyclic burning mimicking actual in-home conditions are being studied.

The sampling system consisted of a filter to collect part-iculates, a cooled condensor followed by an XAD-2 resin trap and finally a solvent-trap in an ice bath were used. No material was detected in the ice bath. Our discussion relates to the materials extracted from the filters and resin traps. Details are discussed in the paper submitted by J.P. Longwell at this conference.

Extraction (Soxhlet) of both the filter and resin trap with methylene chloride yielded the primary samples for chemical analysis and biological assay. Attempts to reduce extraction

samples to dryness for weighing were thwarted by the presence of
an appreciable amount of relatively volatile materials, such as
ethyl benzene, ethynyl benzene, ethenyl benzene, indene, naphthalene
and others of low molecular weight. Our knowledge of amounts are
only approximate at this time as we begin to use appropriate internal
standards for each compounds measurement on gas and liquid chromato-
graph output.

In our studies prior to this year, separations based on gas
chromatography were used in the anticipation that the preponderant
amount of mutagenic material would be found among the nonpolar
polycyclic aromatic hydrocarbons and, thus, permit identification
after GC separation. This may well be true in the case of the
response of human cells to diesel exhaust extract but is clearly
not the case for bacterial mutation.[3]

BACTERIAL MUTATION ASSAY

1 ml cultures of Salmonella typhimurium were treated for 2
hours with the test mixtures dissolved in small volumes of dimethyl
sulfoxide. The fractions of cells surviving treatment and mutating
to resistance to 50 µg/ml 8 azaguanine were determined by simply
spreading cells in appropriate number of agar plates and enumerating
the colonies formed two days later.[4,5]

Two forms of the assay were used. In the first form, bacteria
were present alone so that any catalytic activity modifying the
structure of chemicals in the sample would be due entirely to
enzymes of the bacteria. In the second form an extract of a
homogenate of rat liver was also added to mimic the metabolic
capabilities of mammalian tissue to perform a wide variety of
oxygenation and conjugation reactions with chemicals such as the
polycyclic aromatic hydrocarbons. In general, these bacteria do
not contain enzymes capable of causing reactions of relatively
inert polycyclic aromatic hydrocarbons to form more reactive
and mutagenic derivatives. Thus, the essence of liver is required
in order to observed the mutagenic activity of simple PAH in
bacterial cultures.

OBSERVATIONS

The first question that arises from such a series of physical
manipulations and observations regards reproducibility. For
instance, two samples from the exhaust of the same burners run
under identical conditions of fuel and smoke level should have
identical physical and biological characteristics.

Table 1 presents data which permit considerations of reproducible sample behavior.

The first thing that strikes one about the intended replicate samples from the same burner using identical fuel and nearly identical excess air conditions is that they are not so physically similar as to support an assumption of identity. For instance, the two samples at the lowest smoke level (Table 1) have ratios of extracted material from the filter and resin bed to the particulate weight of 0.6 and 0.4. In Table 1 (High), we note an approximate two-fold variation on this parameter between two samples. One is also interested in the wide variation in purposely duplicate runs in the amount and distribution of extractable material among the filters and the resins traps.

Noting this problem, we can, however, see that the high smoking conditions have produced a low amount of extractable material relative to the amount of particulate produced. What is remarkable is that, under conditions of excess air yielding the lowest smoke level, the yield of extractable material is highest.

Since the condition of high excess air has produced a low smoke level and a high amount of extractable material, particularly on the resin trap, we may ask if this material differs in terms of some gross characteristics, such as ability to mutate bacteria.

The bacterial mutation assays, however, do not reveal any more desirable levels of reproducibility among the replicate runs than did the weighing of extractable material. The expected variation among replicate independent bacterial mutation assays of the same sample is not small, however. Standard deviation around the mean for replicate assays on benzo(a)pyrene (n = 1000) is about 50% of the mean value.

Despite the considerable variation expected of the biological assay, the variation observed between each set of attempted identical burner runs is considerably larger than would be expected by chance. Since, in every case, the material collected on the filter of the first run performed in the engineering laboratory was considerably more mutagenic than in the second run, the conclusion that these runs are non-identical is inescapable.

If, however, we regard each run as different and group the samples from each run in terms of increasing smoke number (samples 7, 8 and 9 of Table 1 for instance), we can see a clear decline in mutational potency of the extracted material as the smoke number increases. This observation combined with the clear decline in the total amount of extractables collected per kg of fuel burned as the smoke number increases leads to the observation that the least sooty operation condition releases the highest amount of mutational activity.

Table 1. Independent Sample Runs of a Flame Retention Burner with No.2 Heating Oil in Continuous Mode, Producing a Smoke Level of 1 (Low), 5 (Intermediate), and 9 (High).

Bach Smoke Level	Sample No.†	Sample wt. (mg)	mg Extract/ mg Particulate	μg Extract/ g fuel	Mutant fractions x 10^5 150 μg/ml x 2 hrs	
					Without Metabolism	With Metabolism
Low 1	7. F	0.5	0.59	3.2	512	235
	X	37			277	159
	11. F	0.8	0.39	2.4	165	21
	X	13			178	26
Inter- mediate 5	8. F	3.1	0.01	0.5	350	140
	X	7.4			No data	NA
	12. F	0.7	0.01	0.6	120	70
	X	3.0			50	NA
High 9	9. F	0.2	0.001	0.5	180	27
	X	1.7			19	NA
	13. F	1.5	0.002	0.8	48	NA
	X	3.9			43	NA

† F, filter; X, XAD-2 resin trap

This observation is paradoxical and the conclusion, in addition to being paradoxical, is possibly premature. However, as we set up a better controlled experiment at MIT, I will take the risk of being wrong and conclude for the purposes of discussion of these two independent runs on the same continuously operated oil burner that (1) operating with excess air led to production of more total extractable material, and (2) that this material on a weight basis was more mutagenic than extractable material produced under operating conditions of less excess air.

Table 2 rearranges the basic data to reflect this way of looking at the preliminary testing results. Clearly, for each run, the observed specific mutagenicity of 150 µg of extract from either the filter or the XAD-2 resin trap decreases with the smoke level of the sample. For instance, the mutation fraction values (x 10^{+5}) for smoke levels of 1, 5 and 9 for run I for the filter extracted materials are 512, 350 and 180, respectively, and 165, 120 and 48 for run II. Similar behavior is seen for the material trapped on the XAD-2 resin.

As Table 2 shows, addition of the metabolic component of rat liver homogenate consistently reduced the mutagenic activity of the extracts. This can be interpreted in a number of ways. For example:

(a) the liver homogenate could inactivate mutagens present by enzymic catalysis or providing substrate for reactions, or

(b) both inactivation of some mutagens and activation of others could be occurring simultaneously.

Our experiences with diesel exhaust have provided evidence that the latter possibility is, in fact, occurring.[3]

In general, compounds active in the absence of a metabolic element will be chemically reactive as are the arene oxides of the PAH or activated by the enzymes of the bacterium as seems to be the case for certain di-nitro PAH derivatives. Those which we would expect to find in these oil burner effluents which are activated by the liver homogenate enzymes would be cyclopenteno (c,d)pyrene, fluoranthene, alkyl phenanthrenes and other simple or alkyl substituted PAH.

Turning to what little we presently know about which chemicals are mutagens in these samples, I should state that it is very little indeed for the practical reason that only a few mg of extract were collected in this pilot study. (For a fairly complete example see Kaden et al., examining a turbulent diffusion flame

Table 2. Comparison of Amount of Extractable Material
Produced and Its Mutagenic Potency Within Runs
When Smoke Level was Changed by Varying Excess
Air.

	Sample No.	mg Extract/ g fuel	Mutagenicity[+] Without Metabol.	With Metabol.
RUN I	7. F	3.2	512	235
	X		277	159
	8. F	0.5	330	140
	X			NA[++]
	9. F	0.5	180	27
	X		19	NA
RUN II	11. F	2.4	165	21
	X		178	26
	12. F	0.6	120	70
	X		50	NA
	13. F	0.8	48	NA
	X		43	NA

[+]Mutagenicity is the mutant fraction observed x 10^{+5} for
2 hour exposure to 150 µg/ml extract.
[++]NA--not active; value less than 17 observed.

or the description of our studies of diesel exhaust by J.P. Longwell
in this conference.)

First, let me point out the fuel itself is not detectably
mutagenic; and, while it contains mutagenic compounds such as
fluoranthene at 50 ppm, such levels are not sufficient to be
detected when the whole fuel is used at a few hundred µg/ml in
the assay. Second, let me be clear in stating that none of the
samples discussed so far were subjected to GC MS analysis. Extracts
of other samples from this series of oil burner trials have shown
the presence of fluoranthene and the alkyl phenanthrenes and
anthracenes in sufficient quantity to suggest that they are responsible for a major amount of the mutation induced in the presence
of the rat liver homogenate.

What compounds are responsible for mutation in the absence of
the liver homogenate is unknown. Some workers have suggested that

mononitro PAH are the culprits. We find pure standards of mono-
nitropyrene to be relatively inactive, while the extremely mutagenic
dinitro compounds have not yet been found in our studies of combustion effluents. Since polar PAH derivatives do not pass through
over GC columns and many polar PAH derivatives have already been
identified in extracts of diesel exhaust, one might suggest the
use of HPLC to separate our extracts for further characterization.

Recently, Ms. Barbara Andon has worked out a micro assay which
permits measuring mutation from each peak of HPLC output. Her
collaborative effort with the analytical chemist Ms. Julie Leary
now permits us to identify the most mutagenic fractions when only
13 μg of a combined filter + XAD-2 extract from our oil burner
series was separated on HPLC. Some forty separate peaks were
eluted, and one was found to contain some 60% of the activity of
the sample with only four others containing significant amounts of
mutagenic activity. We expect this to be fruitful in future studies.

At present, we are setting up a domestic oil burner to produce
the quantities of sample under reproducible conditions to allow us
to identify the most important mutagens in its exhaust under different operating conditions. Perhaps this knowledge will help us
define efficient conditions of use which minimize the mutagenic
properties of oil burner effluents.

Interesting as the analysis of complex mixtures to detect its
mutagens is, toxicologists are acutely aware that finding a mutagen
for bacteria is not necessarily finding a mutagen for humans. In
our laboratory, therefore, conclusions reached using bacterial
assays are compared to observations in which human lymphoblasts are
used directly. This, however, is still not enough. I will briefly
try to outline an approach devised by Thomas R. Skopek, now at
Yale University, and myself to answer two important questions:

1. Are people genetically changed by chemicals in
    their everyday environment?

2. If they are genetically changed, then which
    chemicals are responsible?

The basic idea is that we think we can take a blood sample
and determine the pattern, or spectrum, of mutations which have
taken place among the white blood cells of that sample. We know
the spectra produced in human cell culture for spontaneous change
in the absence of chemicals and can easily distinguish between
that spectrum and those produced by PAH, alkylating agents, UV
light or substituted acridines.

As we develop the technology necessary to test our hypotheses,
we anticipate comparing the actual spectrum of change in human

white blood cells to the spectrum (or spectra) produced by
environmental mutagens which we discover in the complex mixtures
of everyday life, such as oil burner smoke.

    This work has been sponsored principally by the U.S. Department
of Energy and the National Institute of Environmental Health Sciences.

REFERENCES

1. W. G. Thilly and H. L. Liber, Genetic Toxicology, in"Toxico-
        logy: The Basic Science of Poisons", T. Doull, C. D.
        Klassen and M. O. Amdur, Eds., Mac Millan Publishing Co.,
        NY, 2nd edition(1980), pp.139-157
2. D. A. Kaden, R. A. Hites and W. G. Thilly, Mutagenicity
        of soot and associated polycyclic aromatic hydrocarbons
        to Salmonella typhimurium, Cancer Res. 39:4152-4159
        (1979)
3. H. L. Liber, B. M. Andon, R. A. Hites and W. G. Thilly,
        Diesel soot: Mutation measurements in   bacterial and
        human cells, Proceedings at the Conference on Health
        Effects of Diesel Engine Emissions, U.S. Environmental
        Agency, (1979)
4. T. R. Skopek, H. L. Liber, J. J. Krolewski and W. G. Thilly,
        Quantitative forward mutation assay in Salmonella
        typhimurium using 8-azaguanine resistance as a genetic
        marker, Proceedings National Acad. Sci. USA 75:410-414
        (1978)
5. T.R. Skopek, H. L. Liber, D.A. Kaden and W.G. Thilly,
        Relative sensitivities of forward and reverse mutation
        assays in Salmonella typhimurium, Proceedings National
        Acad. Sci. USA 75:4465-4469 (1978)
6. W. G. Thilly, Chemicals, genetic damage and the search for
        truth, Technology Review, Feb/Mar, pp. 37-41 (1981).

DISCUSSION

E. Boyland (London School of Hygiene and Tropical Medicine)

1) If perylene is mutagenic and not carcinogenic it could be
an initiator and would induce cancer if combined with a suitable
promoter.
2) What effect does the presence of lead have on hydrocarbon
production?

Thilly

1) Perylene may be an initiator but we have not detected any
mutagenicity in human cells.
2) I don't know.

C. Stärk (Institut für Chemische Technologie, Darmstadt)

1) In which way do the forward testing systems react on toxic
compounds?
2) Do the forward mutagenic testing systems allow to detect
nitrogenated compounds?

Thilly

1) Forward assays simply present a larger and more heterogeneous
genetic target to test agents than sets of reversion assays.
Thus they facilitate the study of mutation of chemically complex
mixtures by reducing labor and permitting assays with very small
volumes such as HPLC peaks.
2) Forward mutation assays should be sensitive to the same set
of chemicals to which diverse sets of reversion assays are. Yes,
we use our forward assays to study nitro polycyclic aromatic
hydrocarbons.

A. Cavaliere (Laboratorio di Ricerche sulla Combustion, Naples)

     Is there any study on saturation effects of mutagenic test
that indicates the combustion mutagenic products to be in this
range?

Thilly

     We have been careful to study the mutational response at
the lowest concentrations permitted by the need for precision.
None of the PAH studied so far are found in the samples in
concentration which we would expect would saturate the systems
for metabolism, but this may not be true with regard to DNA
repair systems.

D. Rivin (Cabot Corporation)

     Bacterial assays, such as those based on Salmonella
typhimurium may indicate the potential for soot extracts to alter
DNA in the absence of mammalian genetic repair mechanisms.
Do the Ames modifications of S. typhimurium adequately model
human cells as regard to the ability of molecules to penetrate
the cell membrane?

Thilly

     First, the non identity of bacterial and human cells is as
obvious at the level of DNA repair as any where else. However,
the deep rough mutations used in S. typhimurium for mutation assays
are in fact a modification, partial removal of the bacterial poly-
saccharide wall matrix, which would make bacteria more like human
cells with regard to drug uptake through the cell wall/membrane system.

G. W. Smith (General Motors Research Laboratories)

What is the explanation for the fact that perylene has high
mutagenicity but low carcinogenic activity? (Alternatively, is
much progress being made on structural theory of mutagenicity
vis à vis carcinogenicity?)

Thilly

First, perylene is a very potent mutagen for S. typhimurium
but under identical conditions of assay it has not been detected
as a mutagen for human cells.

Secondly, many mutagens will not cause cancer in experiments
such as mouse skin paintings because they do not possess other
qualities necessary for tumor production.

Thirdly, it is not clear that all mutagenic events can initiate
the cancer process.

S. Galant (Société Bertin)

Since you mentioned experiments on both flat flames and
diffusion flames, have you carried out any measurements on mutagenic
activity of current spark ignition engines?

Thilly

Yes, but the chemical analysis has proven difficult due to
our limited sample size. We intend to apply Barbara Andon's
micromutation assay to the output of direct HPLC separation of
particulate extracts. We have no knowledge of the non-particulate
emissions from spark-ignition engines.

J. M. Beér (Massachusetts Institute of Technology)

Prof. Thilly reports higher biological activity of PAH emitted
from non sooting burner flames. PAH and soot samples taken from
Coal Liquid turbulent diffusion flames show no or little extractible
material on soot as long as the soot was held in the sampling train
at a temperature higher than 100°C. It seems that for minimizing
biologically active PAH emission, it would be better to allow soot
to form in the flame and then precipitate it at low temperature
from the flue gas. Do you think that it is worth following this
line of thought?

Thilly

It may be. I think you agree that it is the total emission,
whether on "soot" at the time of collection or not, that we should
seek to analyze.

J. Jagoda (Georgia Institute of Technology)

It seems from what has been said that, under the conditions of sooting, less PAH is present than under non sooting conditions, indicating that some PAH has been transformed into the solid inactive soot. This is also born out from soot mechanism considerations.

It would then almost seem that having soot present would reduce the chance of mutation by PAH, which seems unexpected (at least to me). The answer to this probably lies either in the way in which PAH is assimilated in the human body or in the possible easier burnout of PAH in the gas phase than that adsorbed on soot (i.e., soot acts as an excellent carrier).

Thilly

First, it does not seem to me that the hypothesis that PAH are transformed into soot rather than produced by a concurrent independent process has been proven in our discussions here. The study of PAH exposure routes of humans by any of the kinds of air borne particles has not yet really begun. Yes, burn out in the gas phase might be a reasonable basis for an abatement process.

K. H. Homann (Technische Hochschule, Darmstadt)

How did you make sure that the analytical procedure, such as using columns at elevated temperatures, did not change the contents of the original samples?

Thilly

We are not sure that gas chromatography did not change the samples. We hope to avoid this possibility by using low temperature HPLC systems and checking for total biological activity before and after separation. A set of appropriate reconstruction experiments is planned.

THE TOXICOLOGY OF SOOT

E. Boyland

TUC Centenary Institute of Occupational Health
London School of Hygiene and Tropical Medicine
Keppel Street/Gower Street
London WCIE 7HT England

In 1775 the London surgeon Sir Percival Pott published a
monograph in which he described cancer of the scrotum as an
occupational disease of chimney sweeps. In England chimneys which
contained soot from burning of coal were cleaned by small boys
who were made to climb the flues. In doing this the soot fell
from the chimney and the boys became covered in soot. There
were no facilities for washing or changing of clothes. Pott said
that the disease was due to deposition of soot in the skin of the
scrotum. This was probably the first definition of an external
cause of cancer.

Cancer of the scrotum is no longer a common disease of
chimney sweeps. In the last century, however, skin and scrotal
cancer was seen frequently in men exposed to oil and coal tar.
It still occurs in some engineering works. Early pathologists
in Europe failed to induce cancer experimentally with coal tar, but
Japanese workers showed that the repeated applications of coal
tar to the ears of rabbits induced skin cancer. Later it was
found that the painting of a solution of coal tar on the skin
of mice also induced tumours. This gave a method for the biological estimation of carcinogenic activity.

Using this method Kennaway[1] attempted to identify the carcinogenic compounds, starting with about one ton of coal tar. It was
seen that the active fractions were strongly fluorescent. Using
a simple spectroscope a specific fluorescent spectra was used in
the concentration of the activity. Compounds were also tested
which had similar fluorescent spectra as the active fractions.
As a result of the work, dibenz(a,h)anthracene was found to cause
cancer in mice. This was the first chemical carcinogen to be

13

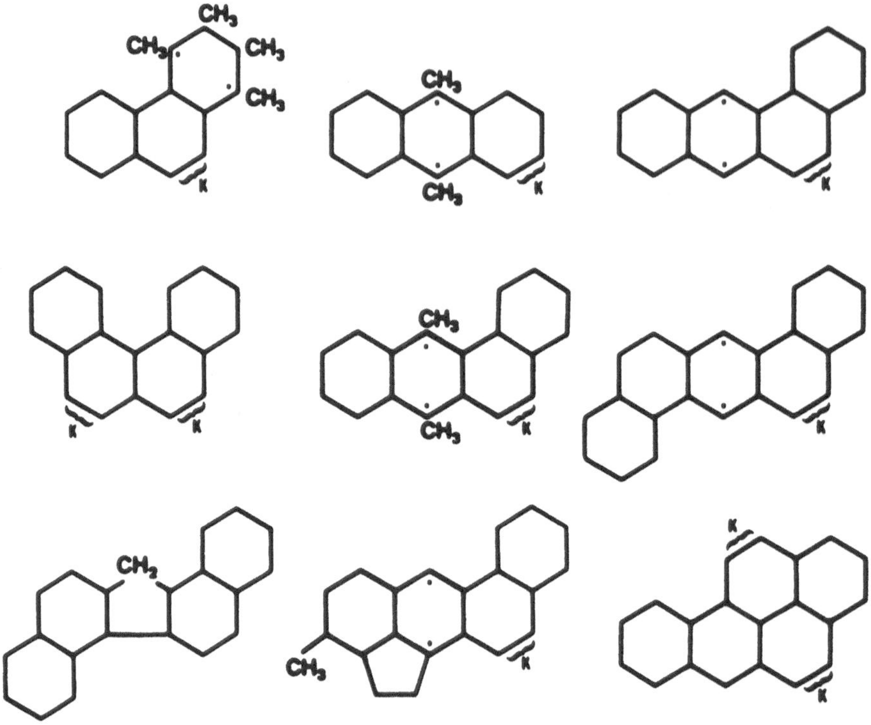

Fig. 1. The K-regions of carcinogenic hydrocarbons.

identified. It is also carcinogenic on injection and when
administered in this way it is still probably the most potent
carcinogenic compound.

Dibenz(a,h)anthracene was not identical with the active
material in the coal tar. Cook et al.[2] isolated an active
compound and showed it to be benzo(a)pyrene. Benzo(a)pyrene
is widely distributed in the environment, because it can be
formed by combustion of any carbonaceous material.

Following these observations Cook and the colleagues in
London and Fieser[3] in Boston synthetised many polycyclic aromatic
hydrocarbons which were tested for carcinogenic activity. By
1940, much knowledge of chemical structure and activity was
available. The active compounds appeared to have a reactive
phenanthrene double bond. Pullman and Daudel[4] called this the
K-region. The role of the K-region in carcinogenesis is still
not understood. (See Fig. 1).

Benzo(a)pyrene and other carcinogenic hydrocarbons are
stable substances, resistant to heat, acid or alkali. Because
of this, in 1932, I thought the activity might depend upon

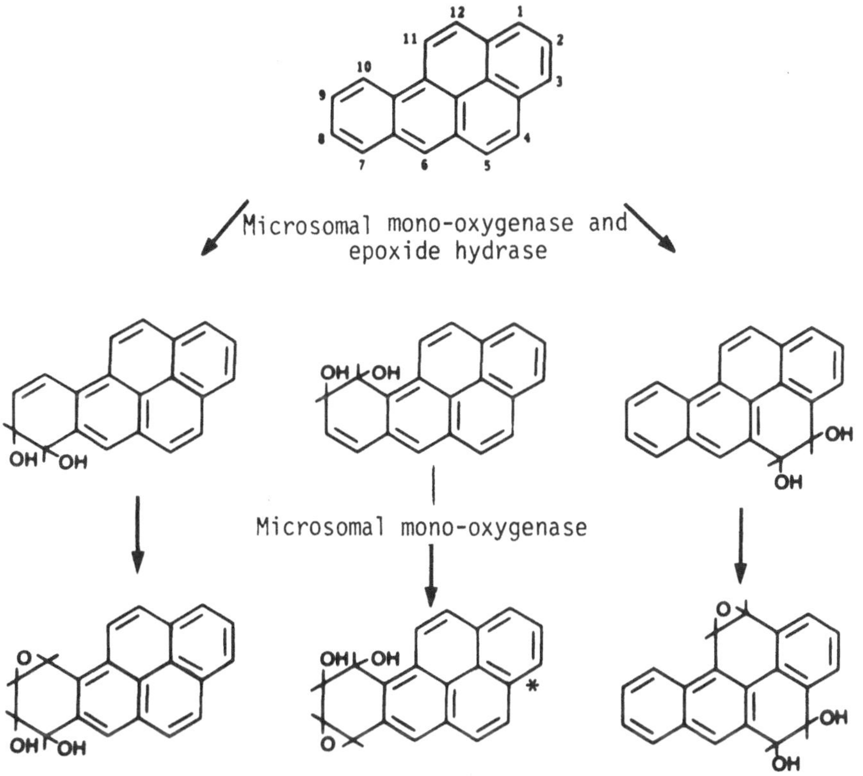

Probable proximate carcinogen (after Sims and Grover[6])

Fig. 2.   Some diol-epoxides derived from benzo(a)pyrene.

metabolic processes.   Very small amounts of benzo(a)pyrene were
available so Boyland and Levi[5] investigated the metabolism of
anthracene in rats and rabbits and identified 1,2-dihydrol, 2-
dihydroxy anthracene and the glucoronide of this as metabolites.
This process of the formation of dihydrodiols was shown to be a
general metabolic process.   By 1950 it was shown that these diols
were formed from epoxides or arene oxides.

There were many known carcinogenic epoxides and it was
thought that the oxides of carcinogens were the reactive species
which reacted with DNA.   Later work particularly that of Sims and
Grover[6] has shown the mechanism is more complicated.   When oxidised
by microsomes all the double bonds of aromatic hydrocarbons are
oxidised to epoxides.   The epoxides can (1) rearrange to form
phenols (2) react with water by the action of the enzyme epoxide
hydrase to yield dihydrodiols (3) react with glutathione by the
action of a glutathione transferase (4) react with protein.

Some of the diols formed in this way have activated double bonds adjacent to the hydroxyl groups. These bonds are then further oxidised to yield diol epoxides. Of the derivatives of this it is the "bay region" diol epoxides which react with guanine residues of DNA and so cause mutations and the first or initiating stage of carcinogenesis (See Fig. 2). This produces latent cancer cells.

Latent cancer will remain as such unless transformed to tumours by the action of promoters. Complete carcinogens such as benzo(a)pyrene have both initiating and promoting activity but there are many different types of promoters, including saccharin, phenobarbitone, benzene valium, duodecane and chloroform. With some promoters there are threshold or safelevels below which they are not hazardous. On the other hand it is probable that there are no safe or threshold levels for initiating agents.

Passey[7] showed that benzene extracts of chimney soot induced cancer when painted on the backs of mice. Generally, however, the carcinogens are bond to the carbon and so inactive. Epidemiological investigations have not shown carbon black to be a human carcinogen.

Carbon black is a soot which is made deliberately and is used in industry. Some forms of black rubber such as used for car tyres may contain as much as 50% carbon black. The recommended TLV (Threshold Limit Value) for carbon black is 3.5 mg per m^3, but there is no evidence of increased disease in men occupied in the manufacture of carbon black.

The polycyclic hydrocarbons which may be carcinogenic are strongly adsorbed on the carbon. They are more readily eluted with toluene or benzene than by cyclohexane but the ease of elution varies with different samples of soot. The hydrocarbons are very slowly eluted with body fluids such as serum. Soot particles would mostly be removed from the lung before carcinogens would be removed. Charcoal is the best known material to absorb and inactivate toxic materials. It is for this reason that given reasonable hygienic standards soot does not seem to present a carcinogenic hazard.

CARBON BLACK

Assessment of possible hazards from carbon black is difficult because different preparations contain variable amounts of many compounds some of which are still not identified. The chemical composition and physical properties differ with methods of preparation and the nature of the starting material and probably from day to day according to environmental and other conditions. A NIOSH report summarised the situation.[8] The committee recommended

Table 1. Physical and Chemical Properties of Various
          Types of Carbon Black.

Property	Furnace Black		Channel Black	Thermal Black
	Oil	Gas		
Composition				
Carbon (%)	98	99.2	88.4-95.2 (avg,91.2)	-
Oxygen (%)	0.8	0.4	3.6-11.2 (avg,7.8)	-
Hydrogen (%)	0.3	0.3	0.4-0.8 (avg,0.6)	-
Ash-Ca, Mg, Na (%)	0.1-1.0	0.1-1.0	0.01	-
Volatile matter (%)	1-2	1-2	5-18	<1
Average particle diameter (μm)	0.018-0.06	0.04-0.08	0.01-0.03	0.14-0.47
Specific surface (m^2/g)	25-200	25-50	100-1,000	7-13
pH	8-9	8-9	3-5	8-9
Benzene extractables (%)	0.05-0.1	0.05-0.15	-	0.03-1.75

a 10 hour TLV limit of 0.1 mg/m^3 of the cyclohexane soluble fraction, and that the material be designated 'suspect carcinogen' if
the cyclohexane extractable material is above 0.1%. The 1968
TLV for carbon black in U.S.A. of 3.5 mg/m^3 remains, but
Troitskaya et al.[9] indicated that the maximum allowable concentration (MAC) in the shopfloor atmosphere in the USSR has been fixed
at 4 mg/m^3.

The Chemistry of Carbon Black

The properties of various types of carbon black are summarized
in Table 1. The most probable carcinogenic hazard is associated
with the 'benzene extractables'. The material extracted from
furnace blacks consists mainly of aromatic hydrocarbons and sulphur
compounds. Some compounds identified by computerised gas chromatographic mass spectrometry and high resolution mass spectrometry
are shown in Fig. 3.[10] The exact structures of compounds marked
c and d are not certain. Some of these compounds such as

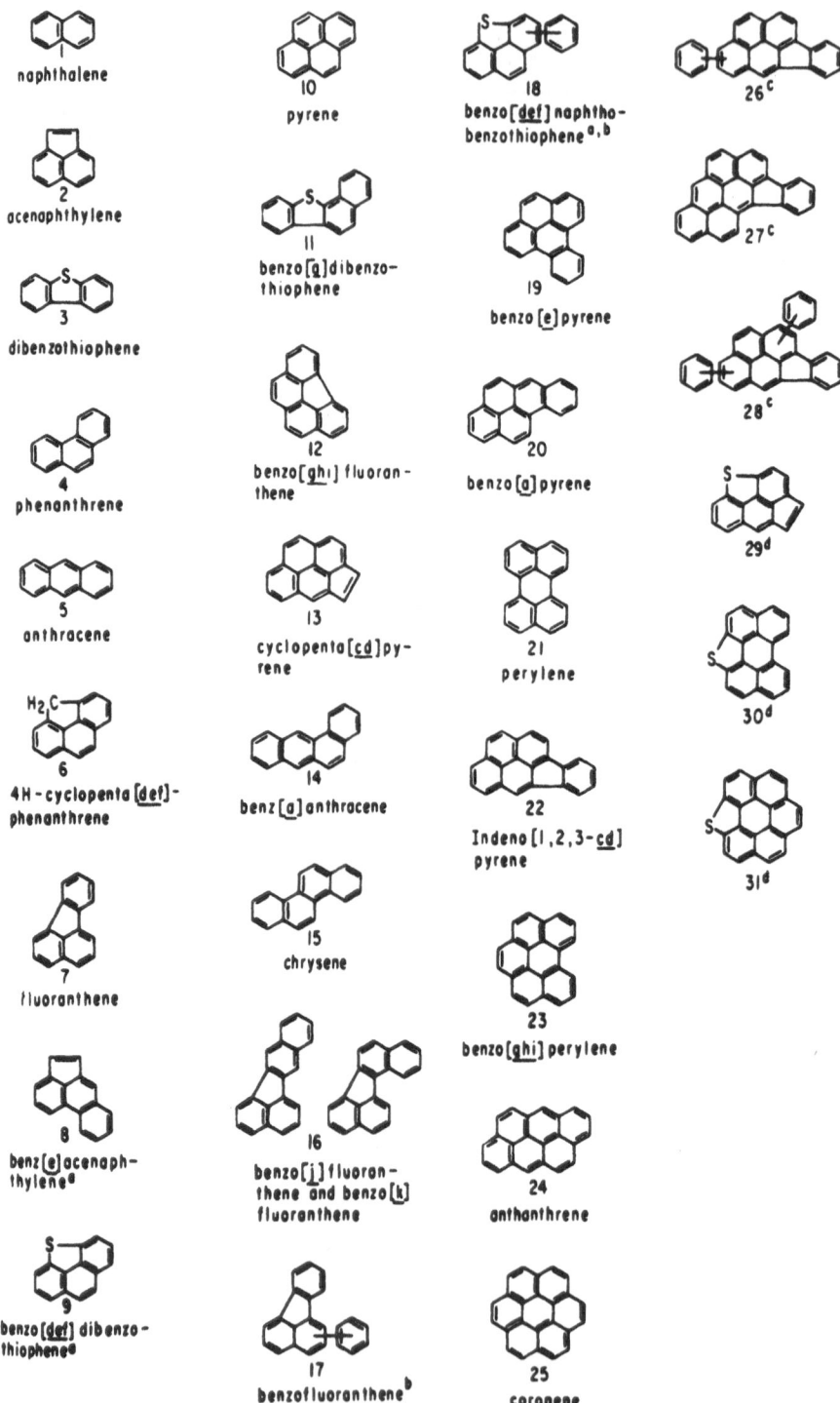

Fig. 3.   Benzene extractables.

benz(a)anthracene and benzo(a)pyrene are known carcinogens.
Chrysene has been described as an initiating agent and pyrene
and fluoranthene as cocarcinogens.[11] Oxygen derivatives[12] and
nitro derivatives[13] of polycyclic compounds have also been found
in extracts of commercial carbon blacks. The amounts of some
cyclic hydrocarbons present in a carbon black are shown in
Table 2.[14]

Russian workers[9] found benzopyrene in benzene extracts of
all types of carbon black examined.

Steiner[15] showed that furnace black but not channel black
suspended in tricaprylin induced sarcomas on injection into mice.
He concluded that the material was carcinogenic because carcinogens
were eluted by tricaprylin. It was also shown that injection
of benzene extracts of furnace black induced sarcomas.

Other workers[16] concluded that carbon blacks were not carcinogenic themselves and that adsorbed carcinogens lost their biological potency.

Elution of compounds from carbon black

The known carcinogens present in carbon black are strongly
adsorbed. Many workers, e.g. Neal et al.[17] and Falk[18], found
that polycyclic hydrocarbons including benzo(a)pyrene were
not eluted by blood plasma, gastric juice or milk. Kutscher
et al.[19], however, found that bovine serum eluted benzo(a)pyrene
from Corax L in 7 hours and from Degussa MT and Degussa 101 in
15 minutes.

The rates of elution by benzene and the amounts of extractable
material from carbon black vary with the type of carbon. Thus
extraction of Vulcan J and 330, which contain over 0.1 % of
extractable material, was almost complete in 150 hours but
extraction of 660 GPF, 330 HAF and Regal 300 which have only
about one third as much extractable material was almost complete
in 50 hours.

It is clear that the known carcinogens present in carbon black
are not easily removed. The fact that they are not easily eluted
by biological fluids indicates that they are unlikely to be
causes of gastric cancer. On the other hand it is probable that
the nicotine present in tobacco smoke could elute carcinogens
from carbon black in the lung. Nicotine is a derivative of
pyridine. Pyridine is known to be active in desorbtion and
nicotine would be expected to act in the same way. It would
be interesting to compare the abilities of cyclohexane, pyridine
and nocotine to elute benzo(a)pyrene from carbon black. Data
on the smoking habits of workers in the carbon black and rubber

Table 2. Polycyclic Aromatic Hydrocarbons Identified in The Carbon Blacks.
(Values expressed as percent of the benzene extracts)
(Mean values and ranges of 5 samples for every black type).

P.A.H.	Sample				
	Vulcan J	Regal 300	HAF 330	GPF 660	339
Phenanthrene and/or anthracene	0.07 (< 0.05–0.1)	absent	< 0.05	absent	0.2 (0.1–0.5)
Fluoranthene	4.1 (3.8–4.3)	3.7 (3.4–3.9)	3.5 (2.9–4.3)	4.1 (3.4–6.4)	4.3 (3.9–5.2)
Benzo(d,e,f)dibenzo-thiofene + benzo(e) acenaphtylene	< 0.05	< 0.3	< 0.3	absent	< 0.05
Pyrene	22.2 (21.3–24.9)	23.2 (21.6–24.0)	16.5 (13.4–18.7)	17.2 (14.3–18.1)	17.2 (14.3–20.1)
Benzo(ghi) fluoranthene	7.2 (5.9–7.6)	6.3 (5.6–7.2)	6.9 (6.5–7.3)	4.6 (4.2–5)	7.8 (7.5–8.1)
Cyclopenta(cd)pyrene	10.2 (5.6–14.7)	< 0.3	7.0 (6.8–7.3)	4.4 (4.2–5)	6.5 (5.2–7.8)
Benzofluoranthenes (total)	0.7 (0.4–1.1)	absent	< 0.3	1.2 (0.9–1.4)	0.6 (0.4–0.9)
Dimethyl Cyclo-penta-pyrene and/or Dimethyl Benzo-fluoranthene	2.5 (2.0–3.5)	0.5 (<0.3–0.7)	0.9 (0.3–1.3)	1.9 (1.7–2.3)	1.4 (1.1–1.7)

P.A.H.	Sample				
	Vulcan J	Regal 300	HAF 330	GPF 660	339
Benzopyrenes (Total)	1.4 (1.1-1.9)	0.5 (≪0.3-0.7)	1.1 (0.4-2.1)	2.6 (1.5-3.2)	2.7 (2.0-3.9)
Indenopyrene	1.7 (1.5-2.2)	0.5 (≪0.3-0.7)	0.1 (abs-0.3)	2.3 (2.1-2.5)	2.9 (2.7-3.1)
Benzo(ghi)perylene	11.7 (10.9-13.8)	6.2 (5.5-8.5)	8.7 (8.2-9.5)	13.5 (12-16.5)	13.6 (12.8-17)
Anthanthrene	3.2 (2.8-3.5)	-	0.8 (<0.3-1.2)	2.3 (1.9-3.3)	3.5 (2.8-5.5)
Benzo-acridine derivative	< 0.05	absent	absent	absent	0.2 (abs-0.5)
Isomer of Coronene	6.3 (5.1-7.2)	1.1 (abs-2.2)	0.1 (abs-0.3)	0.5 (abs-1)	5.1 (4.6-5.7)
Coronene	7.2 (4.1-10)	1.2 (<0.3-3.2)	3.8 (3.5-4.4)	8.9 (8.5-9.8)	11.7 (9.5-13.8)
Total	78.5	43.8	50.05	63.5	77.5

industries are needed. The use of carbon black with less than 0.1% cyclohexane extractable matter might reduce the possible hazards.

As carbon black is often a major constituent of finished rubber products the risks involved in its use need consideration and investigation.

The elution of polycyclic hydrocarbons from carbon blacks is a slow process. It would be unlikely to occur during passage through the stomach. Hydrocarbon could be eluted in the lung where it remains diffused in the tissue for long periods.

REFERENCES

1. E. L. Kennaway, Biochem. 24:497 (1930)
2. J. W. Cook, C. L. Hewet and I. Hieger, J. Chem. Soc.
      395 (1933)
3. L. F. Fieser, Proceedings of the Bicentenial Conference
      on "Production of Cancer by Polynuclear Hydrocarbons"
      (University of Pennsylvania, Philadelphia) (1941)
4. A. Pullman and B. Caudle, Adv. Cancer Res. 3:117 (1955)
5. E. Boyland and A. A. Levi, Biochem. J. 29:2679 (1935)
6. P. Sims and P. Grover, Adv. Cancer Res. 20:165 (1974)
7. R. D. Passey, Brit. Med. J. 2:1112 (1922)
8. NIOSH Report, U. S. Dept. of Health Education and
      Welfare, Center for Disease Control "Criteria
      for a recommended standard - occupational exposure
      to carbon black", Publication n°78-204 (Sept. 1978)
9. N. A. Troitskaya, S. K. Velichkovsky, S. K. Bilmullina,
      T. G. Sazhilla, N. V. Gorodnova and T. D. Andreeva,
      Gig. Truda. Prof. Zabol. 3:32-36 (1975)
10. M. L. Lee and R. A. Hines, Analytical Chemistry 48(13):
       1890 (1976)
11. B. L. VanDuuren,"Chemical Carcinogen" American Chemical
       Society, Monography, 173:24 (1976)
12. A. Gold, Anal. Chem. 47:1469 (1975)
13. W. C. Fitch and D. H. Smith, Envir. Sci. Techn. 13:341
       (1979)
14. G. Locati, A. Fantuzzi, G. Consonni, L. Gotti and G.
       Bonomi, Am. Ind. Hyg. Assoc. 40:644 (1979)
15. P. E. Steiner, Cancer Res. 14:103 (1954)
16. F. Von Haam, H. C. Titus, I. Caplan and G. Y. Shinowara,
       Proc. Soc. Exp. Biol. Med. 98:95 (1958)
17. J. Neal, M. Thornton and Nau, Arch. Environ. Health
       4:598 (1962)
18. H. C. Falk, A. Miller and P. Kotin, Science 127:474
       (1958)
19. W. Kutscher, R. Tomingas and H. P. Weisfeld, Arch. Hyg.
       151:646 (1967)

DISCUSSION

D. Rivin (Cabot Corporation)

The NIOSH Criteria Document described by Prof. Boyland contains major recommendations and conclusions which are not supported by the literature reviewed therein. The document is useful, however, as a general but uncritical survey of the literature on the health effects of carbon black.

Boyland

I have found the NIOSH document extremely useful.

Rivin

Recent work by Lakowitz and Berbn shows that carbon black inhibits the uptake of benzo(a)pyrene (B(a)P) by mammalian liner microsomes in vitro. Similar experiments with asbestos, silica or iron oxide give enhanced uptake of B(a)P. The effect of carbon black is attributed to strong preferential adsorption of B(a)P on the carbon surface.

Boyland

Carbon black is a very strong absorbent and is used to reduce toxicity.

L. M. Appelman (Institute CIVO, The Netherlands)

What is your opinion on the possible classification of carbon black as "suspected carcinogen" in the proposal of a committee, installed by the Dutch Ministery of Health, to establish a classification of carcinogens?

Boyland

There is little evidence that carbon black is a human carcinogen.

M. E. Weill (University de Rouen)

Is it possible to make a comparison between the influence of carbon and of asbestos?

Boyland

No, because the particles are of different size, shape and nature.

L. Le Bouffant (Centre d'Etudes et Recherches des Charbonnages
                de France)

    Concerning the carcinogenicity of asbestos fibres, I am rather
skeptical about the necessity of association with tobacco smoke.
In animal experiments, it is well known that mesotheliomas can
be obtained by intrapleural injection of asbestos fibres alone.
Besides, many lung cancers produced by occupational exposure to
asbestos fibres are observed among non smoker population

Boyland

    Asbestos is carcinogenic in animals because the animal diets
contain tumours initiators. Asbestos is a tumour promoter.

A COMPARATIVE STUDY OF SOOT AND CARBON BLACK

Donald Rivin, Avrom I. Medalia

Cabot Corporation
Concord Road
Billerica, Massachusetts 01821

INTRODUCTION

Soot is a mixture of particulate carbon with various organic and inorganic components. It is generally formed as an unwanted by-product of combustion; thus, soots differ widely according to the type of fuel, the conditions of combustion, and the method of collection. Chimney soot was the first recognized occupational carcinogen. Its carcinogenicity is due to components other than particulate carbon, especially soluble organic compounds which form a significant portion of such soots[1].

In this paper we describe an analytical scheme for the characterization of environmental soots. It is used to compare different kinds of soot with each other and with carbon blacks. Carbon blacks are particulate carbons of high purity and well-defined morphology, manufactured under controlled conditions for a wide range of industrial applications. Emphasis is placed on the forms of particulate carbon in these materials and also the extractable organic compounds and their biological effects.

MORPHOLOGY OF PARTICULATE CARBON

The fundamental colloidal unit of carbon black is an aggregate of spheroidal carbon particles fused together in a random configuration as seen in Figure 1. The aggregates differ in size, but share certain features, including turbostratic arrangement of graphitic layers, colloidal size (i.e. below 1 micron), and a characteristic morphology. We propose to call this unique form of carbon "aciniform", a word of Latin origin meaning "clustered like grapes". Aciniform carbon is defined as particulate carbon

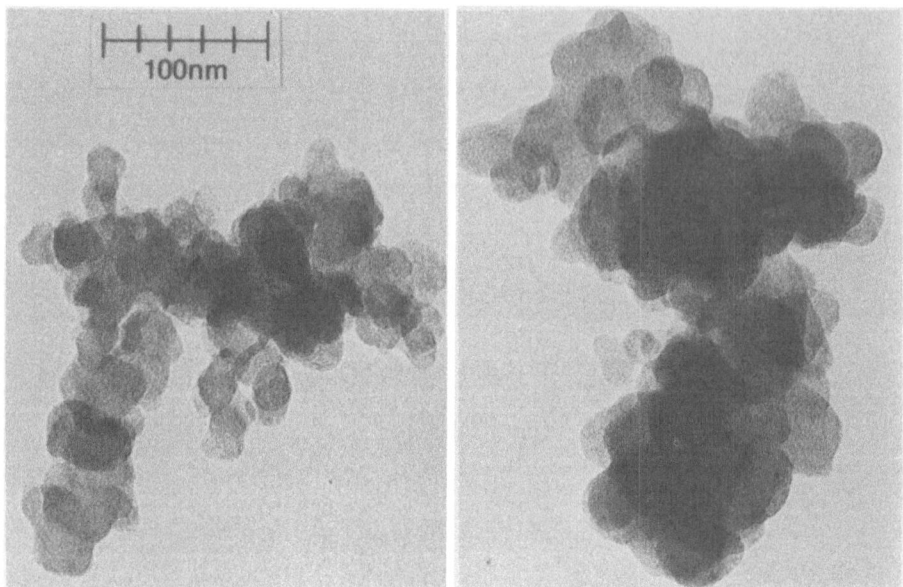

Fig. 1. Aciniform carbon (Furnace process carbon black).

of turbostratic microstructure, formed in the gaseous phase, and
composed of spheroidal particles fused together in aggregates
of colloidal dimensions. Small amounts of other elements may be
within the particles or bound to the surface. Carbon black should
be regarded as a manufactured product, which consists almost
entirely of aciniform carbon. Of course, the size of the particles
and the morphology of the aggregates depend on the grade of carbon
black.

The particulate portion of diesel soot is also exclusively
aciniform carbon. Unlike carbon black, diesel particulate shows
evidence of substantial post-aggregation, in which aggregates of
different particle size, which were formed in different flame
zones, have collided and have been cemented together by subsequent
deposition of a layer of carbon.

Very little aciniform carbon has been observed in soots
recovered from the chimneys of domestic fireplaces in the present
study, although the smoke which escapes from the top of a chimney
may have a high proportion of aciniform carbon[2]. For example, the
chimney soot in Figure 2, collected from a wood-burning fireplace,
contains less than 0.1% by weight of aciniform carbon. About 1/3
of this soot is particulate carbon, consisting mainly of bits of
coke and char. One fifth of the soot is ash, and nearly half of
the soot consists of pyrolyzable material, the greater part of

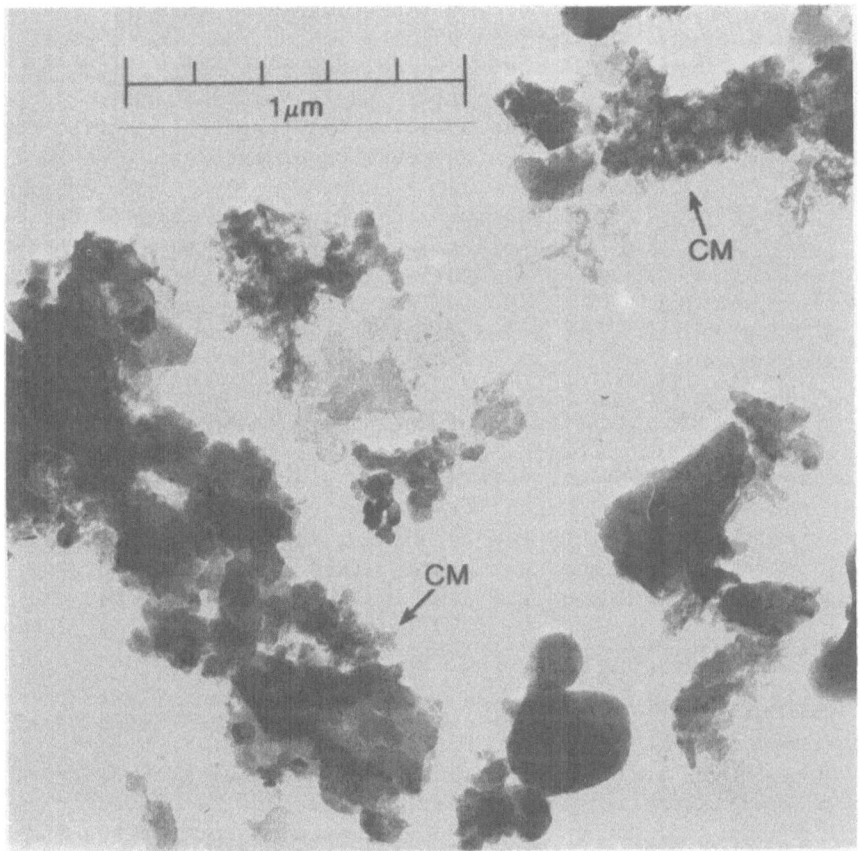

Fig. 2. Chimney soot from wood-burning fireplace after extraction
and wet de-ashing. Note carbonaceous microgel (CM)
particles.

which is apparently resinified, since it is not soluble in boiling
toluene. The micrograph shows many entities which seem to have
been formed by accretion of particles or aggregates of aciniform
carbon, with concomitant deposition and resinification of tars.
This material differs from aciniform carbon in the gel-like morphology, and in the extremely large size (≥ 1 micron) of the entity
relative to that of the particles. This particulate morphology we
call carbonaceous microgel and define as microscopic entities in
which spheroidal carbon particles of colloidal dimensions are
embedded in a carbon or carbonaceous matrix.

ANALYSIS OF CARBON BLACK AND SOOTS

Depending on their origin, environmental soots contain other
particulate materials in addition to aciniform carbon and carbonaceous

microgel. These include coke and char fragments, carbon cenospheres, crosslinked resins, and inorganic salts and oxides. Most soots also have large amounts of adsorbed and occluded components, such as water, sulfuric acid, nitrogen oxides and a Soluble Organic Fraction (SOF) which is mainly composed of solvent-extractable organic compounds formed as by-products of combustion or pyrolysis.

Although complete characterization of soot requires detailed microscopic and chemical analysis, most environmental samples can be adequately described by a more limited analysis of particulate morphology, extract (SOF), and ash content. Some important analytical properties are summarized in Table I for six soots and two carbon blacks.

Carbon blacks are almost pure aciniform carbon whereas most soots have a high ash content but very little aciniform carbon, except for diesel soot which is mostly a mixture of aciniform carbon and SOF. None of the listed soots contain more than fifty percent particulate carbon and as a result of their greater extract, resin, inorganic salt and char content they exhibit much higher hydrogen/carbon and TGA weight loss than the carbon blacks.

ANALYSIS OF EXTRACTS

The SOF contains the most important constituents of soot from an environmental health viewpoint, namely polynuclear aromatic compounds (PNA), a class comprising polynuclear aromatic hydrocarbons (PAH) as well as more polar oxidized and heterocyclic aromatics. A number of these compounds are known to be mutagens and/or carcinogens. It follows that the potential biological effect of an environmental soot is related to its SOF, especially to the amount of PNA not strongly adsorbed on the particle surface. Adsorption capacity of a soot or carbon black can be approximated from its available surface area [(Particulate Carbon)(Specific Surface Area) from Table 1] relative to that of an adsorbed SOF 'molecule'. If the latter is taken at $2500m^2/g$ (i.e. based on an average four ring PNA), then the apparent SOF surface coverage is $< 0.05\Theta$ for the carbon blacks and ranges from 6Θ for DOF to 263Θ for DWF, where Θ is the estimated monolayer coverage. Thus the amount of SOF is well below the monolayer capacity of carbon black but greatly exceeds that which can be adsorbed on the soots.

The presence of non-adsorbed SOF in soots is evident when one compares the amount of material removed with methylene chloride – a relatively poor extractant for aromatic molecules adsorbed on carbon – to that extracted with toluene under conditions known to desorb PNA from carbon black. Methylene chloride extracts forty percent of the SOF from N351 carbon black whereas more than ninety-five percent of the SOF is removed from soots (Table II). The distribution of compound types in the methylene chloride and toluene

Table 1. Analytical Properties of Carbon Blacks and Soots.

Carbon Black or Soot	Ash, %	Extract, % SOF	Extract, % Water	TGAa Weight Loss, %	Atomic Ratio, H/C	Carbon, % Particulateb	Carbon, % Aciniform	Specific Surface Area. m²/gc
Furnace process carbon black (N351)	0.27(.09)d	0.13	0.90	1.5	0.040	99	(99)e	73
Furnace process carbon black (RCF4)	0.54(.27)	0.09	0.87	1.2	0.023	99	(99)	91
Chimney soot from wood-burning fireplace (DWF)	21.8 (20.3)	15.8	14.2	48.0	1.08	50	0.024	3
Blended chimney soot from domestic coal fires (DCF)	24.6 (22.6)	35.6	19.0	52.4	1.21	23	0.36	17
Chimney soot from English coal-burning fireplace (ECF)	45.6 (n.d.)	15.8	14.7	36.4	1.00	–	0.89	< 1
Soot from "soot box" of domestic oilfurnace (DOF)	53.8 (40.7)	0.64	50.7	43.7	n.d.	8	0.83	32
Soot from small diesel engine (DE)	2.2 (0.68)	51.1	3.6	49.2	n.d.	45	50	72
Standard urban dust (NBS SRM-1648)	64.6 (57.7)	2.9	27	36.2	1.67	13	0.47	29

aLoss in weight upon heating to 910°C in nitrogen
bMaterial remaining after subtraction of SOF, water extract and insoluble inorganic matter as estimated by (ash)d.
cN2 BET surface area of sample after extraction with methylene chloride, toluene and water, and deashing with HF and HCl.
dAfter consecutive Soxhlet extraction with methylene chloride, toluene and water. Referred to weight of sample before extraction.
eCarbon black is wholly aciniform carbon. Allowance made for measured extractable and inorganic impurities.

Table II. Extract Composition

	N351		DWF		DCF		ECF		DE	
	MC	TOL	MC	TOL	MC	TOL	MC	TOL	MC	TOL
Total Extract,%	0.5	0.08	15.	0.83	35.	0.62	14.	0.68	51	0.14
Distribution %										
Non-Aromatic	n.d.	n.d.	0.5	0.5	3.4	0.6	1.2	2.3	35	6.2
PAH	70	46	3.9	4.4	11	4.0	3.9	6.9	14	19
Polar PNA	18	45	92	70	73	67	88	95	88	46
Identified PAH Compounds, %	77	72	6.5	3.3	3.9	2.0	16	4.0	8.9	9.2

extracts is similar for the chimney soots but differs for diesel soot DE, in which much of the non-aromatic fraction in the methylene chloride extract appears to be unburned fuel and both extracts have relatively low polar PNA content.

The number of individual compounds present in the SOF from carbon black is much less than in most soot extracts. For example, the PAH fraction from carbon black contains only thermodynamically stable, highly condensed aromatic molecules with no detectable alkylated products, whereas the soots exhibit a broad continuum of structures including alkylated aromatics. Ten unsubstituted molecules identified by HPLC, account for seventy-five percent of the PAH extracted from N351 carbon black but only for four to fifteen percent of the PAH fraction from soot.

The distribution of major identified PAH's given in Table III offers an additional basis for comparing soot and carbon black extracts. Note particularly the effect of adsorption on the composition of the extracts from carbon black. PAH with less than five rings are preferentially desorbed with methylene chloride whereas the larger, more strongly adsorbed PAH are extracted with toluene at a higher temperature. No comparable effect is observed with soots since most of the SOF is admixed, rather than adsorbed.

MUTAGENICITY

In so far as most of the SOF constituents of soots are un-identified, the potential biological activity of these materials cannot be estimated from analytical data but must be measured directly. Mutagenicities of extracts and extract fractions using Ames Salmonella typhimurium tester strain TA98 are summarized in Table IV.

Table III.   PAH Distribution

PAH, ppm[a]	N351 MC	N351 TOL	DWF MC	DWF TOL	DCF MC	DCF TOL	ECF MC	ECF TOL	DE MC	DE TOL
Anthracene	–	–	–	–	400	1.	n.d.	n.d.	200	8
Phenanthrene	–	–	51	2	500	1.	250	–	2400	5
Fluoranthene	30	6.	120	n.d.	n.d.	n.d.	n.d.	n.d.	1750	n.d.
Pyrene	182	13	57	3.4	390	1.5	311	11	1325	2.8
Benzanthracene	0.5	0.2	24	n.d.	n.d.	n.d.	120	n.d.	230	n.d.
Cyclopenta-pyrene	52	29	72	n.d.	n.d.	n.d.	n.d.	n.d.	<20	n.d.
Benzo(a)pyrene	0.7	4.4	22	2.6	120	0.8	57	1.9	33	1.3
Indeno-pyrene	0.2	8.8	16	2.1	33	0.7	26	1.0	165	2.9
1,12 Benzperylene	4.4	91	14	1.7	53	<0.5	122	1.2	190	2.3
Coronene	<0.02	92	3	<0.5	19	<0.5	13	0.4	61	<0.5

[a]Parts of PAH per million parts of original carbon black or soot.

Most of the mutagenic components in soot are extracted with
methylene chloride. They are mainly polar PNA's which are direct
acting mutagens. The small amount of adsorbed SOF on carbon black
yields a low mutagen content, predominantly in the toluene extract.
Almost all of these extracted molecules are unsubstituted PAH which
are only weakly mutagenic in the absence of enzymatic activation.

Carbon black and soot extracts may have comparable mutagenicity
per unit weight of extract but due to their much greater SOF content,
soots have up to three orders of magnitude more mutagens per unit
weight of sample.

This difference in mutagen content may be enhanced in vivo.
Body fluids elute less than 1/10,000 of the PAH adsorbed on carbon
black[3] whereas the large unadsorbed polar PNA fraction in soots
should be more readily removed. Nevertheless, certain high-SOF
soots, such as diesel particulate, have not been found to be toxic
or carcinogenic to humans or laboratory animals.[4]

EXPERIMENTAL SECTION

Materials

This report gives properties of eight representative materials -
two carbon blacks and six environmental soots - out of a total of
six commercial carbon blacks and eighteen soots which were examined.
The soots include samples taken from the following sources:
residential fireplaces and stove chimneys(9), industrial and power
plant furnaces and boilers(5), diesel engines(3), and an urban
atmosphere(1). Those samples explicitly covered in this report are
described below.

Table IV. Salmonella   (TA98) Mutagenicity Assay of
          Extracts.

SAMPLE	EXTRACT	Net Revertants			
		per µg extract		per mg samples	
		(−S9)	(+S9)	(−S9)	(+S9)
N351	MC	4	9	2	4
	Toluene	3	20	2	12
	Tol(PAH)	2	17	1	7
RCF	MC	−	−	< 1	< 1
	Toluene	−	−	< 1	< 1
DWF	MC	0.6	0.4	90	59
	MC(PAH)	1	2	6	11
	MC(POLAR)	0.5	0.3	71	39
	Toluene	1	0.6	8	5
DCF	MC	1.7	0.8	600	280
	MC(PAH)	2.2	3.6	80	134
	MC(POLAR)	1	0.4	260	100
ECF	MC	8.5	4.8	1200	670
	MC(PAH)	7.3	17	40	95
	MC(POLAR)	13	5.7	1600	700
DE	MC	4.3	1.8	2200	930
	MC(PAH)	1.3	1.3	90	90
	MC(POLAR)	8.5	6.1	1000	715

Furnace process carbon blacks N351 and RCF4 are typical of
grades used in rubber and ink applications, respectively.

Soot DWF was taken from the chimney of a domestic hardwood-
burning fireplace. It is a dry, granular black powder with irregular flakes ≤ 1 mm. Electron microscopy shows thin flakes and gel
fragments ≤ 2 µm. Soot DCF is a blend obtained by vacuuming domestic
flues in London, mostly based on coal fires using Welsh Boiler Nut
smokeless coal, Derbyshire and Staffordshire coals. It is an oily
black powder with 50-100 µm irregular particles. Electron microscopy shows mainly carbon gel fragments ≤ 2 µm. Soot ECF was taken
from the "shelf" in the flue of an open fireplace burning several
varieties of coal. It is fine brownish-black powder having some
aciniform carbon aggregates of 50-120 nm in the presence of carbon
gel and flakes of inorganic material. Soot DOF was taken from the
soot box of a domestic oil furnace. It is a brownish-black powder
with an acrid odor and contains a large quantity of crystalline
inorganic flakes 100-250 µm in size and also some carbon gel and

coke fragments ≤ 0.5 μm. Soot DE was collected by electrostatic precipitation from the exhaust of a Petter 4.5 hp single cylinder diesel engine operated on no. 2 diesel fuel. It is a low density, black powder with no flakes or lumps > 0.5 mm. Standard urban dust is a Standard Reference Material (SRM 1648) from the National Bureau of Standards collected over one year by continuous filtration of air in the vicinity of St. Louis, Missouri. It is a fine, dark grey powder having mainly crystalline inorganic particles ≤ 40 μm and appreciable 40-120 nm carbon gel.

Methods

Determination of ash content and hydrogen/carbon elemental analyses are by standard combustion methods on dried samples. Consecutive extractions with methylene chloride (4 hrs.), toluene (48 hrs.) and water (48 hrs.) are carried out in a Soxhlet apparatus. Soluble organic fraction (SOF) is defined here as the sum of material extracted with methylene chloride and toluene, although in several cases the aqueous extract also contains considerable organic material. Thermal Gravimetric Analysis is performed in nitrogen and air using a Perkin-Elmer TGS-2 apparatus with a weight loss precision of 1 % absolute. Aciniform carbon is estimated by dispersing the sample in xylene, centrifuging off particles coarser then 1 μm, then determining the amount of colloidal (aciniform) carbon by optical absorbance at 800 nm.

SOF components are separated into fractions by a five-step chromatographic procedure:

1) Pass concentrated extract through a hydrated silica gel column, elute with methylene chloride or toluene.
2) Back-flush column with methylene chloride/methanol (50/50) to obtain 'very polar' fraction.
3) Dissolve eluate from (1) in hexane/ethyl acetate (98/2) and introduce into a bonded phase guard column followed by a silica gel column. The non-aromatic fraction elutes after a short time.
4) Back-flush silica gel with same solvent to obtain PAH fraction.
5) Back-flush bonded column with ethyl acetate to isolate 'moderately polar' fraction.

Eluates from steps (2) and (5) are combined and reported as a polar PNA fraction. This assignment is supported by proton magnetic resonance analyses that indicate that this fraction is rich in heteroaromatic molecules and by ultra-violet absorption measurements which show a high content of polynuclear aromatics.

Individual PAH compounds were analysed by reverse phase HPLC (Zorbax ODS column, 75% acetonitrile/water to acetonitrile gradient) with both fluorescence and ultraviolet detectors.

   Mutagenicity assays were conducted at Arthur D. Little Inc.
using Ames Salmonella typhimurium tester strain TA98 with and
without rat liver S-9 microsomal fraction. Extract mutagenicity
is calculated from the slope of the initial linear portion of
the dose-response curve. Mutagenicity is also reported per unit
weight of the original soot or carbon black sample using gravimetric extract data.

Acknowledgements

   We thank D. R. Sanders for the extract composition analyses
and H. M. Cole, J. E. Connolly, F. A. Heckman, D. McLaughlin,
D. L. Peterson and J. E. Steeper for their experimental contributions to this study. We are also indebted to J. O'Gieblyn, S. A.
Hoenig, J. B. Horn, J. J. Johnson and H. Marsh for the collection
and careful identification of environmental soots. The support
of Cabot Corporation is appreciated. This paper was presented
on behalf of the Environmental Health Association of the Carbon
Black Industry.

REFERENCES

1. R. D. Passey, Brit. Med. J. 2:1112 (1922)
2. J. L. Muhlbaier, paper to International Conference on
        Residential Solid Fuels, Portland, Oregon, June 1-4,
        1981.
3. F. Buddingh, M. J. Bailey, B. Wells and J. Haesemeyer,
        Am. Ind. Hyg. Assoc. J. 42:503 (1981)
4. Health Effects of Exposure to Diesel Exhaust, National
        Research Council, Washington, DC, National Academy
        Press, 1981.

DISCUSSION

J. M. Beér (Massachusetts Institute of Technology)

   It seems to me that the classification of Dr. Rivin offered
for soots is not fully representative of the soot formed in the
gas phase in combustion processes. The high ash concentrations
reported for samples collected in chimneys will most likely
contain fuel particle or droplet fragments which are the results
of liquid or solid phase cracking and polymerisation rather than
gas phase soot formation. The soot formed in the gas phase in
flames of hydrocarbon fuels bears close resemblance to that in the
carbon black process.

Rivin

   Soots are complex mixtures of extractable and residue components in addition to a particulate carbonaceous fraction.

Aciniform carbon is the major particulate carbon from high temperature combustion and pyrolysis systems. Environmental soots, however,
contain appreciable amounts of other carbonaceous particulates
depending on fuel and the conditions of formation and collection.
For example, carbonaceous microgel probably forms at relatively
low temperatures by condensed phase carbonisation.

Meaningful comparison of the physical, chemical and biological
properties of collected soots requires suitable identification and
characterization of the components of these mixtures.

POLYCYCLIC AROMATIC HYDROCARBONS AND SOOT

FROM PRACTICAL COMBUSTION SYSTEMS

John P. Longwell

Massachusetts Institute of Technology
Cambridge, Massachusetts

INTRODUCTION

The increasing study of formation and control of soot and mutagenic compounds in combustion systems is motivated by public concern and regulatory activity on suspected or known sources of carcinogens. Combustion of hydrocarbon fuels produces widely varying amounts of known carcinogens which are frequently found adsorbed on soot. While the effect on public health is unclear, more knowledge of the chemistry and physics of formation and destruction of these materials is clearly needed. Increased use of solid fuels, low hydrogen content liquids and the automotive diesel engine has the potential for increased emission of these substances. Gasoline engine and oil burning furnaces will continue to be major fuel consumers and research relevant to these devices will be of equal importance.

Mutagens are found in high boiling temperature hydrocarbon fuels and are also synthesized by high temperature pyrolysis or in rich mixture flames from all fuels containing C-H bonds. The balance of these sources depends on both the fuel and the combustion system. While it is not possible, at this time, to determine the relative importance of these sources, there is some evidence, based on chemical analysis and biological assay which can serve as a guide to ongoing and future research. In this paper some results from field sampling and laboratory research will be examined for major trends and characteristics. In order to achieve some consistency, most of this discussion will be based on work at MIT where similar sampling, chemical analytical and bio-assay techniques were used in the several programs. Major emphasis will be placed on PAH since emission of these substances motivates much of the particulates related public concern.

37

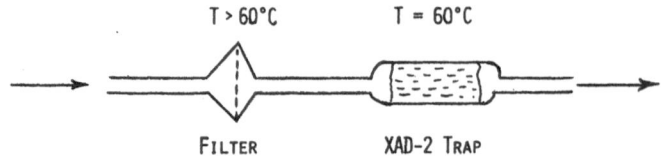

Figure 1.   PAH sampling technique.

Sampling and Analysis

A diagrammatic sketch of typical sampling apparatus is shown
in Figure 1. Sample gasses are drawn into a probe and then through
a filter. The sampling, filter and tubing temperature are maintained at a temperature above 60°C and preferably around 200°C to
prevent PAH condensation. The mixture after passing through the
glass fiber filter where soot is removed, is cooled to 60°C and
passed through an XAD-2 (polystyrene) trap which removes the
normally liquid hydrocarbons. Both filter and XAD-2 trap are
extracted with methylene chloride for further analysis and study.

The fraction of soluble material recovered on the filter is a
strong function of filter temperature. Table 1 illustrates this
effect on specific PAH found in the products of laminar premixed
sooting combustion of benzene[1].

At 85°C only 4% of naphthalene is collected on the filter
along with the soot. At 200°C, it is clear that a substantial
fraction of these PAH will be lost if the absorbent trap is not
used or if the filter temperature is not held below 85°C.

A practical example of this problem is found in devices
designed to remove particulates from diesel exhaust.

Some tests[2] of partitioning of soot and PAH by a particulate
trap using the sampling train described above were carried out.
The particulate trap temperature in these tests was above 165°C
and it is seen that, while approximately 90% of insoluble particulates are removed, the soluble material is removed much less

Table 1.   PAH Retention on Filter. Fraction on Filter at:

Compound	40°C	85°C	200°C
Naphthalene	0.6	0.04	0.001
Anthracene	0.9	0.7	0.05
Fluoranthene	1.0	0.8	0.4

Table 2. Diesel Exhaust Filtration

Oldsmobile 350 Diesel, Corning Glass Works Filter,
Standard Diesel Fuel, 100 Ft-lb load, Filter Temp. > 165°C

RPM	Corning Filter	Particulates mg/m^3	Extract mg/m^3
1000	No	72	7
	Yes	2.4	5
2000	No	59	~10
	Yes	7.7	7

efficiently (~ 50%). The material passing through the trap exhibited mutagenic activity at about the same level as that collected
without the trap.

After methylene chloride extraction and concentration, the
samples of soluble material are analyzed by chromatographic and
mass spectrometric techniques. Part of the sample can also be
evaluated for mutagenicity by single cell tests. The assays reported
here were developed and carried out in Prof. W.A. Thilly's laboratory
at M.I.T.[3,4]. In Table 3 the mutagenicity of some compounds found
in combustion products and possessing activity, is shown for forward
mutation of bacterial cells (Salmonella Typhimurium) and for human
lymphoblast cells.

The ratio of activity to the benzo(a)pyrene at 80 μmolar
concentration is shown for both tests. Of special interest are
the methyl anthracenes and methyl phenanthrenes. The compounds
are quite active and are found in petroleum and coal liquids and
in diesel combustion products. They are not generally a major

Table 3. Relative Mutagenicity of Soot Extract Components.

Compound	Relative Mutagenicity (a)	
	Bacteria	Human Cells
Benzo(a)pyrene	1.0	1.0
Phenanthrene	< .013	0.01
1 Methyl phenanthrene	0.5	0.2
2 Methyl phenanthrene	0.3	< .005
9 Methyl phenanthrene	0.05	0.25
Fluoranthene	1.0	0.5
Cyclopenteno (cd) pyrene	1.51	0.14
2 Methyl anthracene	0.15	0.05
9 Methyl anthracene	0.05	1.0

(a) Activities compared at 80 μmolar.
    PMS activation used in all cases.

Table 4. Mutagenicity of Cyclopenteno(cd)pyrene and
          its Epoxide.

| | Concentration Required for Significant Mutant Fraction μM | | | |
| | Bacterial Test | | Human Cell Test | |
Compound	PMS Activation	No Activation	PMS Activation	No Activation
Cyclopenteno(cd) pyrene	6	NA 40	7	NA 90
Cyclopenteno(cd) pyrene epoxide	6	0.7	NA 20	0.4

NA = Not Active

component in the combustion products of laboratory pre-mix flames.
The two isomers 9 methyl anthracene and 9 methyl phenanthrene are
notable for their higher activity in human cell assays while the
2 methyl isomers are almost inactive. Fluoranthene is a major
component of all PAH from hydrocarbon combustion and is active in
both bacterial and human cell tests. Cyclopenteno(cd)pyrene is
also found in most laboratory combustion produced PAH and can
account for a major part of activity in bacterial tests. It is
less active in human cell tests where most of the activity can
usually be accounted for by other components. Human cell tests are
relatively new and application has, so far, been limited to diesel
soot extracts. In both bacterial and human cell tests, PAH show
activity only in the presence of oxidative enzyme activator (Aroclor
PMS). Some oxygenated compounds, however, are equally or more active
in the absence of PMS. An example is shown in Table 4.

In the absence of PMS activation cyclopenteno(cd)pyrene is
seen to be inactive at 40 μM conc. for the bacterial test and at
90 μM for the human cell test while, as shown in Table 3, it is
quite active when PMS is used. When the olefinic bond is epoxidized
it is extremely active in the absence of PMS for both bacterial and
human cell tests and inactive at 20 μM for human cells with PMS
activation. While the extracted PAH from laboratory flames
require PMS activation for significant mutation, extracts from
diesel and domestic oil burner combustion gasses show very high
activitity without PMS activation and also contain polar compounds
including oxygenates. Identification and testing of these polar
compounds is an active area of research.

Significant mutation in the absence of PMS, while not
well understood, is taken as an indication of non-flame partial
oxidation within or downstream from the combustor since these
compounds have not been observed in studies of laboratory flames
where all fuel passes through the flame zone.

Table 5. Mutagenic Activity of Eluate Fraction
          of Methylene Chloride Diesel Particulate
          Extract.

1978 Oldsmobile 350 Hot Start Federal
Test Procedure

Eluate Fraction	Wt %	Activity (a)
Hexane	27	0
1:1 Hexane/toluene	6.5	40
Toluene	6.5	30
Methylene chloride	14	15
2:1 Methylene chloride/methanol	38	15
1:2 Methylene chloride/methanol	4	0
Methanol	4	0

(a) Bacterial tests with PMS activation

Studies of Combustion Systems

The projected increase in automotive diesel transportation
has triggered extensive studies of the composition and health
effects of diesel exhaust. A summary and analysis of the current
state of knowledge in this area has been published by the National
Research Council Diesel Impacts Study Committee[7]. Production
of particulates and vaporizable hydrocarbons depends on engine
design and operating conditions and on fuel and lubricant
composition. Particulates collected in a dilution tunnel account
for around 0.3 wt % of the fuel consumed with a variation of
approximately a factor of two for normal operating conditions.
The soluble fraction is typically 20 wt % of the total particulate
weight, again with approximately a factor of two variation.
This soluble material is a complex mixture of both fuel and
lubricant derived compounds which includes raw fuel, cracked
and oxidized fuel and flame produced PAH. Because of their
complexity, these mixtures must be fractioned before analysis.
Table 5 shows the results of one such fractionation using a
silica acid column.[5]

The hexane fraction contains paraffinic components and
also is responsible for 40% of the bacterial mutation activity.
The relatively small hexane/toluene fraction (6.5%) contains
most of the PAH and accounts for 30% of the mutagenic activity.
The more polar materials, while large in volume (56 % by weight)
only account for 15% of the activity. These results are with
PMS activation. Without PMS the activity of the polar fraction
would have been dominant. The PAH containing hexane/toluene

Table 6. Individual Mutagenic Contributions
          Identified in Hexane/Toluene
          Fraction.

	Wt % of Total Extract	Mutagenic Contribution (Mutant Fraction x 10^6) (a)
Benzo(a)pyrene	< 0.3	< 0.3
Fluoranthene	0.2	1.3
Methyl phenanthrene	0.24	0.6
Component Contribution		1.9-2.2
Total Methylene Chloride Extract		4.3

(a) Human Lymphoblasts with PMS Activation

fraction was analyzed by gas chromatograph-mass spectrometer.
Table 6 lists the major identified mutagens and displays an
estimate of their contribution to the total mutagenic activity,
using PMS activated human cells.

Approximately 40% of the total extract mutagenicity is accounted
for by these hydrocarbons. It is probable that inclusion of methyl-
anthracenes and PAH found in the two adjacent fractions will account
for the bulk of the mutagenicity and that, in this case, fluoranthene will be the major contributor. Cyclopenteno(cd) pyrene is
present and is a major contributor to bacterial mutagenicity but,
because of its relatively low human cell activity (see Table 3) was
not an important contributor in this sample. Continuation and
extension of the analysis and bio-assay studies to the other fractions should, in time, provide additional insight into the compounds
responsible for mutagenic activity in human lymphoblasts.

The residential oil burner consumes a fuel similar to, and
frequently interchangeable with, automotive diesel fuel. During
1979, in the United States, diesel powered highway transportation,
consumed 810,000 barrels per day of fuel, and domestic space heating
consumed the same amount, 810,000. Commercial heating, which used
similar but larger equipment, consumed an additional 430,000 B/D.
Consumption for domestic and commercial heating is concentrated in
urban and suburban areas while a major fraction of diesel fuel is
consumed by trucks in intercity highway transportation. Relatively
little attention has been given to emissions from domestic oil
burners; however, some perspective can be gained from available
information obtained using the sampling and bio-assay techniques
previously discussed and reported in (2).

Table 7. Residential Oil Burner Tests
(Flame Retention Head)

Bacharach Smoke No.	%Excess Air	Soot mg/g Fuel	Extract mg/g Fuel[1]	Bacterial Mutagenic Activity % of BaP[2]
1	24	0.006	0.003	6
5	17	0.06	0.0005	2
9	13	0.5	0.0007	0.05
5 (Cyclic)	-	0.11	0.01	1.0
Diesel	-	3.0	0.6	∿1.0

(1) The extractable fraction contains oxygenated hydrocarbons.
(2) With Aroclor (PMS). Extracts were considerably more active
     without activation.

In these studies a No. 2 heating oil containing 31% aromatics
(FIA analysis) was burned. The burner was a commercially available
(Beckett Model A1) flame retention head system which consumed 1
gallon per hour of fuel. Results of this work are summarized in
Table 7.

Soot emission from these burners is controlled by increasing
excess air to reduce the Bacharach smoke number. In the field, most
burners operate in a smoke number range of 1-4. As shown in Table 7
decreasing excess air from 24 to 17% increases the smoke number
from 1 to 5 and increases the soot production from .006 mg/g fuel
to .06, a factor of 10. A further decrease to 13% excess air
increases soot output by another factor of 10. Production of
methylene chloride soluble extract, however, decreases when excess
air is decreased below 24% and, of considerable interest, the
mutagenicity decreases drastically as soot production increases so
that adjusting a mildly smoking burner (smoke number = 5) to a
smoke number of 1 greatly increases the atmospheric loading of
mutagenic material. These measurements were made under steady state
conditions; however, the normal mode is cyclic. Results of a 5
minutes on - 10 minutes off series of tests where excess air was
adjusted for a steady state smoke number of 5 are also shown. The
particulates and extractables loadings are significantly increased
over those found for steady state operation and the specific
mutagenic activity was increased by a factor of 5.

The bacterial tests reported employed PMS activation; however,
as in the case of extracts from diesel exhaust, higher activity was
observed in the absence of the activator. This observation is
consistent with the observed presence of oxygenated hydrocarbons.

The activity shown is in the same range as diesel exhaust extract; however, the amount is lower by a factor of ∿60. Since in urban and suburban areas, especially in winter, considerably more oil is consumed for space heating than is consumed in diesel transportation, human exposure to combustion produced mutagens from oil burners is not necessarily less than from diesel engines and it is believed that this type of combustion system deserves considerably more study.

Sources of PAH and Soot

The composition of the soot and XAD-2 trap extracts indicates that these materials are a mixture of fuel and fuel derived species that have had quite different histories in their passage through the combustion system. A listing of major pathways is:

	Process	Soluble Components	Soot Formation
I	Flame	PAH	Low to moderate
II	High Temp. Pyrolysis	PAH	Moderate to very high
III	Low Temp. Pyrolysis	PAH, oxygenates	Low
IV	Vaporized Fuel and Lube	Fuel and Lube Components	None

The flame process where all fuel components pass through an active flame zone produces PAH and soot from fuel fragments. As equivalence ratio increases beyond stoichiometric, PAH are found downstream from the flame zone. Soot is formed at richer mixtures and increases rapidly at mixture ratios approaching the rich stability limit. At atmospheric pressure, soot production does not generally exceed 1% of the fuel carbon.

If hot products from rich mixture combustion are mixed with additional fuel, soot and PAH are formed by pyrolysis. In this process over 50% of the fuel carbon can be converted to soot. Under high temperature conditions relatively little PAH are formed. A similar process, of course is found in the pre-flame zone of laminar diffusion flames. In this case soot and PAH can be consumed in the flame. At lower temperatures, corresponding to combustion products that have lost heat by radiation or by mixing with cool gasses, pyrolysis and partial oxidation can produce mutagenic material without production of soot.

Finally, the presence of raw fuel or engine lubricant is frequently detected. Distillate fuels frequently contain small amounts of PAH; however, the concentration is generally low compared with the concentration found in extracts. Some products from coal

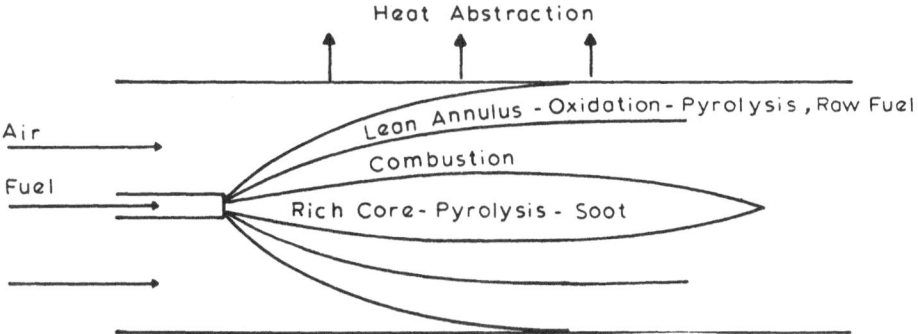

Fig. 2. Turbulent Diffusion Flame. Sources of soot and extractable
         material.

liquefaction, however, can contain large PAH concentrations.

     Figure 2 is a diagrammatic sketch of a turbulent diffusion flame
showing the zones where the processes discussed above are believed
to occur. A fuel rich core is shown which rises in temperature by
mixing with the hot combustion products surrounding this core. A
zone where the fuel-air mixture is too lean to support a stable flame
is also shown. In this zone, mixing with hot combustion products can
result in oxidative pyrolysis or, at sufficiently low temperature,
can result in escape unmodified fuel molecules. Some further
discussion of processes I and II of gasoline and non-vaporizable
fuel burning follows.

Pre-mixed Flames

     Combustion of homogeneous fuel-air mixtures, of special import-
ance in the spark ignition automobile engine and in the laboratory,
is being studied in the laminar flat flame burner and in the strongly
back-mixed jet-stirred reactor. These devices are illustrated in
Figure 3.

     Studies of composition within low pressure flat laminar flames
provide a wealth of information on the formation of PAH and soot[8,9].
Composition profiles for a 20 Torr benzene/oxygen/argon flame of
equivalence ration = 1 are shown in Figure 4.

     The species shown go through a maximum in the flame zone and
are quickly consumed before 6 mm in this non-sooting stoichiometric
flame. Higher molecular weight species also form and are consumed.
At higher equivalence ratios, hydrocarbon species and soot survive
and are found downstream from the flame zone. Of special interest
is the zone from 0 to 2 mm from the porous disk source of fuel/
oxidant mixture. Acetylene, formaldehyde and phenyl are indicated
to reach the burner surface by diffusion counter to the flowing
stream. Highly oxygenated deposits are formed on the cool burner
face. Higher molecular weight species are also formed in this cool

Stirred-Plug Flow Reactor

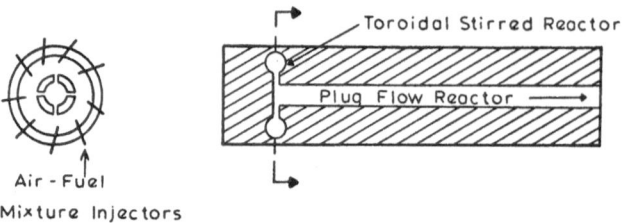

Laminar Flat Flame

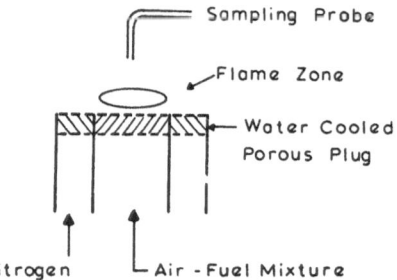

Fig. 3.   Burners for premixed flame studies.

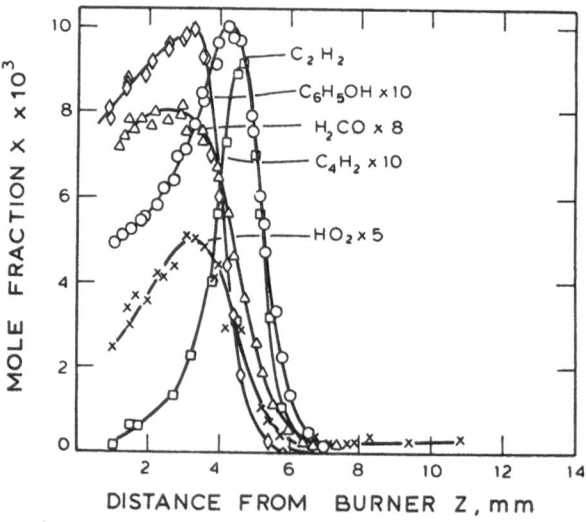

Fig. 4.   Concentration profiles.   Benzene – O_2 – 30% argon.
Unburned gas velocity = 50 cm/s.   Equivalence ratio = 1.0
Pressure = 20 Torr.

boundary zone. It has been suggested that this diffusion of flame intermediates is responsible for the solid deposits formed on cylinder walls and piston heads in spark ignition engines[10].

The spark ignition engine also produces significant concentration of PAH. Since these engines operate at approximately stoichiometric equivalence ratios and with nominally premixed and prevaporized fuel-air mixtures, laboratory flat flame studies would not predict appreciable PAH concentration in the material that passed through the flame.

The following observations from gasoline engine tests are found in [11]:

1. Fluoranthene emission is characteristically around 0.5 µg/g fuel and benzo(a) pyrene around 0.004 µg/g fuel.

2. A large amount (10 X) is found in the lubricating oil.

3. Adding BaP to the fuel increases BaP emissions but PAH are emitted when absent in the fuel.

4. Very lean and very rich operation increases PAH emissions.

It appears that, while PAH in the original fuel can contribute to exhaust emissions, PAH are also synthesized. Local rich mixtures can, of course, be responsible as in the case of the diesel engine or the residential burner. It also appears that quenching near cold surfaces could be responsible.

Fluoranthene production per gram of fuel was around a factor of 30 lower than for the diesel engine; however, the amount of gasoline burned in the U.S. in 1979 was 7.1 million barrels per day while diesel fuel used for transportation was 0.81 MB/D. This source, therefore, can not be considered insignificant for high traffic areas. Use of catalytic converters, for CO and hydrocarbon destructions, in U.S. automobiles can also greatly reduce PAH emissions when in good working order.

Samples have also been taken from a toluene-air flat flame.[2] Results from this work are compared with stirred reactor results in Table 8.

For the flat flame, a sharp maximum in PAH concentration occurs at 6 ms. The concentration after 10-20 ms is relatively constant and is lower by 2-3 orders of magnitude. The well-stirred reactor at 5 ms residual time produces PAH concentrations approximately equivalent to those found at 120 ms in the flat flame systems. Intense back mixing, therefore, gives much lower PAH than simple averaging of the concentrations found in the flat flame system.

Table 8.   PAH Production in Pre-Mixed Toluene-Air
Flames. Equivalence ratio : 1.65.
Pressure: 1 atmosphere.
PAH: μg/g fuel.

	Flat Flame at		WSR at
	2 mm (6 ms)	40 mm (120 ms)	5 ms
Phenol	300	1	1
Naphthalene	1200	4	2
Phenanthrene	100	5	3
Fluoranthene	40	3	–
Cyclopenteno(cd)pyrene	50	1	2

This is believed to be primarily the effect of lower reactant
concentrations; inherent in a WSR.
but it is also clear that quenching, pyrolysis and partial oxidation must be taken into consideration if practical flames are
to be understood.

Non-Flame Processes

    High temperature pyrolysis of hydrocarbons is the major
process for manufacturers of carbon blacks. In these processes
fuel is mixed with hot combustion products in a continuous flow

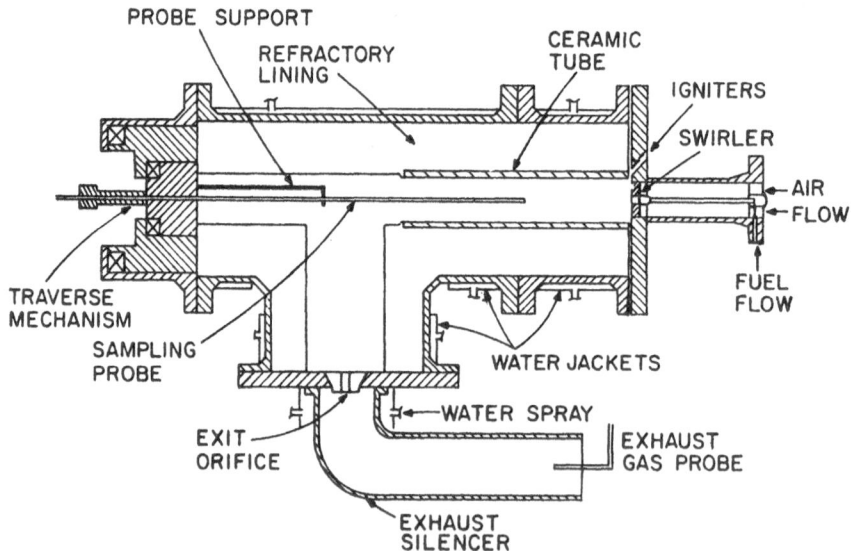

Fig. 5. Schematic cross section of experimental burner.

system. Demonstration that this process can be important in turbulent diffusion flame combustion systems can be found in [12,13]. The apparatus employed in [12] is shown diagrammatically in Figure 5.

Air entered the system through swirl vanes and diesel fuel was injected at the axis of the burner through a standard swirl type pressure atomizing nozzle. Samples were taken on the axis for determination of composition. While the overall equivalence ratio of the system was less than one, fuel concentration was much higher along the axis. The effect of changing overall fuel rate on stoichiometry and soot and PAH production at a point three diameters from the fuel injector is shown in Figure 6.

While GC/MS analysis of the methylene chloride soluble material from these laboratory premix flames shows a fairly large number of individual PAH compounds, no high molecular weight oxygenated compounds have been found. It is also found that these extracts have no significant mutagenic activity in the absence of PMS. It can therefore be concluded that the mutagenic activity observed with extracts from oil burner and diesel exhaust are caused by compounds produced by non-flame pyrolysis and partial oxidation.

With PMS activation the extracts from these laboratory flames are quite active because of their high content of mutagens (Table 9).

In this case cyclopenteno(cd)pyrene is a major contributor because of its high concentration and because of its high activity in bacterial tests (1.5 x BaP). In human cell tests

Table 9. Predicted and Measure Mutagenicity of Extract From Laminar Toluene Heptane Flame.

S. Typhimurium with PMS activation
Equivalence ratio = 1.9.

Compound	Conc. in Extract wt %	Calculated Mutagenic Contribution
Cyclopenteno(cd)pyrene	14	35×10^{-5}
Benzo(a)pyrene	4	15×10^{-5}
Fluoranthene	4	10×10^{-5}
	Total	60×10^{-5}
Measured Mutagenicity		90×10^{-5}

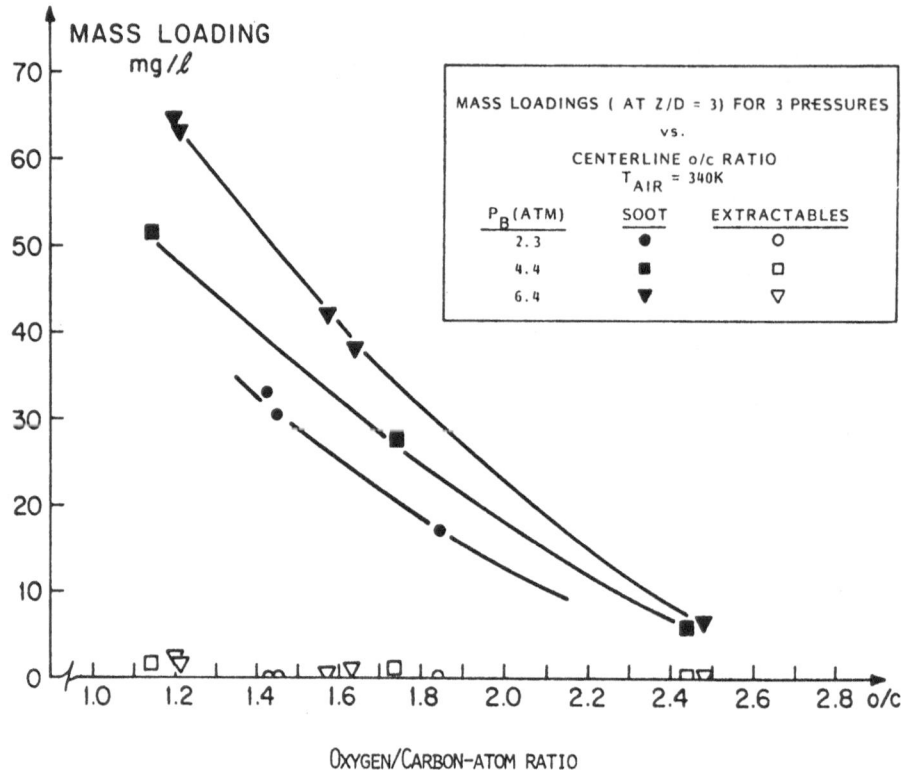

Fig. 6. Centerline soot and extractables.

it would be less important because of its lower activity (0.15 x
BaP). The three compounds shown account, within experimental
error, for the measured activity of the total mixture. Inclusion
of methyl anthracenes and methyl phenanthrenes would increase
the calculated activity. It is expected that continued research
on pre-mix flames will allow modeling and prediction of the mutagenic activity of the combustion products from these systems,

    In this figure oxygen to carbon ratio includes all species
(there was very little free oxygen). Up to $Z/D = 5$, soot
concentration along the axis increases somewhat as distance
from the fuel inlet increased, indicating that oxidation was not
rapid. Soot loadings are much higher than found in stable
pre-mix flames where conversion of 1.0% of the fuel to soot is
seldom exceeded; the maximum of 65 mg/l for this system corresponds
to 50% conversion of fuel carbon to soot. Increasing pressure
from 2.3 to 6.4 atmospheres causes only a moderate increase in
soot production. Extractables are less than 1.0% of soot
formation; however, other studies at different gas flow rates
have given extractables in the 2-10% range at low values of

Z/D. Fluoranthene production for example was in the range 500-
2000 μg/g while a total diesel exhaust extract produces 1.0 μg/g.
This is roughly in proportion to the greater soot production.
(50% vs 0.3%). It therefore appears that a relatively small
fraction of the total fuel subjected to this type of process
could contribute a major fraction of the soot and PAH.

Bacterial tests of these extracts gave no significant
activity without PMS activation, but high activity was observed
with PMS. The activity could be accounted for by summing the
activity of the identified PAH, with fluoranthene and benzo(a)
pyrene making the major contributions.

Oxygenated compounds were not found so that the formation
of oxygenated compounds, and the corresponding bacterial

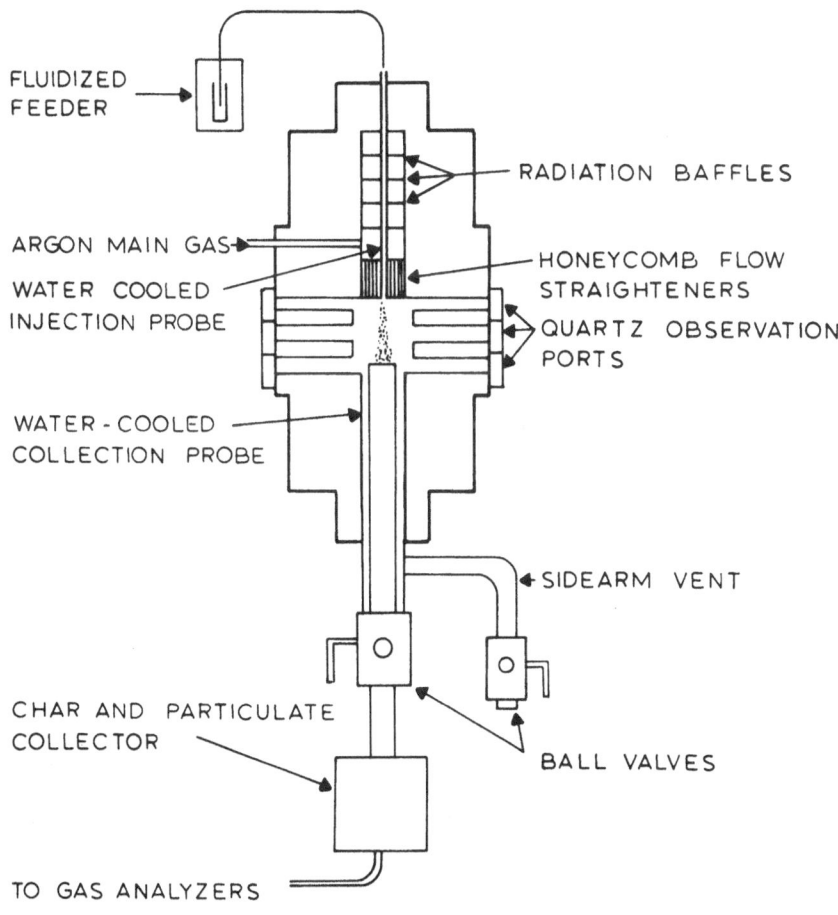

Fig. 7. Laminar flow pyrolysis furnace.

Table 10.   Single Particle Pyrolyses

Fuel	Temperature K	PAH µg/g Fuel	
		Fluoranthene	BaP + BeP
Montana Lignite	1300	900	
	1400	130	
	1760	50	
	1943	5	
Bituminous Coal	1250	450	1800
SRC-II Liquid	1273	170	70
Core of High Pressure Diesel Fuel Burner	2000+	500-2000	300-800

mutation activity in the absence of PMS, does not appear to occur in the core of this turbulent diffusion flame.

Studies of non-flame pyrolysis of coal particles and oil drops are being carried out.[2] A diagrammatic sketch of the apparatus used is shown in Figure 7.

Particles are fed to a 2 inch I.D. heated section and collected in a probe in which cold quench gas enters through a porous wall to prevent wall deposition. Char or droplet residues are removed by inertail separation, and soot and tars are captured by a filter-XAD-2 trap combination.

Table 10 presents a few results from this work which was carried out in the absence of oxygen.

PAH yields were maximized at around 1250 K and the results with Montana Lignite particles show a pronounced decrease in PAH yields as temperature is increased to approach typical combustion temperatures. Bituminous coal was found to produce comparable

Table 11.   Bacterial Mutagenicity of PAH Evolved During Pyrolysis Experiments.

Pyrolysis of:	Mutant Fraction x 10^5	
	Expected	Observed
1. SRC-II (2.9:1), 1273 K	85	47
2. SRC-II (2.9:1), 1273 K, 5% O_2	80	45
3. Montana Lignite Coal, 1300 K	38	55

Table 12. Sample PAH Yields From Practical Systems

Source	PAH μg/g Fuel	
	Fluoranthene	BaP + BeP
MIT Experimental Furnace-SRC-II	0.1	0.2
Coal Fired Pilot Plant	0.002	0.003
Chain Grate Stoker (Commercial)	0.5	0.2
Residential Stoker	90	30
Gasoline Engine without Catalytic Converter	0.05	0.004
Diesel Engine	1.0	0.2

amounts of fluoranthene and much larger amounts of the benzo pyrenes. For lignite the benzo pyrenes were not found in easily measured quantity. SRC-II liquid was obtained from the SRC-II pilot plant which liquefied bituminous coal by treatment with hydrogen in the presence of a solvent. Pyrolysis of 150 μm drops produced somewhat less PAH than were found from coal pyrolysis. For comparison, results from the previously discussed high pressure diesel fuel burner are shown. The yields of PAH from this burner are comparable to those found with coal. In this case maximum yields are shown since, further downstream, pyrolysis rapidly destroys PAH.

Determination of PAH yields from these single particle studies allowed estimation of mutagenicity of the mixture. Table 11 compares these estimates with measurement in bacterial mutation (with PMS) and demonstrates agreement within experimental error.

Comparison of Practical Systems

Some additional information on PAH emissions from practical systems is shown in Table 12.

The SRC-II fuel studied in droplet pyrolysis experiments was also burned in the MIT 3 MW experimental furnace[2]. In these experiments stage combustion for nitric oxide control was used and the sample was taken at the downstream end of the second stage. In this case, fluoranthene yield is lower by a factor of 10^4 and benzopyrene yield is lower by a factor of 10^3. In one example of a full scale coal fired power plant[14] emission of fluoranthene was reported to be 0.002 μg/g coal, a factor of 50 lower than the experimental furnace and a factor of ~ 10^5 lower than the particle pyrolysis yields.

Coal burning stokers, however, produce much larger yields of PAH[15]. As in fireplaces and hand stoked coal and wood burning stoves, large amounts of pyrolysis products escape the solid bed.

Combustion above the bed can consume these products; however, substantial emissions are generally observed in small scale equipment and for the residential stoker emissions are only about one order of magnitude less than the laboratory particle pyrolysis experiments. The gasoline engine and diesel engine examples are shown again for comparison.

General Conclusions and Observations

1. PAH in combustion products can come from the fuel, fuel non-flame pyrolysis or from quenching of rich mixture flames.

2. In heterogeneous combustion systems, high temperature non-flame pyrolysis may be an important contributor.

3. A limited number of PAH account for most of the observed bacterial and human cell mutagenicity when PMS activation is employed.

4. Techniques for distinguishing between the various sources would be useful for diagnosis and correction of combustion emission problems.

5. Modeling PAH formation in practical flames requires much more quantitative knowledge of PAH sources and post-flame reactions.

ACKNOWLEDGEMENTS

Much of the MIT work discussed in this paper is not fully published and represents a large number of individual contributions.

GC/MS Analysis and Sample Workup
    Prof. Biemann, Dr. W.A. Peters, Ms. J. Leary, Mr. E. Kruzel
Bio-Assay
    Prof. W.A. Thilly, Ms. B.M. Andon
Premixed Flame Studies
    Prof. J.B. Howard, Prof. J.P. Longwell, Dr. G. Prado,
    Dr. J. Bittner, Mr. P. Westmoreland, Mr. J. Nenninger
Coal Particle and Droplet Studies
    Prof. J.M. Beér, Dr. M. Jacques, Mr. M. Togan
Turbulent Burner Studies
    Prof. J. P. Longwell, Dr. J. Rife, Mr. W. Hall
Field Studies of Practical Systems
    Prof. J. P. Longwell, Dr. G. Prado, Mr. E. Kruzel.

    Support from the National Institute of Health Sciences, Center Grant No 5P30ESO2109, The Department of Energy, ED-1-COZ-77, and the General Motors Co. is gratefully acknowledged.

REFERENCES

1. C. Ruoff and D. Swanson, 1979, Temperature Effects
    on gas-solid partitioning of polynuclear hydro-
    carbons in the exhaust gas of benzene/air flames,
    B.S. Thesis Massachusetts Institute of Technology

2. Center for Health Effects of Fossil Fuel Utilization
    2nd Annual Progress Report, Massachusetts Institute
    of Technology, 1980

3. T.R. Skopek, H.L. Liber, J.J. Krolewski and W.G. Thilly,
    Quantitative forward mutation assay in Salmonella
    typhimurium using 8- azaguanine resistance as a
    genetic marker, Proc. Natl. Acad. Sci. U.S.A.,
    75:410-414 (1978)

4. W.G. Thilly, J.G. DeLuca, E.E. Furth, H. IV. Hoppe
    D.A. Kaden, J.J. Krolewski, H.L. Liber, T.R. Spoker,
    S.A. Slapikoff, R.J. Tizard and B.W. Penman,
    Gene-locus mutation assays in diploid human lymphoblast
    lines, in: "Chemical Mutagens", F. J. de Serres and
    A. Hollaender, Eds., Plenum Publishing Corporation,
    New York, 6:331-364, 1980

5. T.R. Barknecht, M. Yu, R.A. Hites and W.G. Thilly
    Mutagenicity of Diesel Soot Components to Human Cells
    Manuscript in preparation, Massachusetts Institute
    of Technology, 1981

6. T.R. Barfknecht, B.M. Andon, E.L. Cavalieri and W.A.
    Thilly, Mutagenicity of Cyclopenteno(cd)pyrene
    and Derivatives to Bacteria and Diploid human
    Lymphoblasts
    Manuscript in preparation, M.I.T., 1981

7. Health Effect of Exposure to Diesel Exhaust, National
    Academy Press 1981

8. J.D. Bittner, 1981, A Molecular Beam Mass Spectrometer
    Study of Fuel Rich and Sooting Benzene/Oxygen/Argon
    Flames, Sc.D. Thesis, M.I.T.

9. K.H. Homann and H.Gg. Wagner, Some new aspects
    of the mechanism of carbon formation in premixed
    flames, Eleventh Symposium (International) on
    Combustion, The Combustion Institute, 1967
    p.371-379

10. J.D. Bittner, S.M. Faist, J.B. Howard and J.P. Longwell
    Deposit Formation by Diffusion of Flame Intermediates
    to a Cold Surface, M.I.T. 1981 (to be published)

11. P.S. Perderson, J. Ingwersen, T. Nielsen and E. Larsen
    Effects of Fuel, Lubricant and Engine Operating
    Parameters on the Emission of Polycyclic
    Hydrocarbons, Environmental Science and Technology,
    14(1) (1980)

12. W.C. Hall, Soot formation in a diesel fuel spray
    flame, M.I.T., M.S. Thesis, 1981

13. G. Prado, M.L. Lee, R.A. Hites, D.P. Hoult and J.B. Howard,
     Soot formation in a turbulent diffusion flame,
     Sixteenth Symposium (International) on Combustion,
     The Combustion Institute, 1975
14. R.L. Bennet, K.T. Knapp, P.W. Jones, J.E. Wilkinson
     and P.E. Strup, Measurement of polynuclear
     aromatic hydrocarbons in stack gases. Polynuclear
     aromatic hydrocarbon, Ann Arbor Science, 1979
15. R.P. Giammar, R.B. Engdahl and R.E. Barrett,
     Emission from Residential and Small Commercial
     Stoker-Coal-Fired Boilers Under Smokeless
     Operation, EPA-600/7-76-029 (1976).

DISCUSSION

P. Cadman (The University College of Wales)

     Are there any studies on the effect of hydrogen or hydrogen
compounds such as water on the production of PAHs and soot/carbon
black in combustion?

Longwell

     It has been observed that dilution by water vapour is some-
what more effective for soot reduction than dilution by CO_2 or N_2.
Increasing the hydrogen content of a fuel reduces soot formation
for a given equivalence ratio.

S. Galant (Société Bertin)

     What about pressure effects on PAH formation rates?

Longwell

     Reliable experimental data are not available. However,
there is a general indication that conversion of fuel to soot and
PAH increases with pressure.

STRUCTURE OF SOOTING FLAMES

Jack B. Howard and James D. Bittner

Department of Chemical Engineering
Massachusetts Institute of Technology
Cambridge, Massachusetts 02139 USA

INTRODUCTION

Studies of the structure of sooting flames have provided valuable information about chemical and physical processes occurring prior to and during soot formation. These studies have included both laminar and turbulent premixed and diffusion flames, thus permitting sooting behavior to be compared and contrasted for various conditions of practical and theoretical interest. The wide range of available information is covered in recent reviews.[1-3]

The most extensive information on structure of sooting flames is that from flat low-pressure premixed flames of the type used in the pioneering work by Homann, Wagner, and coworkers.[4-9] The relatively thick and steady reaction zones of such flames combined with the simplicity of one-dimensional geometry allows the flame structure to be resolved in considerable detail, thereby providing insight into the events associated with soot formation. The focus of this paper is the occurrence of high molecular weight species prior to and during the first appearance and subsequent growth of soot particles. The molecular weights of these species extend far beyond those ($\sim$ 300 amu or less) of polycyclic aromatic hydrocarbons observable by use of conventional sampling probes and gas chromatography.[10] Such species are of course not unexpected since soot formation is a growth process, and they are intermediate in size between the fuel molecules and the smallest soot particles that have been counted under the electron microscope ($\sim$ 1.5 nm diameter or 2000 amu). However, the question remains as to whether the growth of soot occurs by (a) small species adding directly to the soot particles while in parallel forming high molecular weight byproducts,

(b) small species forming high molecular weight material, some of which serves as the reactants for soot growth while some is thermally or oxidatively decomposed and the remainder survives as byproducts, or (c) a combination of the above. Clarification of the role of the high molecular weight material would shed light on the formation mechanisms of both soot and the associated conventionally measured polycyclic aromatic hydrocarbons.

PREVIOUS OBSERVATIONS

Homann and Wagner[8] observed heavy hydrocarbons in premixed flat C_2H_2/O_2 and C_6H_6/O_2 flames burning at 2.67 kPa using both mass spectrometric analysis of gases withdrawn from the flame with a molecular beam sampling system and in situ optical absorption and emission measurements. The same species were also observed to evaporate from soot samples under high vacuum conditions in the mass spectrometer. The observed species include molecular weights up to 550 amu (the limit of the mass spectrometer used) and they first appear prior to onset of soot formation. The concentration of species larger than about 300 amu reaches a maximum close to the position where soot formation begins, which is just behind the main oxidation zone in C_2H_2/O_2 flames but within the oxidation zone of C_6H_6/O_2 flames. The heavy species then disappear soon after the onset of soot formation, in the zone where most of the soot mass is formed. These species, whose individual mole fractions were estimated to be about 10^{-7} (about 10^{10} molecules/cm^3), were considered to be intermediates or "nuclei" for carbon particles.[8]

Tompkins and Long[11] studied heavy hydrocarbons in premixed flat C_2H_2/O_2 flames at 2.67 kPa by solvent (chloroform) extraction and other analyses of material collected isokinetically through a water-cooled funnel. The mass fluxes of total polymeric material (total sample condensed in a water-cooled filter), soluble material, insoluble material, and polycyclic aromatic hydrocarbons each reaches a maximum early in the flame just beyond the blue-green zone and then undergoes an initially rapid decline. The flux of insoluble material then becomes constant while those of soluble material and polycyclic aromatics increase with increasing distance from the burner. Based on the physical appearance of the collected material, Tompkins and Long suggest that a polymer-like material is first formed which then undergoes carbonization and dehydrogenation higher in the flame. Based on studies of soot particle formation under these conditions[7,12] and recognizing the uncertainty in the effective location of the flame position sampled by the funnel of Tompkins and Long, the location of the above flux maxima is apparently close to that of the onset of soot formation. Although the formation of some polymeric material within the sampling funnel cannot be excluded, the occurrence of the strong flux maxima indicates that the amount of heavy hydrocarbon material prior to and during the early stage of soot formation exceeds the amount of soot eventually formed.

The use of optical measurements in low-pressure flat flames provides additional evidence of a system of heavy hydrocarbons molecules with a concentration maximum in the region of the flame where soot first appears. In both C_6H_6/O_2 flames[9] and a C_2H_2/O_2 flame[13] the absorption profile, in the region of the flame beginning just before the onset of soot formation and extending through the subsequent zone where most of the soot growth occurs, is found to exceed the absorption attributable to the soot particles alone. The difference is absorption by heavy hydrocarbon species and the magnitude of this excess absorption at different heights above the burner provides an approximate measure of the relative concentration of these species. By assuming the absorptivity of the heavy hydrocarbon material to be the same as that of the young soot particles, and by measuring the latter at room temperature using particles deposited on glass slides in a molecular beam sampling apparatus, Wersborg et al.[13] estimated the absolute concentration profile of the heavy hydrocarbons. While the results are only an approximation owing to the assumptions made, they do indicate that the peak volume concentration in that case is roughly comparable to the volume concentration of soot eventually formed.

Additional evidence of the heavy hydrocarbon species comes from studies of ionic effects in soot formation, a subject reviewed recently by Calcote[14]. Using molecular beam sampling with on-line mass spectrometry, Wersborg et al.[15], Homann[16], Delfau et al.[17], and Olson and Calcote[18] measured concentration profiles (relative values in the last study) of heavy ions in fuel-rich and sooting low-pressure premixed flat flames. The maximum concentration of the heavy ions occurs at about the same flame position as that of the neutral species of the same range of mass numbers, which is consistent with the picture that the ions, though only a fraction of the heavy hydrocarbons, reflect qualitative features of the latter. The concentration of the heavy ions increases strongly as equivalence ratio is increased from nonsooting to sooting values, and their mean molecular weight increases with distance from the burner in a given flame. These ions grow in the zone of soot formation to a size of several thousand mass units, the equivalent diameters being comparable to those of young soot particles. While the role of these species in soot formation is not well established, the occurrence of their peak concentration near the onset of soot formation is another example of the association of heavy hydrocarbons with the sooting process.

PRESENT STUDY

We have recently reported[19,20] measurements of species concentrations

Table 1.   Experimental Flame Conditions [pressure, 2.67 kPa (20
           torr); cold gas velocity, 0.5 m/s (298 K and 2.67 kPa)]

Equiv-alence Ratio, ϕ	C/O Mol Ratio	Inlet Conditions				
		Composition, mol %			Flux, 10^{-6} mol/cm^{2}s	
		C_6H_6	O_2	Ar	C_6H_6	O_2
1.8	0.72	13.5	56.5	30	7.26	30.4
2.0	0.80	14.7	55.3	30	7.91	29.8

in low-pressure premixed flat C_6H_6/O_2 flames under the near-sooting
(equivalence ratio ϕ = 1.8) and sooting (ϕ=2.0) conditions given in
Table 1.   The data, obtained with a molecular beam sampling apparatus equipped with an on-line quadrapole mass spectrometer, include
concentration profiles for individual species of molecular weight up
to 202 amu.   As can be seen in the accompanying discussion of this
work[21], numerous hydrocarbon species are observed which exhibit concentration maxima indicative of intermediates in the overall reaction
scheme leading from fuel to final products.   Comparison of the different mole fraction profiles reveals a general progression to
larger species having smaller mole fractions that peak farther from
the burner.

     In order to discern if this type of progression or sequential
growth might lead to the soot in sooting flames, a better understanding is required of the heavy hydrocarbons discussed above which
are in the mass range between the species whose individual concentration profiles are known and the first observable soot particles.   Of
particular interest is whether the progression to larger molecular
weights and smaller mole fractions also occurs in the mass range of
the heavy hydrocarbons, and to what extent the magnitude and rate of
change of the total mass fraction of heavy species are commensurate
with those of the initial soot.   These questions are addressed presently with data obtained by extending the use of the molecular beam
mass spectrometer to species larger than 200 amu, using the technique
outlined below.   The apparatus is described elsewhere[19,20].

     Many species larger than 200 amu are detectable under the con-
ditions of Table 1, but in the near-sooting (ϕ = 1.8) flame, most
of the signals are too low to make quantitative measurements in the
normal mode of operation.   In order to study the high-mass material
the mass spectrometer was operated without the DC voltage on the
quadrapole and thus as a high-pass filter that transmits all and
only ions with mass/charge ratio greater than a specified value.
The signal is effectively the sum of the ion currents of species
with molecular weights, M, greater than a specified low-mass cutoff,
m.   The sensitivity and mass discrimination effects in this high-
mass mode are difficult to quantify (see below), but the results are

adequate for the study of trends. By varying m within the mass
range of the high molecular weight species, trends in the shifting
concentration and molecular weight distributions can be observed as
parameters such as distance from the burner and equivalence ratio
are varied.

The mass scale for the high-mass work was calibrated up to 671
amu by observing the fragment ions from perfluorotributylamine (PTA)
$[(C_4F_9)_3N]$ at an electron energy of 25.0 eV (nominal). The high-
mass signals in the flames were measured with an electron energy of
17.0 eV (referenced to argon) to prevent fragmentation. At this
energy, the only significan fragmentation was the formation of M-1
ions from species with methyl and methylene groups. This did not
distort the relationship between the molecular weight distribution of
the neutrals from the flame and that of the ions formed by electron
impact.

DATA AND ANALYSIS

Near-Sooting Benzene Flame

High mass signals, $I_{M>m}$, from the near-sooting flame for three
values of the cutoff mass (m = 200, 450, and 700 amu) are plotted
in Fig. 1 as intensities relative to that of argon. The signal from
species larger than the cutoff mass maximizes at a certain height
above the burner. As the cutoff mass increases, the signal maxima
decrease and move further from the burner, i.e., the total number of
species above the cutoff mass decreases while the molecular weight
of the most abundant species increase. These observations alone
would be consistent with either of the following two concepts, or
both: that the heavy hydrocarbons, beginning perhaps with individually measurable polycyclic aromatics between 100 and 200 amu, react
among themselves so as to grow in size while decreasing in number;
or that the heavy hydrocarbons are destroyed by oxidation, with the
smaller species being preferentially consumed. These concepts are
further discussed later.

The signal from species larger than 200 amu, $I_{M>200}$, maximizes
at 8.4 mm. This is about the same location where the mole fraction
of mass 202 ($C_{16}H_{10}$) maximizes[21], thus implying that much of the
M>200 material is in a narrow mass range beginning at 200 amu. The
difference between the signals from species with masses greater than
200 amu, $I_{M>200}$, and those greater than 450 amu, $I_{M>450}$, is a relative measure of the number concentration of species between 200 and
450 amu. Comparison of this difference with $I_{M>450}$ and $I_{M>700}$ in
Fig. 1 reveals that, in the region between the burner surface and
about 8 mm, the material with molecular masses above 450 amu is a
small molar fraction (less than 5%) of the material between 200 and
450 amu, and the material with masses above 700 amu is even less.

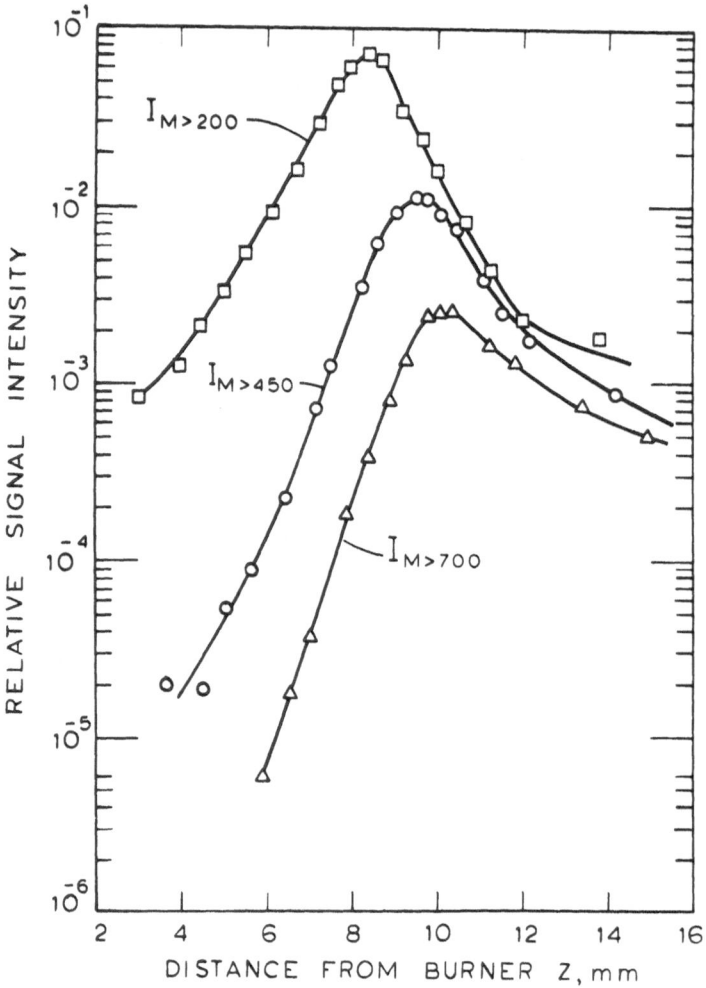

Fig. 1.   Relative Intensity of High-Mass Signals as Function of
          Distance from Burner for Different Cutoff Mass Numbers
          in Near-Sooting Benzene/Oxygen Flame [Equivalence
          ratio, 1.8. Other conditions, see Table 1.]

However, at distances beyond 10 mm, material with masses above 450 amu is a significant molar fraction (about 70%) of the total material with masses greater than 200 amu.

The distribution of high molecular weight species was investigated more extensively at several distances z from the burner by measuring the signal $I_{M>m}$ while increasing the low-mass cutoff in 50 amu increments between 200 and 750 amu. The results are 12 profiles of the type, and including those, shown in Fig. 1. The various $I_{M>m}$ values at each of six distances from the burner between 7.2 and 12.1 mm have been normalized by dividing by $I_{M>200}$ at each

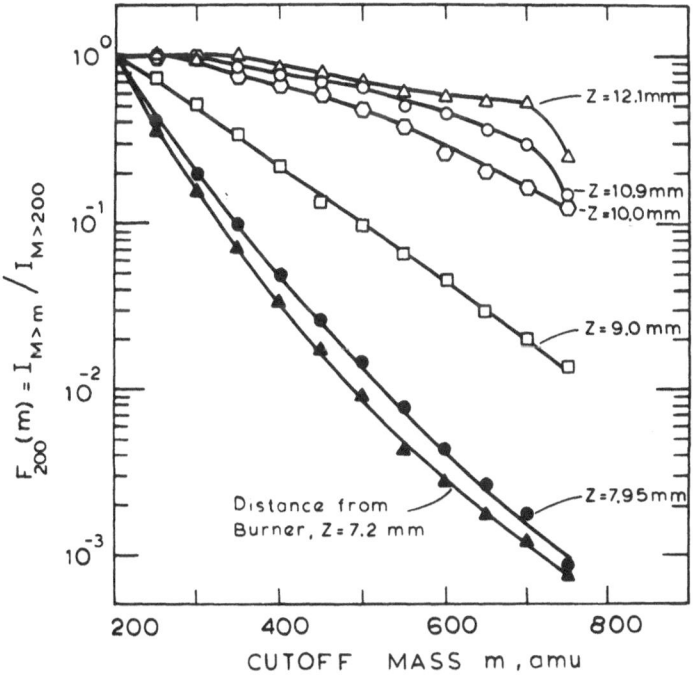

Fig. 2.   Cumulative Distribution Function of High-Mass Signal as
           Function of Cutoff Mass Number at Different Positions in
           Near-Sooting Benzene/Oxygen Flame [Equivalence ratio,
           1.8.   Other conditions, see Table 1.]

distance, and are shown in Fig. 2. To the extent that the relative
signal intensity is a measure of the relative number concentration,
the curves in Fig. 2 are cumulative distributions of the molecular
mass m for species larger than 200 amu. Thus the value of the cumulative distribution function

$$F_{200}(m) = \frac{I_{M>m}}{I_{M>200}} \tag{1}$$

at a point on one of the curves represents the number of species
larger than mass m expressed as a fraction of the total number of
species larger than 200 amu. As distance from the burner increases,
the fraction of the M>200 amu material that is from the higher
masses (e.g., >700 amu) increases several orders of magnitude.

The derivative with respect to mass of the cumulative distribution function gives the probability density function (p.d.f.)

$$f_{200}(m) = \frac{-dF_{200}(m)}{dm} \tag{2}$$

Accordingly, $f_{200}(m)dm$ represents the total number of M>200 amu
species that is contributed by species of mass m, and

$$\int_{200\ amu}^{\infty} f_{200}(m)dm = 1 \tag{3}$$

The probability density functions for five distances from the
burner are shown in Fig. 3. As expected from the cumulative distribution functions, the probability density function at 7.95 mm
is heavily weighted toward the mass range 200-300 amu. As the distance from the burner increases, the p.d.f. increases at the higher
masses and decreases at the lower masses. At distances above 10.0
mm, the signal between 200 and 300 amu is a small to negligible
fraction of the total, and the distributions are relatively uniform in the range 300-600 amu. At 10.9 and 12.1 mm, the distributions turn upward at the far right indicative of a faster growth
of the higher-mass species, but it is not known if this apparent
behavior is real or an artifact of the data at this high-mass extreme of the mass range studied.

These observations show that although the number density of
high molecular weight species (M>200 amu) is decreasing with distance from the burner beyond 8.4 mm (Fig. 1), the mean molecular
weight is increasing. In order to determine how the total mass of
heavy species changes as the number decreases, the p.d.f. has been
multiplied by m to convert to a mass basis and by

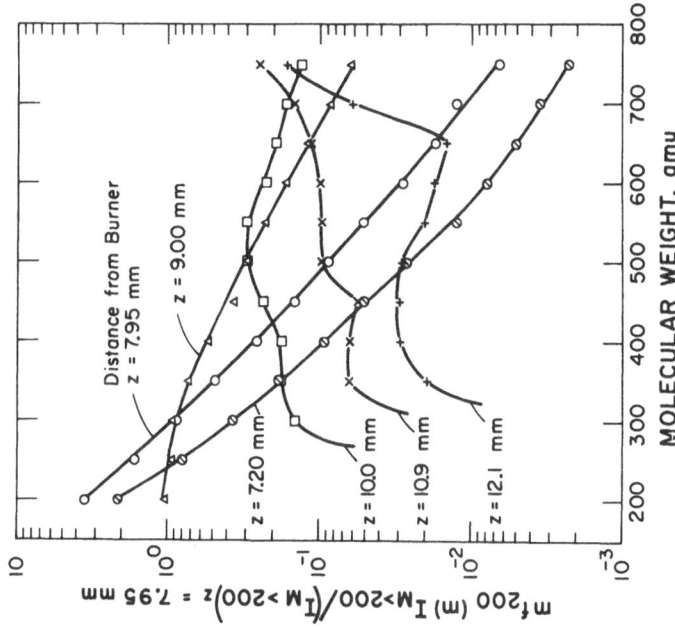

Fig. 4. Mass of Material as Function of Molecular Weight at Different Heights above Burner, Normalized with Number of Moles of M > 200 amu Material at 7.95 mm above Burner [Benzene/Oxygen flame. Equivalence ratio, 1.8. Other conditions, see Table 1].

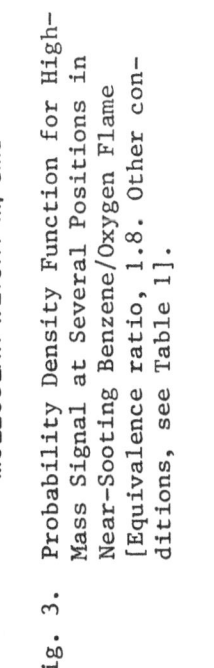

Fig. 3. Probability Density Function for High-Mass Signal at Several Positions in Near-Sooting Benzene/Oxygen Flame [Equivalence ratio, 1.8. Other conditions, see Table 1].

$$I_{M>200}/(I_{M>200})_{z=7.95mm}$$

to normalize the mass with the constant number of moles of heavy
material existing at an arbitrarily fixed distance (z = 7.95 mm)
from the burner. The resulting function,

$$mf_{200}(m)\, I_{M>200}/(I_{M>200})_{z=7.95mm}$$

is shown in Fig. 4 for different distances from the burner. A point
on one of the curves represents the mass of species of the indicated
molecular weight existing at the indicated distance from the burner,
expressed as a ratio to the total number of moles of M>200 amu material existing at 7.95 mm. Figure 5, which is a cross plot of Fig.
4, shows how the same function varies with distance from the burner
for different molecular weights.

It can be seen in Fig. 5 that the mass of material at a given
molecular weight exhibits a peak at a height above burner which increases with increasing mass number. The shapes of the curves and
the location of the peaks is the same as would be seen if number of
moles instead of mass were plotted, since the number curves could be
produced by vertical displacement of each mass curve in accordance
with division by the constant molecular weight indicated thereon.
In comparison with mole fraction profiles for individual species of
m less than or equal to 200 amu, the curves in Fig. 5 constitute a
continuation of the progression already seen for the individual
hydrocarbon species, namely molar concentration profiles peaking at
larger distances from the burner and with smaller peak values as m
increases. Figure 5 shows that the peak mass concentrations also
decrease as m increases.

Figure 4 shows that the mass distribution of the heavy material
shifts from being strongly biased toward the smaller species in the
interval 7.20-7.95 mm to being essentially depleted of species
smaller than about 300 amu at 10.0 mm and beyond. The mass distribution becomes relatively uniform over a wide but decreasing range
of molecular weights, e.g., 300-750 amu at 10.0 mm and 350-650 amu
at 12.1 mm. The level of the approximately uniform portion of the
distributions decreases with increasing distance from the burner,
meaning that the heavy species in this molecular weight range are
being consumed, and with a net mass rate that is almost independent
of molecular weight. However, at the high molecular weight end of
the distributions the curves turn upward at 10.9 and 12.1 mm, apparently reflecting a slower net rate of mass consumption of the
largest species ($\sim$ 650 amu and larger).

An apparent kink in the curves at about 400, 450, and 650 amu
for 10.0, 10.9, and 12.1 mm, respectively, and the emergence of the
upward slope in the region of the largest mass numbers at the high-

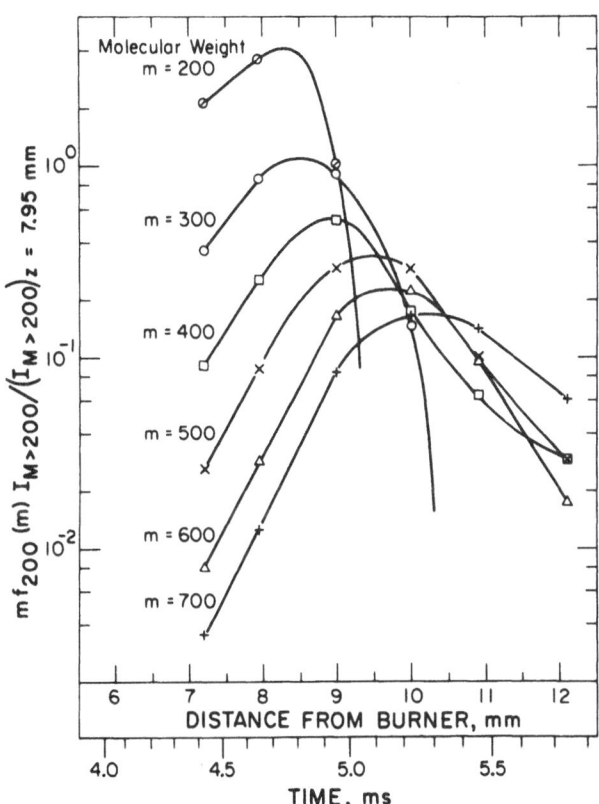

Fig. 5. Variation with Distance from Burner of Mass of Material
in Molecular Weight Range m to m + dm at Different Molecular Weights, Normalized with Number of Moles of M > 200
amu Material at 7.95 mm above Burner [Benzene/Oxygen
flame. Equivalence ratio, 1.8. Other conditions, see
Table 1.]

est heights above the burner give the appearance of a bimodal distribution. Thus, one could draw in Fig. 4, for example at 12.1 mm,
two distributions with maxima at about 450 amu and 750 amu or beyond,
the sum of which would give the observed curve. Because of experimental uncertainties, we do not know if this bimodal behavior is
real. Since the heavy material is being destroyed in a region of the
flame where growth can also occur, the observed change in the distribution with distances from the burner is presumably the net effect
of destruction and growth, and the apparent evolution of the two modes
may reflect different relative contributions of these opposing processes within the two different regions of the molecular weight
spectrum. Accordingly, the change with height above the burner of
the first (lower-mass number) mode of the distribution might reflect
primarily destruction while the behavior of the second mode might be
strongly influenced by growth processes. To speculate further, the
relatively slow net consumption of the highest molecular weight material in this nonsooting flame might under sooting conditions become
net production, with the second (higher-mass number) mode of the
distribution being an early stage of what would become the soot particle system if the equivalence ratio were high enough.

The mean molecular weight and overall relative mass and number
concentrations of the high-mass material are calculated as follows.
Since $mf_{200}(m)dm$ is the total mass of species of molecular mass m,
per mole of all species exceeding 200 amu, the mass mean molecular
weight of the M>200 amu material is

$$\bar{M}_{M>200} = \int_{200amu}^{\infty} mf_{200}(m)dm \tag{4}$$

Using, from above, $I_{M>200}/(I_{M>200})_{z=7.95mm}$ as the ratio of the moles
of M>200 amu material at any distance z from the burner to that at
7.95 mm, the molar and mass concentrations of all the M>200 amu
material at any z, relative to those at 7.95 mm, are given as follows:

$$\frac{C}{C_{z=7.95mm}}(molar) = \frac{I_{M>200}}{\left[I_{M>200}\right]_{z=7.95mm}} \tag{5}$$

$$\frac{C}{C_{z=7.95mm}}(mass) = \frac{\bar{M}_{M>200}}{\left[\bar{M}_{M>200}\right]_{z=7.95mm}} \left[\frac{C}{C_{z=7.95mm}}(molar)\right] \tag{6}$$

The results calculated with Eqs. 4-6 are shown in Fig. 6. As
distance from the burner increases over the interval studied here,

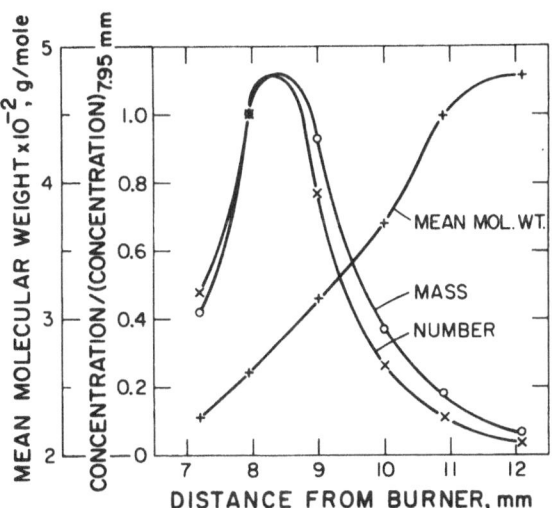

Fig. 6.   Variation of Mean Molecular Weight and Mass and Number Con-
centrations of M > 200 amu Material with Distance from
Burner [Benzene/Oxygen flame. Equivalence ratio, 1.8.
Other conditions, see Table 1].

the total number of moles and the total mass of the heavy hydrocarbon material proceed through maximum values at about 8.3 and 8.5 mm,
respectively, and then decrease to values at the largest distance
(12.1 mm) that are factors of about 30 and 20, respectively, smaller
than those at the peaks. At the same time the mean molecular weight
$\bar{M}_{M>200}$ increases from about 230 amu at 7.20 mm to about 480 amu
at 12.1 mm. Whether the maximum $\bar{M}_{M>200}$ is truly located as depicted
in Fig. 6, or slightly before or beyond 12.1 mm cannot be discerned
from these data.

Sooting Benzene Flame

     The behavior of the high mass signal was studied as the fuel
equivalence ratio was increased beyond the sooting limit of $\phi = 1.9$
while the burner chamber pressure, cold gas velocity, and mole percent argon in the unburned mixture were maintained constant at the
values used in the near-sooting $\phi = 1.8$ flame. The high mass signal for species larger than 700 amu, $I_{M>700}$, at $\phi = 2.0$ (a sooting
flame; conditions in Table 1) is plotted as a function of the distance from the burner in Fig. 7. For comparison, the high-mass
signals $I_{M>200}$ and $I_{M>700}$ from the near-sooting flame are also
shown. The $I_{M>700}$ signal for $\phi = 2.0$ increased by a factor of 10^4
over the distance of 6 mm to a maximum at 12.7 mm from the burner.
This corresponds to the boundary (Fig. 7) between the blue zone and
the orange zone that is characteristic of the continuum emission
from soot. This maximum is 100-times higher than the $I_{M>700}$ signal
maximum for the $\phi = 1.8$ flame, and even larger than the $I_{M>200}$ signal maximum for the $\phi = 1.8$ flame.

     To investigate the effects of the increased fuel equivalence
ratio on the flame structure, the mole fraction profiles of O_2, C_2H_2,
and C_6H_6 were measured in the sooting flame. These profiles and
those from the near-sooting flame are shown in Fig. 8. The dotted
lines are the locations of the $I_{M>700}$ maxima of Fig. 7. As was the
case in the near-sooting flame, the mole fractions of low molecular
weight aromatics in the sooting flame are expected to disappear as
that of benzene does. Since the polyacetylenes are equilibrated
with H_2 and C_2H_2 in the near-sooting flame[19,20], their mole fraction
profiles are expected to follow the C_2H_2 mole fraction profile
through a similar relationship in the sooting flame. Because of the
competition between the chain branching reactions

$$H + O_2 \; \overset{\rightarrow}{\underset{\leftarrow}{}} \; OH + O \tag{R1}$$

$$O + H_2 \; \overset{\rightarrow}{\underset{\leftarrow}{}} \; OH + H \tag{R2}$$

and the hydrocarbon oxidation reactions that consume radicals, the
relative amounts of O_2 and C_2H_2 are indicative of the concentrations
of the oxidizing species O and OH.

     At any fixed distance from the burner, the O_2 mole fraction in
the $\phi = 2.0$ flame is higher than that in the $\phi = 1.8$ flame because
the flame front has broadened and moved away from the burner at the
increased ϕ. In the region just prior to the maximum of $I_{M>700}$ (8
to 10 mm for $\phi = 1.8$; 10 to 13 mm for $\phi = 2.0$), the mole fraction
of benzene is the most sensitive of the three species to changes
in ϕ. Therefore, the exact shape of the benzene profile is difficult to measure in this region. At 10 mm in the $\phi = 1.8$ flame a
change in benzene flow rate of 0.7% resulted in a 43% increase in
the benzene signal.

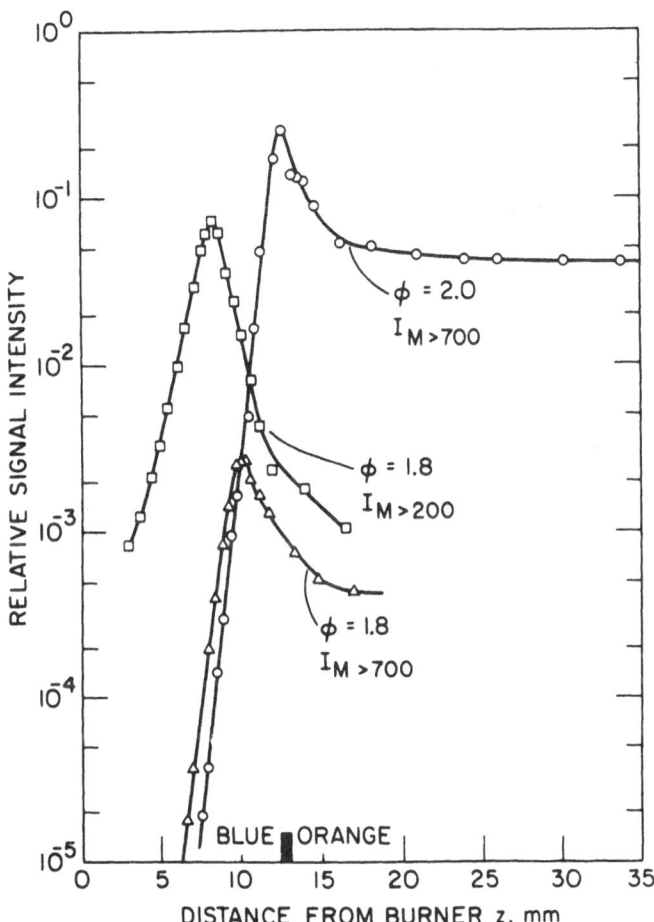

Fig. 7. Relative Intensities of High-Mass Signals as Function
of Distance from Burner in Near-Sooting ($\phi = 1.8$) and
Sooting ($\phi = 2.0$) Benzene/Oxygen Flames [Conditions, see
Table 1.]

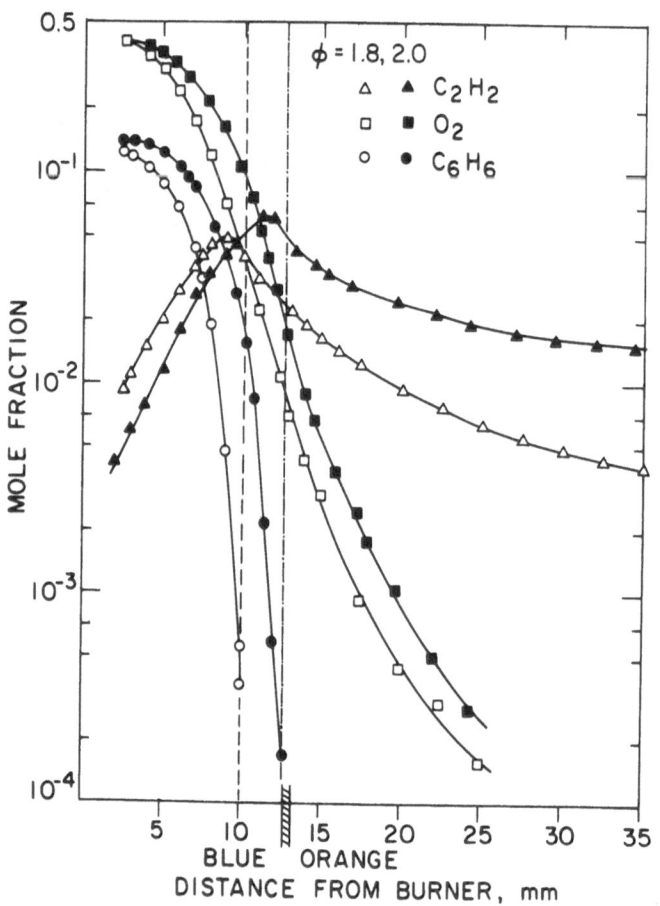

Fig. 8. Mole Fractions of C_6H_6, O_2, and C_2H_2 vs. Distance from Burner in Near-Sooting (ϕ = 1.8) and Sooting (ϕ = 2.0) Benzene/Oxygen Flames [Conditions, see Table 1.]

There is much similarity in the shapes of the $I_{M>700}$ curves
in Fig. 7. The curve for ϕ = 2.0 relative to the one for ϕ = 1.8 is
higher, indicative of about 100-times larger number concentrations
of M > 700 amu species at equivalent flame positions, and displaced
downstream as a consequence of decreasing flame speed with increasing
ϕ under fuel rich conditions. (The temperature profile and time
scale change little as ϕ is increased from 1.8 to 2.0 since the
effect of lowering the adiabatic flame temperature is offset by a
smaller heat loss to the burner.) Both curves exhibit a strong peak
soon after a linear upstream segment which has essentially the same
slope in both cases. Downstream of the peaks the signals decrease
rapidly over a distance of 3 or 4 mm and only slowly thereafter.

DISCUSSION

Diffusion and Reaction of Heavy Species

The linear segments of the $I_{M>m}$ curves immediately upstream
of the peaks are consistent with the appropriate interpretation that
the heavy species are not reacting in that region, i.e., their diffusion toward the burner is balanced by convection, the net flux
being zero. Accordingly, if C and D are the molar concentration and
effective diffusion coefficient of the heavy species in a certain
range of molecular weights, then the slope of the linear segment
immediately upsteam of the peak, assuming C is proportional to the
signal intensity, is

$$d \ln C/dz = v/D \qquad (7)$$

where v is the bulk gas velocity. Thus, neglecting the relatively
small effects on v and D of changes in temperature, number of moles
and gas composition in the region of interest here, the slope should
increase with decreasing D and hence with increasing molecular
weight. Such behavior is indeed seen in Figs. 1, 5, and 7. Furthermore, the calculated values of D for the different molecular
weights represented in Fig. 5 are in good agreement with estimations
using the semiempirical correlation of Fuller et al.[23] with the
temperature dependence recommended by Sherwood et al.[24], and assuming
the species to be polycyclic aromatic hydrocarbons. For example,
using the data and Eq. 7, the value for 200 amu is 0.07 cm^2/s (1 atm
and 300 K), while the estimated value for pyrene (202 amu) is 0.06
cm^2/s. At the higher molecular weights the observed values of D
vary approximately as $M^{-2/3}$ in agreement with the trend estimated
from the semiempirical correlations. Therefore, as discussed above,
the heavy species are not reacting in the linear region of the
$I_{M>m}$ signals upstream of the maxima. This conclusion apparently
applies not only to the ϕ = 1.8 data which are the primary basis for
the above analysis but also to the ϕ = 2.0 profile in Fig. 7 which
agrees in both degree of linearity and slope with the equivalent

profile ($I_{M>700}$) for the near-sooting case.

In the region of rapid consumption immediately downstream of the concentration maximum, inspection of Fig. 7 shows that the $I_{M>700}$ curves decrease with approximately constant slope for a considerable portion of the consumption, and that this slope is approximately the same for both equivalence ratios even though the concentrations are about 100-times larger in the sooting flame. In addition to the slope, D and v are also approximately constant in this region and their respective values in the two flames differ little. Under these conditions the implied reaction rate $\dot{R}$ (moles/cm^{3}s) per unit concentration C (moles/cm^3) is approximately constant and given by

$$\dot{R}/C = (v - D \ d \ \ln \ C/dz)(d \ \ln \ C/dz) \qquad (8)$$

Thus the approximately constant slope (d ln C/dz) implies a net consumption process approximately first-order in the concentration of species being consumed.

Therefore after building up to a maximum concentration the M>700 amu material then undergoes approximately first-order net decomposition with only a small difference in the fractional rate $\dot{R}/C$, s^{-1}) under the greatly different concentration conditions of the near-sooting and sooting flames. Inspection of the other $I_{M>m}$ profiles in Figs. 1 and 7 reveals for each cutoff mass an approximately linear segment downstream of the peak similar to the foregoing for M = 700 amu. However, the slopes are seen to decrease as m increases. Therefore, according to the above discussion and Eq. 8, the heavy material defined by each cutoff mass seems to undergo downstream of the concentration maximum a net consumption that is approximately first-order in the species consumed, with a fractional rate that decreases as the cutoff mass increases. This apparent effect of molecular weight is suppressed by the mass cutoff data which represent the lumped behavior of material having a wide range of molecular weights. The profiles of normalized differential mass concentrations in Fig. 5 reveal a much stronger decrease of fractional consumption rate with increasing molecular weight than is obvious from the $I_{M>m}$ profiles.

The approximate first-order consumption undergone by the high-mass material would seem to exclude second-order growth reactions of these species with each other as being a major part of the observed behavior not only because of the difference in order but also because the observed consumption involves the decrease of total mass, not just number concentration. However, this type of growth may contribute to the decrease in number being somewhat faster than the decrease in mass, i.e., the increase of average molecular weight (Fig. 6). If one considered only the heavy species data, the consumption behavior might be attributed to first-order thermal decomposition or pyrolysis, i.e., fragmentation of the carbon network and

production of low molecular weight species such as C_2H_2. Such a process might include H-atom addition to a ring and ring fragmentation,
or H-atom abstraction of H with the product radical decomposing to
lower molecular weight carbon fragments. For example, Tompkins and
Long[11] attributed to carbonization and dehydrogenation the decreasing flux of total polymeric material and chloroform insoluble material which they observed as mentioned above. However, pyrolysis alone
cannot account for the present observations. The only manner by
which pyrolysis can remove mass from the M>200 amu category would be
the formation of M<200 amu products, but the total mass of such species does not increase as the heavy species are consumed in the ϕ =
1.8 flame. In fact, inspection of Fig. 5 in the region around 10 mm
where the M>700 material peaks shows that m=200 amu and m=300 amu
species are rapidly disappearing beyond 10 mm (data points not shown
at 10.0 and 10.9 mm are off scale). A similar rapid decrease can be
seen for C_6H_6 in Fig. 8 and for polycyclic aromatic hydrocarbons
smaller than 202 amu whose individual mole fractions have been measured[19-22]. Nevertheless, a contribution by pyrolysis or oxidative
pyrolysis to the heavy species consumption cannot be excluded since
the M<200 amu fraction of the products from such decomposition would
quickly be destroyed by the same, presumably oxidation, processes
that account for the rapidly decreasing concentrations of the benzene
and other species less than 300 amu mentioned above. Further considerations described below suggest that oxidation alone could account
for all of the observed consumption of heavy species.

Oxidative Consumption of High-Mass Material

As discussed above, the peak concentration and rapid net consumption of M>700 amu species begins at the start of the orange luminous zone in the sooting flame and at the location of the impending
start of the orange luminous zone in the near-sooting flame. Since
the onset of soot formation is often located near the start of the
orange or yellow luminous zone and is often referred to as occurring
after the oxidation zone in premixed flames, the question may arise
as to whether the amount of oxidation occurring in the zone of rapid
net consumption of M>700 amu material is large enough to be significant in this consumption. Homann et al.[4] found that a considerable
fraction of the initial oxygen remains as O_2 at the start of the
yellow-orange luminous zone of a low-pressure sooting C_6H_6/O_2 flame.
In the present study, fluxes of O_2, C_2H_2, and C_6H_6 as given in Table
2 were calculated using the quasi one-dimensional flame equation[25]
and measured profiles of the species mole fractions, temperature,
and area expansion ratio[22]. The effective diffusion coefficients in
the mixture were calculated for the individually identified species
using the equation given by Fairbanks and Wilke[26] with binary diffusion coefficients calculated from Lennard-Jones potential parameters
as described by Fristrom and Westenberg[27]. Effective diffusion coefficients for the high-mass material were obtained from the measured
signal profiles in the region upstream of the peak signal, where

Table 2. Fluxes of Oxygen and Carbon as Different Species at Beginning and End of Zone of Rapid Consumption of M > 700
amu Species in Nearly Sooting and Sooting Benzene/Oxygen
Flames [Beginning (a) and end (b) of zone arbitrarily
defined as location of (a) peak signal intensity and (b)
transition from relatively fast to relatively slow decrease of signal intensity (see Figure 7 and distances
(mm) from burner indicated in table)].

Species	Flux of Oxygen or Carbon, % of Oxygen or Carbon Fed					
	Nearly Sooting (φ=1.8)			Sooting (φ=2.0)		
	Begin. (10.1mm)	End (17.0mm)	Decrease (6.9mm= 1.74ms)	Begin. (12.6mm)	End (19.6mm)	Decrease (7.0mm= 1.84ms)
Oxygen as O_2	37.2	1.1	36.1	18.3	0.69	17.6
Carbon as M > 700amu Species	0.006	0.001	0.005	0.6	0.1	0.5
Carbon as C_2H_2	29.8	6.8	23.0	32.1	9.8	22.3
Carbon as C_6H_6	2.6	0	2.6	1.0	0	1.0

essentially only diffusion occurs as described above. The profiles
of temperature and area expansion ratio measured for the φ = 1.8
flame were assumed to apply also for the φ = 2.0 flame when the profiles of the latter are transposed together so as to align the peak
of the M>700 amu signal, and hence (see Fig. 8) the location of
C_6H_6 disappearance and the peak C_2H_2 mole fraction, with those of
the φ = 1.8 flame.

The fluxes of M>700 amu material were calculated in a similar
manner after converting the high-mass signal intensities ($I_{M>700}$)
which are relative to argon, to mole fraction ($X_{M>700}$) using the
measured argon mole fraction and the conversion $X_{M>700} = 8x10^{-4}$
$X_{Ar}I_{M>700}$. The coefficient $8x10^{-4}$ was obtained by calibrating the
high-mass signal with the data on individual species, including
species mole fraction and the first derivative of cumulative mole
fraction with respect to molecular weight, at 202 amu, where the
two different types of measurements overlap. This number repre-
sents $1/\alpha_{i,Ar}S_{i,Ar}$, where $\alpha_{i,Ar}$ is the mass discrimination factor
of the sampling instrument and $S_{i,Ar}$ is the sensitivity of the mass
spectrometer, both for species i relative to argon. From measurements of this coefficient at several heights above the burner, the
above value is reproducible to within a factor of 1.5. Like the

accuracy of the individual species data[22] used in the calibration, the above value is accurate to within a factor of 2 at the lower end of the range of mass numbers studied (200-750 amu). The accuracy at higher mass numbers is only approximately known because the mass discrimination of the filter in the high-mass mode is not well established. However, the main factor is the decreasing gain of the electron multiplier as molecular weight increases, the effect of which is offset by beam discrimination. Therefore the effects of increasing mass number on $\alpha_{i,Ar}$ and $S_{i,Ar}$ are partially compensating. The above coefficient is estimated to be accurate to within a factor of about 3 at mass numbers up to 750 amu.

It is seen from the flux values in Table 2 that the fraction of the initial oxygen remaining as O_2 at the beginning of the zone of rapid consumption of M>700 amu material is 37.2% and 18.3% for the sooting and near-sooting flames, respectively. The occurrence of these large fluxes with only small O_2 mole fractions (Fig. 8) reflects rapid diffusion associated with the steep mole fraction gradients. Since the O_2 remaining at the end of the zone of rapid consumption of M>700 amu material is about 1% of the initial value in both flames, the amount of oxidation occurring within this zone is indeed substantial, thus allowing the possibility that oxidation plays a role in the destruction of the high-mass material or its pyrolytic decomposition products.

Overall Stoichiometry up to High-Mass Peak

Calculation of other species fluxes at the location of the peak M>700 amu signal reveals additional information about the overall stoichiometry of the reactions occurring up to that position in the two flames. Thus at 10.1 mm in the near-sooting flame, the fractions of the initial oxygen represented by the fluxes of the major oxidation products are: CO, 0.35; CO_2, 0.10; and H_2O, 0.20. The slight excess of the sum of these values over the oxygen consumption of 0.63 represents experimental error. The next-largest flux of oxygen (-0.003) is that carried by OH. Although minor in the mass balance, this flux is noteworthy in that it is directed upstream, presumably reflecting OH consumption by hydrocarbons relative to its generation by branching reactions (R1 and R2) being larger upstream of 10.1 mm where about 97% of the fuel is consumed. The fractions of the original carbon represented by species fluxes at 10.1 mm are: CO, 0.49; CO_2, 0.07; C_2H_2, 0.30; C_4H_2, 0.08; and C_6H_6, 0.03. All other species each carry less than 0.006 of the carbon, the main ones being C_3H_3, CH_4, and CH_3, and together represent about 0.02 of the carbon. The carbon flux as all aromatics except C_6H_6, including the M>200 amu material, is only 3×10^{-4} of the total carbon.

At the equivalent position (12.6 mm) in the sooting flame, the more limited data permit the direct calculation of only the fluxes shown in Table 2. Nevertheless, some conclusions about the distri-

bution of oxygen and carbon among the species that dominate the
material balances can be drawn as follows, by consideration of con-
straints imposed by these few known fluxes together with trends
observed in the fluxes of other species when ϕ is increased from
1.0 to 1.8[22]. Given the O_2 flux of about 0.18 of the original oxygen, the fraction of the oxygen carried by CO, CO_2 and H_2O must be
about 0.82, since oxygen in OH and other species is negligible in
these balances. Similarly, since the sum of the fractions of the
original carbon carried by C_2H_2 (0.32), C_6H_6 (0.01), and M>700 amu
species (0.006) is about 0.34, the total carbon in CO, CO_2, and the
other hydrocarbons must be about 0.66. The part of this remaining
carbon that can be attributed to the other hydrocarbons is constrained by the fraction of the carbon in CO and CO_2 which in turn
is constrained by the fraction of oxygen in CO and CO_2 and the CO/
CO_2 ratio. From a hydrogen balance using the fluxes of C_2H_2 and
C_6H_6, the fraction of the total hydrogen in H_2O, H_2, and other
hydrocarbons is 0.67. In the ϕ = 1.8 flame at 10.1 mm, the H_2 flux
is 0.2-times that of C_2H_2, and as ϕ increases from 1.0 to 1.8 the
H_2 mole fraction increases approximately 6-fold at similar stages
of fuel consumption in these two flames. The C_2H_2 flux at the
position of the M>700 amu peak increases by only a few percent in
going from ϕ = 1.8 to ϕ = 2.0 (Table 2). Therefore, the H_2 flux
at 12.6 mm in the ϕ = 2.0 flame should be at least 0.2-times that of
C_2H_2, which means that H_2 carries at least 0.06 of the total hydrogen, leaving no more than 0.61 for H_2O and other hydrocarbons. Since
the H/O feed ratio is 0.8, the foregoing means that the fraction of
the total oxygen in H_2O must be less than 0.8x0.61/2 or 0.24. This
fraction from 0.82 leaves 0.58 or more for the fraction of the total
oxygen in CO and CO_2. In order to place a limiting value on the CO/
CO_2 flux ratio, we note that this quantity increases approximately
10-fold at similar stages of fuel consumption as ϕ increases from
1.0 to 1.8. Since the ratio is 7 at 10.1 mm and ϕ = 1.8, we assume
the value at 12.6 mm and ϕ = 2.0 is at least 7 and probably larger.
Accordingly, the above 0.58 or more of the total oxygen as CO and
CO_2 would be equivalent to 0.64 or more of the carbon as CO and CO_2,
leaving 0.02 or less of the carbon for hydrocarbons other than C_2H_2,
C_6H_6, and M>700 amu material.

The 0.02 figure would seem to be conservatively large based on
the assumptions employed about the CO/CO_2 and H_2/C_2H_2 flux ratios.
On the other hand, the figure would seem to limit the C_4H_2 flux to
too small a value based on the value observed at ϕ = 1.8. This
discrepancy may reflect error accumulated by taking flux differences,
which may be several percent of the carbon. Therefore we only conclude that the hydrocarbons other than C_2H_2, C_6H_6, and M>700 amu
material at the M>700 amu peak concentration appear to represent no
more than a small fraction of the carbon fed. Definite specification
of the upper limit will require a more complete set of species flux
data for this sooting flame.

If the ratio of the mass flux of M>700 amu species to that of
the M>200 amu species at the M>700 amu concentration peak is the
same in both the ϕ=1.8 and ϕ=2.0 flames, then from the measured
values of the ratio at ϕ = 1.8 (0.2) and the M>700 amu flux at ϕ =
2.0 (0.006 of the total carbon), the estimated flux of M>200 material at the M>700 amu concentration peak (12.6 mm) for ϕ = 2.0
is 0.03 of the total carbon. Accordingly, the total flux of species
larger than 200 amu but less than 700 amu would be 0.024 of the
total carbon. Although this result is only an approximation owing
to the above assumption and the uncertainty in the calculated mole
fraction, and hence flux, of the high mass material, its close comparison with the estimated upper limit for all hydrocarbons other
than C_2H_2, C_6H_6, and M>700 amu material indicates that these other
hydrocarbons are mainly in the mass range 200-700 amu and (as discussed below) presumably of aromatic structure.

Therefore it appears that the only hydrocarbons other than
C_2H_2 and C_6H_6 (and possibly C_4H_2, see above) remaining with large
enough flux to be significant in the mass balance at the M>700 amu
concentration peak in the sooting flame are aromatics with mass
numbers above 200 amu, and these comprise about 0.03 of the total
carbon. In contrast, at the M>700 amu peak in the near-sooting
flame, the hydrocarbons other than C_6H_6 and C_2H_2 represent about
three-times more of the total carbon (0.10) than in the sooting
flame but the fraction of the total carbon in aromatic species
other than C_6H_6 is 100-times smaller (3×10^{-4}) than in the sooting
flame.

The total fraction of the carbon as hydrocarbons at the M>700
amu concentration peak is considerably less in the sooting case
(0.36 vs. 0.43), though not as different as the fractions of the
initial O_2 consumed (0.82 at ϕ = 2.0 vs. 0.63 at ϕ = 1.8). Up to
this position in the flames, the carbon atoms oxidized to CO or CO_2
per oxygen atom consumed from O_2 is almost the same in both cases
(0.64 for ϕ = 1.8; 0.62 for ϕ = 2.0). Therefore in approaching the
M>700 amu peak the ratio of the flux of carbon as hydrocarbons to
that of oxygen as O_2 increases from the feed values of 0.72 and
0.80 to 0.84 and 1.6 for ϕ = 1.8 and 2.0, respectively. Thus, in
terms of only the fuel and O_2 remaining, the flame gases in the
region upstream of the M>700 amu peak become more fuel rich as oxidation consumes carbon and oxygen in a smaller C/O ratio than the
feed value, and the enrichment increases with increasing equivalence
ratio.

Again considering the overall reaction occurring in the region
upstream of the M>700 amu concentration peak, where most of the fuel
is consumed, slightly less than one (0.90 at ϕ = 1.8; 0.96 at ϕ =
2.0) mole of C_2H_2 is produced per mole of C_6H_6 consumed. The C_2H_2,
being much less reactive than C_6H_6 in the oxidation process, builds

up a substantial flux which maximizes (slightly downstream of the maximum C_2H_2 concentration) at essentially the same location (to within 0.1 mm) as the maximum concentration of M>700 amu material. At the end of this region, the species comprising the 2/3 of the carbon not represented by C_2H_2 include CO, CO_2, and M > 200 amu material, reflecting oxidation and growth reactions, and small hydrocarbon intermediates such as C_4H_2. The growth reactions are in a competition with oxidation (and possibly also pyrolysis) which in both flames allows net production of M>700 amu material only until the C_6H_6 is almost gone. The C_6H_6 decomposition products apparently include important reactants, especially phenyl radical, that initiate growth of aromatic structures from the abundant acetylenic species (e.g., C_2H_2, C_4H_2, C_4H_4, etc.). Also the mole fractions of OH and O, which are produced from O_2 relatively slowly by the branching reactions, are held down by relatively fast reactions with C_6H_6, and it is not until C_6H_6 concentration becomes low that more OH and O are available for oxidation of the growing polyaromatic species. Therefore as C_6H_6 becomes depleted, the combined effects of decreasing rate of mass addition to the heavy species and increasing rate of oxidation of these species cause termination of their net production.

Since the heavy species can also grow by addition reactions among themselves, the increase in flux of species of a given mass number does not respond exclusively to the availability of C_6H_6 decomposition products. Accordingly, referring to Fig. 5 and recognizing that the maximum flux of a species occurs slightly downstream of its maximum concentration, it is seen that the location of the maximum flux, and hence the termination of net production of that species, increases with increasing mass number. For example, while the maximum rate of C_6H_6 consumption occurs at 8.2 mm in the $\phi = 1.8$ flame, the maximum fluxes of 200 amu and 700 amu species occur somewhat downstream of 8.3 mm and 10.3 mm, respectively. Similarly, the net production of M>200 amu and M>700 amu material ceases at 8.7 mm and 10.5 mm, respectively. Clearly our emphasis above on the occurrence of the peak concentration of M>700 amu material where the C_6H_6 is almost (97.4%) gone is an arbitrary choice from among many symptoms that could be cited to illustrate the competition between production and growth. Also, if concentration data were available for M>1000 amu species, then the peak concentration would be found where C_6H_6 is somewhat greater than 97.4% depleted. However, because of the very steep C_6H_6 mole fraction profile (Fig. 8), the location of depletion is relatively insensitive to the nominal value chosen to represent depletion as long as it is less than a few percent remaining.

Distinction of Sooting Condition

We now consider the nature of the competition between production and consumption of high mass species in terms of changes occurring when the equivalence ratio is increased past the sooting limit. In the ϕ = 1.8 flame, the competition allows the concentration of M > 700 amu species to increase only to a stage of oxidation where 0.63 of the O_2 has been consumed, at which point 0.56 of the carbon has been oxidized to CO and CO_2, 0.10 appears as small hydrocarbons other than C_2H_2 and only 3 x 10^{-4} has grown to species larger than 200 amu (within which 6 x 10^{-5} of the carbon is in species larger than 700 amu). Under the more fuel-rich conditions of the sooting flame, the growth reactions are more competitive and the concentration of M>700 amu material increases until 0.82 of the O_2 is consumed while 0.64 of the carbon is oxidized to CO and CO_2, 100-times more of the carbon than at ϕ = 1.8 grows into heavy species, and essentially all of the light hydrocarbons excluding C_2H_2 (and associated polyacetylenes) are consumed by oxidation or growth. Therefore at this position where soot first begins to form as ϕ is increased past the critical value of 1.9, the sooting condition differs from that of near-sooting by less O_2 being left for continued OH and O generation and a larger increase having occurred in the ratio of total hydrocarbons to O_2. However these differences per se are not large nor do they occur abruptly as ϕ is increased, and substantial O_2 does remain at the critical position even in the sooting flame. Also, differences between the sooting potentials of the two flames represented by the different C_2H_2, C_4H_2, and C_6H_6 concentrations at the critical position are probably not important even though these species are noteworthy as possible contributions in growth reactions. The amount of carbon available as C_2H_2 plus C_4H_2 is relatively large and not very different in the two flames, and these species are somewhat equivalent since at this flame position, at least in the ϕ = 1.8 flame, they are in a partial equilibrium with H_2 as mentioned above. Furthermore, the amount of C_6H_6 present at the critical position cannot be an important distinction between the sooting potentials since it is less abundant in the sooting flame.

The important difference in the two flames is the amount of heavy material generated up to the critical position, which is 100-times larger in the sooting flame. This large difference, which occurs rapidly as ϕ is increased, is of course associated with the apparently small and gradually occurring differences noted above, and it is a consequence of the different balances between oxidation and growth in the two flames.

Not only does the amount of high mass material at the critical position increase 100-fold as ϕ is increased from 1.8 to 2.0, but the ratio of carbon as M > 200 amu material to that remaining as C_6H_6 increases from 0.01 to 2. Thus the mass of heavy species in the sooting flame exceeds that of C_6H_6 before the latter becomes negligible. Therefore it might appear that an important distinction between the two flames at the critical position could be availability of OH and O in the consumption of high-mass material immediately

downstream of the maximum concentration of this material. Data from
the near-sooting flame show that OH and H (and presumably O) begin
increasing substantially as C_6H_6 is depleted, implying that mole
fractions are held down by reaction with C_6H_6 as long as the latter
is available. Since in the sooting flame the amount of high-mass
material builds up and exceeds the amount of C_6H_6 present before
the latter becomes depleted, OH and O may be held down by reaction
with the high-mass material beyond the zone of C_6H_6 consumption.
Being aromatic, the high-mass material may have a high reactivity
to O and OH more like that of C_6H_6 than the considerably lower value
of C_2H_2. Accordingly, the concentration of oxidizing species in the
zone of net consumption of the high-mass material may be lower under
sooting than near-sooting conditions. However, this distinction
does not appear to be the critical factor since the fractional rate
of net consumption of M>700 amu material downstream of the peak concentration is not very different in the two flames, as noted above.
Once again it appears that the important distinction is the amount
of high-mass material built up before the net consumption of this
material begins. The capacities of the two flames at the high-mass
peak for further oxidation (and perhaps pyrolysis) are sufficiently
similar that about the same fraction of the peak quantity of high-mass
material is destroyed in the zone of rapid consumption downstream
of concentration peak (see below). Whether the soot mass comes
from the residual of the high-mass material or from addition reactions between a certain class of these heavy aromatic species and
small compounds in the relatively abundant C_2H_2 and C_4H_2 system,
the distinguishing feature of the sooting condition seems to be the
peak amount of heavy aromatic species produced.

Overall Stoichiometry after High-Mass Peak

Downstream of the peak flux of high mass material where the
consumption rate exceeds the growth rate, Table 2 shows that the
remaining C_6H_6 is completely consumed and O_2 and C_2H_2 are reduced
to similar levels in both flames. The amount of high-mass material
is reduced about six-fold in both flames before the rate of consumption becomes considerably smaller at the position where the
amount of O_2 remaining has been reduced to about 1% of the feed
value. The mean molecular weight of the high-mass material continues to increase during this stage of net consumption (Fig. 6),
which could indicate preferential consumption of smaller species,
addition reactions between heavy species and decomposition products
of the high-mass species or of C_2H_2, or both.

As can be seen in Table 2, C_2H_2 is by far the largest source
of carbon for oxidation in the zone of rapid net consumption of
heavy hydrocarbons. As discussed above, the C_2H_2 flux increases
with distance from the burner until C_6H_6 is essentially gone. Thus
the C_2H_2 production rate exceeds the consumption rate as long as

the rate of C_6H_6 decomposition, which is the main source of C_2H_2, is high enough. Acetylene, being much less reactive than C_6H_6 and the other aromatic compounds under these conditions, is abundant throughout the region of production and consumption of increasingly large aromatic hydrocarbons. As mentioned above, acetylene with polyacetylenes and H_2 approach a partial equilibrium which is observed to be well established at about 10 mm and beyond in the near-sooting flame. Some of the species such as C_2H_2, C_4H_2, and C_4H_4 in this acetylene system are of considerable interest as reactants for the growth of aromatic material[19-22]. The addition to the growing aromatics of only a minor fraction of the carbon passing through the C_2H_2 channel would make a major contribution to the total mass of heavy hydrocarbons. For example, at the peak concentration of M>700 amu species, the ratio of carbon flux as C_2H_2 to that as M>700 amu species is 5000 at $\phi = 1.8$ and 50 at $\phi = 2.0$.

Thus some of the C_2H_2 decay occurring downstream of the peak concentration of M>700 amu species could be consumption in growth reactions, but the main mode of C_2H_2 decay would appear to be oxidation, for the following reasons. First, comparison of the data (Table 2) from the two flames shows that a somewhat larger fraction of the C_2H_2 available at the beginning of this region is consumed in the near-sooting flame than in the sooting flame although the amount of heavy material with which carbon from the C_2H_2 system might react is 100-times less in the near-sooting case. Second, the amount of C_2H_2 lost in both cases greatly exceeds the amount of aromatic material. Although this point alone would not exclude the possibility of a large part of the C_2H_2 proceeding through the heavy aromatics which are being oxidized faster than they grow, this possibility is rejected in view of the first point.

Acetylene is the dominant, though not the only, hydrocarbon being oxidized in the zone of rapid consumption of M > 700 amu material. As ϕ is increased from 1.8 to 2.0, the moles of O_2 consumed per mole of C_2H_2 destroyed in this zone decreases from 2.2 to 1.0, and the atomic ratio of oxygen consumed as O_2 to carbon lost from all hydrocarbons decreases from 1.4 to 0.93. This behavior indicates that hydrocarbons oxidation in this zone, presumably by OH and O, proceeds mainly to CO in the sooting flame while some oxidation of the CO to CO_2, presumably by R3, occurs in the near-sooting flame.

$$CO + OH \rightleftarrows CO_2 + H \hspace{4cm} (R3)$$

The different extents of CO oxidation in the two flames is reasonable since OH mole fraction increases substantially as C_6H_6 becomes depleted at $\phi = 1.8$[22], whereas at $\phi = 2.0$ the amount of carbon as heavy aromatics builds up and exceeds the amount as C_6H_6 before the latter becomes small enough to allow the substantial increase in OH mole fraction. We did not measure the OH mole fraction profile for $\phi = 2.0$; the implied assumption that the reactivity of OH to the

heavy aromatics is relatively large, like that of OH to C_6H_6, is
supported by the fact that OH is quite reactive even to soot particles[28].

The postulated difference in OH mole fraction in the two flames
is consistent with the observation that the maximum fractional net
consumption rate of M > 700 amu material is somewhat larger at $\phi = 1.8$
than at $\phi = 2.0$ (although the difference is small as noted above). The
O-atom concentration, like that of OH, is probably suppressed by
reaction with C_6H_6 or heavy aromatics when either of them is sufficiently abundant. The importance of O-atom is indicated by this
species accounting for about one-half of the C_2H_2 decay in this
region of the $\phi = 1.8$ flame.

Chemistry of Formation of the High Mass Species

Possible mechanisms for the formation of high mass species in
these flames are discussed elsewhere[19-22]. As discussed above, the
trends in the high mass signals in the near-sooting flame lead us to
suggest that individually measurable species with masses between 110
and 210 amu that maximize between 8 and 10 mm continue to grow to
species with masses greater than 700 amu. Since the mass numbers at
which signals are observable between 200 and 300 amu correspond to
aromatic structures, we presume the high mass material also to be
aromatic. The rapid growth of large aromatic species seems to use
one or more products of C_6H_6 decomposition since, as discussed above,
the net production of heavy species in a given mass range ceases at a
certain late stage of C_6H_6 depletion. Many of the lower molecular
weight polycyclic aromatics that are formed in the initial steps of
this rapid growth have structures that indicate they are the products
of reactions of aromatic species with C_2 and C_4 species. The mole
fractions of the more hydrogenated species, e.g., C_2H_4, C_4H_4, and
C_4H_6, maximize between 7 and 8 mm where the mole fractions of the
lower molecular weight polycyclic aromatics are increasing rapidly.
Thus the growth that leads to heavy aromatics may begin with C_2 and
C_4 products of ring decomposition reacting with rings that have
remained intact.

The data are consistent with the picture that not all of the
initial C_6H_6 rings are opened, in that the moles of C_2H_2 produced is
slightly less than the moles of C_6H_6 consumed. The difference between the amount of carbon in C_2H_2 and one-third of the carbon in
C_6H_6 consumed up to the M>700 amu peak in the sooting flame is about
one-sixth of the amount of carbon in the M>200 amu material at that
location (the latter being about 6% of the carbon fed). This observation supports the concept that the growth involves addition reactions between a small fraction of the initial C_6H_6 rings that are
not opened and some of the abundant C_2 and C_4 products of ring destruction. The same picture holds qualitatively for the near-sooting
flame where growth is less competitive and a much larger fraction

of the aromatic intermediates are subsequently destroyed by oxidation.

For the near-sooting flame it is possible to calculate net reaction rates of several species pertinent here using the measured mole fraction profiles[22]. It is found that the net production of phenyl radical (C_6H_5) is maximum at the same location (8.2 mm) where the net rate of C_6H_6 consumption is maximum, and the former is about 5×10^{-3} of the latter. Also, the maximum net production rates of both C_7H_8 (toluene) and C_7H_7 (benzyl radical) occur near 8.2 mm and are about 6×10^{-3} and 9×10^{-4}, respectively, of the maximum net rate of C_6H_6 consumption. Since these maximum net rates are the result of both production and consumption, the true rates are of course larger than these values. While these rates are small enough to indicate that most of the C_6H_6 rings are destroyed, they are clearly large enough to support a role in the growth process for aromatic intermediates of C_6H_6 destruction. Thus it appears that a mixture of monocyclic aromatic species, e.g. benzene, phenyl radical, toluene and benzyl radical, and non-aromatics such as C_2H_2, C_4H_4, C_4H_3, C_2H_4, etc. results in the rapid production of higher molecular weight aromatics. The role of the initial aromatic ring in this process may be to provide a structure that is capable of stabilizing the radical adduct formed by addition of the non-aromatic species.

Role of High-Mass Species in Soot Formation

As discussed above, the orange luminosity that begins at the location of the peak concentration of M>700 amu material in the sooting flame is indicative of the onset of soot formation. Therefore the beginning of the visually observed luminous zone in the ϕ = 2.0 flame can be taken as the approximate location of the first appearance of soot particles although precise specification of this location would require resolution of the continuous emission into soot-particle and molecular contributions. Data on emission and absorption from soot and its intermediates in flat premixed C_6H_6/O_2 flames at 2.67 kPa are given by Homann et al.[9].

The onset of soot formation could of course be located by time-resolved measurements of soot particle concentration but such data are not available for the present flames. We have measured the $I_{M>1000}$ profile in a premixed flat C_2H_2/O_2 flame[20] for which detailed soot particle data are available from work with an earlier version of the present molecular beam sampling instrument[12]. This C_2H_2/O_2 flame was operated at the same pressure and burner velocity as the present C_6H_6/O_2 flame but at ϕ = 3.0 (C/O = 1.2) and without the argon dilution. Both flames produce about the same final mass concentration of soot. In the C_2H_2/O_2 flame, the peak concentration of M>1000 amu material and the beginning of orange luminosity both occur at the same location as the onset of soot formation, i.e., the first appearance of the smallest observable soot particles (1.5 nm diameter, or about 2000 amu).

As discussed above, the concentration of M>700 amu material in the C_6H_6/O_2 flame at ϕ = 1.8 exhibits a rapid decrease immediately downstream of the peak (Fig. 7), after which the concentration decreases much more slowly. The same behavior is seen for the M>1000 amu material in the C_2H_2/O_2 flame mentioned above. Since the beginning of this zone of rapid consumption is associated with the onset of soot formation and it is the concentration of the high-mass species being consumed that distinguishes the sooting condition, the question arises as to what role the high-mass species play in the soot formation processes. Although the available information is inadequate for a complete answer, it is interesting to compare the quantity of high-mass material available where the soot begins to form with the amount of soot eventually formed, and to compare the rate of consumption of high-mass material with the rate of soot formation. These comparisons are discussed below.

Homann et al.[9] measured the concentration of soot at 70 mm above the burner in flat premixed C_6H_6/O_2 flames at 2.67 kPa, 0.4 m/s burner velocity and a range of equivalence ratios. Although the argon dilution and higher burner velocity of the present study are expected to decrease the soot yield relative to that reported by Homann et al., the latter data are useful as an approximate basis for comparison. The interpolated value for ϕ = 2.0 is 6 x 10^{-9} g soot/cm^3, and being at 70 mm this value is clearly for the post-flame zone. When the fluxes (times flame area) of both the post-flame soot and high-mass material at different positions are expressed in terms of the fraction of the carbon fed that is carried by these species, it is found that the ratio of M>700 amu material to soot eventually made is 1.7 and 0.3 at 12.6 mm and 19.6 mm, respectively. As shown in Table 2, these positions define the beginning and end of the zone of rapid consumption of the M>700 amu material. This ratio is 2.8 at 12.7 mm, the approximate location of the maximum flux of M>700 amu material. In a similar manner the ratio of M>200 amu material to the post-flame soot is 16, 31, and 1.8 at 12.6 mm, 12.7 mm, and 19.6 mm, respectively. Therefore it appears that the amount of high-mass material at the onset of soot formation (12.6 mm) is significantly larger than the (overestimated) amount of soot eventually formed, which would support the picture that a major fraction of the soot mass could come from these large aromatic species. However, as discussed above, the rapid consumption of high-mass material that occurs in the region 12.6-19.6 mm immediately after the onset of soot formation is accompanied by extensive oxidation and C_2H_2 consumption (Table 2). The amount of high-mass material remaining at the end of the zone of rapid consumption appears to be comparable to the amount of soot eventually formed, but this conclusion is tentative owing to the uncertainties in the calculation.

Since a substantial flux of carbon as C_2H_2 and M>200 amu species remains at 19.6 mm where less than 1% of the original O_2 is left, it is reasonable to assume that only a small further loss of mass from

the M>700 amu material occurs and that the average molecular weight
increases through growth reactions. Such behavior would be consistent with, but an extension of, the trends seen in Figs. 4 and 5 at
the larger distances from the burner and for the higher molecular
weights. The evidence (discussed above) of species growth in the
near-sooting flame that can be seen in these figures would presumably be even stronger in the sooting flame where, in the zone of
rapid consumption of M>700 amu material (see Table 2), less oxidation occurs and the concentration of high-mass species is much
greater. Thus the system of decomposing heavy aromatics in the
presence of C_2H_2 and related acetylenic species offers the ingredi-
ents of growth, and growth indeed occurs, the smaller of the high-
mass species being consumed while the mass flux of larger species
increases. Although the mass flux of the M>200 amu material remaining at the end of the zone of rapid consumption of M>700 amu
species is large enough to account for the soot eventually formed,
the mass flux of C_2H_2 is much larger and the relative extents to
which acetylenic species and high-mass aromatics contribute to the
growth of larger species and hence of soot is not established. However, it is clear that the heavy aromatics are the critical ingredi-
ent since it is their flux and not that of acetylenic species that
differs between the near-sooting and sooting flames.

     In order to compare the rate of consumption of high-mass species
in the zone of rapid consumption with the rate of soot production in
the same zone we again refer to the optical absorption data of Homann
et al.[9] for the benzene flames mentioned above. These workers tentatively resolved the total absorption profile into components due to
different groups of species, one of which was soot. While the
results are not intended for precise quantitative conclusions, they
are the only information we have on the approximate rate of soot
formation in benzene flames under conditions similar to those of the
present study. Judging from the tentative resolution and the trends
in the absorption profiles with changes in the C/O feed ratio, it
appears that about 0.2 to 0.3 of the final soot mass is formed in
the 1.8 ms interval corresponding to the zone of rapid consumption
of M>700 amu material in the present ϕ = 2.0 flame (Table 2). The
remaining, larger fraction of the soot formation occurs downstream
of the zone of rapid consumption of M>700 amu material in a time
interval that is almost 10-times larger than that for the rapid
consumption of the M>700 amu species. Analogous behavior can be
seen in Fig. 5 where, for example, the characteristic time for the
decay of 200 amu species is considerably shorter than that for the
production of 700 amu species. The difference here is not as large
as that between the soot and M>700 amu species, but neither is the
difference between 700 amu and 200 amu as large as that between
the soot particle mass and the average molecular weight of M>700 amu
species in the zone of their rapid consumption (which is perhaps 800
to 900 amu according to the trend observed in the near-sooting flame).
Thus in the evolution of the system of heavy hydrocarbons to larger

and larger species of smaller and smaller peak concentrations, the
time required for the production of the larger species within the
transient distribution increases as the growth proceeds. Assuming
the agents of growth are the heavy aromatics, their decomposition
products, and C_2H_2 and related species, the increasing growth times
might reflect preferential consumption and hence faster depletion
of the more reactive species, namely aromatics. Accordingly, the
formation of a substantial fraction of the soot mass subsequent to
and much more slowly than the rapid consumption of M>700 amu species
would imply that a considerable fraction of the soot is grown by
addition of C_2H_2, C_4H_2, C_4H_4, etc. to the heavy aromatics. Thus the
growth of the system of heavy aromatics leads to soot, the growth
agents being both heavy aromatics and acetylene species, but shifting increasingly to the latter as the aromatics become depleted.

Qualitatively similar behavior is observed in the sooting
acetylene flame mentioned above, but there are some quantitative
differences. The high-mass (M>1000 amu) signal rapidly decays for
about 1.1 ms immediately downstream of the peak value and then
exhibits a much slower decline, the entire pattern resembling the
M>700 amu behavior in the sooting benzene flame. However, the time
duration of the zone of rapid high-mass consumption is 40% shorter
and the fractional decay of the high-mass concentration in this
zone is 30% smaller than in the benzene flame. The fraction of the
original O_2 remaining at the high-mass peak is about 10-times
smaller in the acetylene flame, but the remaining fractions of O_2
differ little at the end of the zone of rapid consumption of high--
mass material. Therefore oxidation and hence the consumption of
hydrocarbons, mainly C_2H_2, in this zone are much more extensive in
the benzene flame. However, the C_2H_2 consumption per mole of O_2
consumed in this zone of the acetylene flame is substantially
larger than the value of 1.0 found in the benzene flame, implying
that the portion of the C_2H_2 consumption attributable to growth
reactions in this early stage of soot formation is larger in the
acetylene flame than in the benzene flame. Furthermore the absolute
amount of C_2H_2 consumed in excess of that which would be required to
convert to CO all the O_2 consumed in this zone is much larger in the
acetylene flame. Therefore the amount of C_2H_2 consumed in growth
reactions in the zone of rapid consumption of high-mass species is
apparently larger in the acetylene flame even though the total C_2H_2
consumption, including growth and oxidation, is of order 10-times
larger in this zone of the benzene flame.

We have not determined the conversion factor between signal
intensity and mole fraction for the heavy hydrocarbons in the acetylene flame, nor have we measured mole fractions of individual
polycyclic aromatics in the sooting benzene flame. However, we have
observed that the peak mole fractions of individually measured polycyclic aromatics are 30- to 100-times smaller in the acetylene flame
than in the near-sooting benzene flame. Therefore the peak concen-

trations of heavy hydrocarbons, i.e., heavy aromatics, are presumably
much smaller in the acetylene flame than in the sooting benzene flame.
The concentration of C_6H_6 at the peak concentration of the heavy aromatics is 40-times smaller in the acetylene flame than in the sooting
benzene flame.

Thus it appears that the onset and earliest stage of soot formation in the acetylene flame, in comparison with the same stages in
the benzene flame, occur in the presence of much lower concentrations
of aromatics and much less oxidation, and with a larger C_2H_2 consumption in the growth reactions. Because of the much smaller concentrations of aromatics in the primary reaction zone of the acetylene
flame, the growth rate of heavy species does not match their destruction rate until the O_2 is almost gone. Therefore the occurrence of
the peak concentration of heavy hydrocarbons and hence the onset of
soot formation are delayed almost to the end of the oxidation zone.
The soot formation begins, as in the benzene flame, in a mixture of
heavy aromatics and C_2H_2, and of course many related species such as
C_4H_2 and C_4H_4. Addition reactions among these aromatics and acetylenic species would appear to be as important for the growth leading
to soot in this flame as in the benzene flame. Since the flux of
aromatics available at the onset of soot formation is much less in
the acetylene flame, it is plausible that a larger portion of the
final soot mass may be added as acetylenic species in this case.

Once soot formation begins, the time required for the growth
of the final particles in the acetylene flame is not greatly different from that in the benzene flame. What is strikingly different is the fraction of the soot formed during the zone of rapid
consumption of high-mass species. In the acetylene flame, only
8×10^{-4} of the final soot mass is formed up to 18 mm[12] where the
rapid consumption of M>1000 amu species ends. Furthermore, with
soot nucleation defined as the appearance of 1.5 nm diameter (M =
2000 amu) particles, only about 0.02 of the final cumulative number of nuclei have been formed at 18 mm[29]. Therefore, in the
acetylene flame, most of the soot particles first become observable
downstream of the zone of rapid consumption of the heavy species
that distinguish the sooting condition, and almost all of the final
soot mass is formed downstream of this zone, while the concentrations
of heavy hydrocarbons and C_2H_2 continue to decrease. The above
difference notwithstanding, a common feature of the acetylene and
benzene flames is the emergence during the rapid consumption of high-
mass material of the heavy species that grow to become the first
observable soot particles.

ACKNOWLEDGEMENT

This work was supported in part by the U.S. Environmental
Protection Agency, Grant No. R803242, and the National Institute
of Environmental Health Science, Center Grant No. 5 P30 ES02109-02.

REFERENCES

1. Lahaye, J. and Prado, G., in Chemistry and Physics of Carbon
    (P.L. Walker, Jr. and P.A. Thrower, Eds.), Marcel Dekker,
    New York, 1978, Vol. 14, p. 167.
2. Wagner, H. Gg., Seventeenth Symposium (International) on Com-
    bustion, The Combustion Institute, Pittsburgh, 1979, p. 3.
3. Haynes, B.S. and Wagner, H. Gg., Prog. Energy Combust. Sci.,
    7: 229-273 (1981).
4. Homann, K.H., Mochizuki, M., and Wagner, H. Gg., Z. Phys.
    Chem. N.F., 37: 299-313 (1963).
5. Homann, K.H. and Wagner, H. Gg., Ber. Bunsenges. Phys. Chem.
    69: 20-35 (1965).
6. Bonne, U. and Wagner, H. Gg., Ber. Bunsenges. Phys. Chem., 69:
    35-48 (1965).
7. Bonne, U., Homann, K.H., and Wagner, H. Gg., Tenth Symposium
    (International) on Combustion, The Combustion Institute,
    Pittsburgh, 1965, p. 503.
8. Homann, H.G. and Wagner, H. Gg., Eleventh Symposium (Inter-
    national) on Combustion, The Combustion Institute,
    Pittsburgh, 1967, p. 371.
9. Homann, K.H., Morgeneyer, W., and Wagner, H. Gg., Combustion
    Institute European Symposium (F.J. Weinberg, Ed.), Academic
    Press, London, 1973, p. 394.
10. Prado, G.P., Lee, M.L., Hites, R.A., Hoult, D.P., and Howard,
    J.B., Sixteenth Symposium (International) on Combustion,
    The Combustion Institute, Pittsburgh, 1977, p. 649.
11. Tompkins, E.E. and Long, R., Twelfth Symposium (International)
    on Combustion, The Combustion Institute, Pittsburgh, 1969,
    p. 625.
12. Wersborg, B.L., Howard, J.B., and Williams, G.C., Fourteenth
    Symposium (International) on Combustion, The Combustion
    Institute, 1973, p. 929.
13. Wersborg, B.L., Fox, L.K., and Howard, J.B., Combustion and
    Flame, 24: 1-10 (1975).
14. Calcote, H.F., Combustion and Flame, 42: 215-242 (1981).
15. Wersborg, B.L., Yeung, A.C., and Howard, J.B., Fifteenth
    Symposium (International) on Combustion, The Combustion
    Institute, Pittsburgh, 1975, p. 1439.
16. Homann, K.H., Ber. Bunsenges. Phys. Chem., 83: 738-745 (1979).
17. Delfau, J.L., Michaud, P., and Barassin, A., Combust. Sci.
    Technol., 20: 165-177 (1979).
18. Olson, D.B. and Calcote, H.F., in Particulate Carbon Formation
    During Combustion (D.C. Siegla and G.W. Smith, Eds.),
    Plenum Press, New York, 1981. p. 177.
19. Bittner, J.D. and Howard, J.B., Eighteenth Symposium (Inter-
    national) on Combustion, The Combustion Institute,
    Pittsburgh, 1981. p. 1105.
20. Bittner, J.D. and Howard, J.B., in Particulate Carbon Forma-
    tion During Combustion (D.C. Siegla and G.W. Smith, Eds.),

Plenum Press, New York, 1981, p. 109.

21. Bittner, J.D., Howard, J.B., and Palmer, H.B., Chemistry of Intermediate Species in the Rich Combustion of Benzene, (this volume).

22. Bittner, J.D., A Molecular Beam Mass Spectrometer Study of Fuel-Rich and Sooting Benzene-Oxygen Flames, Sc.D. Thesis, Department of Chemical Engineering, M.I.T., Cambridge, Mass., 1981.

23. Fuller, E.N., Schettler, P.D., and Giddings, J.C., Ind. Eng. Chem., 58(5): 19-27 (1966).

24. Sherwood, T.K., Pigford, R.L., and Wilke, C.R., Mass Transfer, McGraw-Hill, New York, 1975, p. 18.

25. Fristrom, R.M. and Westenberg, A.A., Flame Structure, McGraw-Hill, New York, 1965, pp. 74-80.

26. Fairbanks, D. and Wilke, C.R., Ind. Eng. Chem., 42: 471-475 (1950).

27. Fristrom, R.M. and Westenberg, A.A., ibid., p. 276.

28. Neoh, K.G., Howard, J.B., and Sarofim, A.F., in Particulate Carbon Formation During Combustion (D.C. Siegla and G.W. Smith, Eds.), Plenum Press, New York, 1981. p. 261.

29. Howard, J.B., Wersborg, B.L., and Williams, G.C., in Faraday Symposia of the Chemical Society, 7: 109-119 (1973).

DISCUSSION

H.F. Calcote (AeroChem Research Laboratories, Inc.)

You point out that there is not much difference in the flame structure when going from non-sooting to sooting flames. How then do you account for the fact that the critical concentration at which soot is formed occurs at a very distinct equivalence ratio and is very reproducible from one laboratory to the next? Does this not indicate that a critical change is occurring in chemistry of the system similar to what occurs at an explosion limit?

Howard

The so-called onset of sooting is associated with only small changes in the profiles of temperature and concentration of the species that carry most of the material fed. The critical change, as we pointed out, is in the peak concentration of heavy hydrocarbons, but this change reflects only a shift in the relative rates of formation and destruction, and hence of the availability of these species, not an initial appearance of new reaction or new species. Similarly, the crossing of an explosion limit involves only a shift in relative rates of reactions and availability of key

species. At both the explosion and sooting limit, the critical
reactions were operating and the key species were present before
the limit was reached.

J. Vandooren (Laboratoire de Physico-Chimie de la Combustion,
              Louvain)

    In the determination of the structures of rich acetylene and
benzene flames, did you monitor CH and CH_2 radicals?

Howard

    In the rich benzene flame (ϕ = 1.8) we looked for CH, CH_2, and
C_2H but were unable to find either of these species. Considering
the detectability limits of the instrument, these species apparently
were not present at the following or higher mole fractions: CH and
CH_2, 10^{-5}; C_2H, 5 x 10^{-6}. The limits stated here are conservatively
high since radicals such as C_6H_3 and C_2H_3 were measured at these or
lower mole fractions. Although our work in rich acetylene flames
is incomplete, we have detected CH_2 but not measured its concentration and we have found C_2H to be absent at the above stated detection
limit.

J. Beér (Massachusetts Institute of Technology)

    How can one explain the breakthrough of oxygen or oxidising
compounds such as OH in a premixed flame of ϕ = 1.8?

Howard

    The conversion of oxygen and the breakthrough of oxygen and
other oxidizing compounds is kinetically controlled in the premixed
flames discussed in the presentation.

R. Delbourgo (C.N.R.S., Rouen)

    Are the concentration profiles shown corrected for diffusion?
If they are not, correction may shift the profiles one relatively
to another and influence their interpretation in terms of relative
consumption or production.

Howard

    The concentration profiles reflect the coupled effects of
diffusion, convection and reactions. No adjustments have been
made to the concentrations to remove the contribution of diffusion.
However, diffusion is accounted for in the flux profiles and in
the calculated net reaction rates.

K. H. Homann (Technische Hochschule, Darmstadt)

Could you please define what you mean by surface growth of soot particles?

Howard

Surface growth as I defined it is the addition to the particle of species smaller than or equal to the smallest counted particles.

S. Galant (Société Bertin)

Have you studied various fuels and the incidence of temperature profiles upon final soot mass loadings and/or intermediate nuclei?

Howard

We have studied and are still studying different fuels, but we have not studied the effects of temperature as an independent variable.

A. F. Sarofim (Massachusetts Institute of Technology)

Is there sufficient data on a given flame system to relate the surface growth of soot, as determined from particle size distribution, to the rate of consumption of high molecular weight compounds?

Howard

Not yet. We need composition profiles for several more cutoff masses for the high-mass species in the ϕ = 3 C_2H_2/O_2 flame. In the ϕ = 2.0 C_6H_6/O_2 flame we need particle size and number concentration profiles.

K. H. Homann (Technische Hochschule, Darmstadt)

I would like to point out that sampling a low pressure flame from a height of 2-3 mm above the burner can lead to a gross disturbance of the oxidation zone of the flame. Therefore, measurements of concentrations by means of a molecular beam nozzle from that height might not give realistic results.

Howard

The zone of soot and PAH formation begins at distances above the burner greater than 2 to 3 mm. For example, in the case of the high-mass species discussed in the presentation, the lowest height above the burner employed in the analysis was 7.2 mm.

CHEMISTRY OF INTERMEDIATE SPECIES

IN THE RICH COMBUSTION OF BENZENE

J. D. Bittner, J. B. Howard and H. B. Palmer[*]

Department of Chemical Engineering
Massachusetts Institute of Technology
Cambridge, Massachusetts 02139
[*]Fuel Station Section, Steidle Building
Pennsylvania State University
University Park, Pennsylvania 16802

INTRODUCTION

The large potential importance of diesel engines and of synthetic
fuels has led to a surge of interest in the associated problem of
soot formation. It is well known[1] that the formation of soot and
polycyclic aromatic hydrocarbons (PAH) is strongly influenced by
fuel type. Studies in gas turbine-type combustors[2-4] and simple
laboratory systems such as laminar diffusion flames[5-9], premixed
flames[10-11] and well-stirred reactors[12-13], have shown that aromatic
fuels have a high propensity to form soot. Recent studies of soot
formation in premixed flames using light scattering techniques[14-15]
have demonstrated the application of the laws of physical coagulation
to the later stages of soot formation in flames of different fuels.
These studies show that the important difference between fuels,
and hence more generally a dominant factor in soot formation, is
the mass of carbonaceous material that enters this coagulating
system. Thus, it is important to understand the early stages of
the chemistry of the fuel oxidation and pyrolysis, the pre-particle
chemistry. Since previous studies suggest that the intact aromatic
ring, not fragments thereof, is largely responsible for the marked
propensity of aromatics to form soot[16-19], a detailed study of the
preservation or destruction of the ring in rich flames of benzene
seems worthwhile.

95

     This paper reports such a study. Mole fraction and flux profiles
of stable and free radical species in a benzene-oxygen-argon flame
near the sooting limit are presented, and the production of C_1 - C_5
hydrocarbons from the aromatic ring and the role of the aromatic
ring in PAH and soot formation are discussed.

EXPERIMENTAL

     The molecular beam mass spectrometer system (Figure 1) that
has been developed to measure stable and radical species in fuel-rich
and sooting flames, the calibration procedures and other experimental
techniques have been described elsewhere[20],[21]. The flames were produced at a burner chamber pressure of 2.67 kPa and a cold gas velocity
of 0.5 ms^{-1} at 298K. The unburned gas composition was 13.5m % C_6H_6,
56.6m % O_2 and 30.0m % Ar. This corresponds to a fuel equivalence
ratio Ø, of 1.8; the sooting limit is Ø = 1.9.

PRODUCTION OF C_1 - C_5 HYDROCARBONS FROM BENZENE

     The apparent importance of the intact aromatic ring in promoting
soot formation and the need for information on its destruction process
were mentioned above. The distribution of non-aromatic hydrocarbons
produced by the destruction of the aromatic ring may also be of
primary importance in the soot and PAH formation process. From
structural considerations, many of the PAH observed in sooting
flames cannot be formed by condensation reactions of intact aromatic
rings. Results presented below suggest that in flames of aromatic
fuels, much of the carbon that ends up as soot may come from non-
aromatic hydrocarbon intermediates. Specifically, it is very
probable that many of the elementary steps in soot formation are
bimolecular addition reactions of aromatic radicals to unsaturated
aliphatic molecules.

     Mole fraction profiles of the major stable and radical species
observed in the Ø = 1.8 benzene flame are shown in Figure 2. Mole
fraction profiles of forty other species with masses up to 202 amu
have been presented elsewhere[21]. The following discussion concen-
trates on the region between the burner surface and 16mm above the
burner, where the fuel is consumed and intermediate hydrocarbon
species, PAH, and soot are produced. Insight into the chemical
processes occurring in the flame was enhanced by calculating the
molar fluxes (per unit area of the burner surface), F_i, for eighteen
species using the equation

$$F_i = X_i \frac{PA}{RT} \left(v - \frac{D_{i,mix}}{X_i} \frac{dX_i}{dz} \right) \qquad (Eq.1)$$

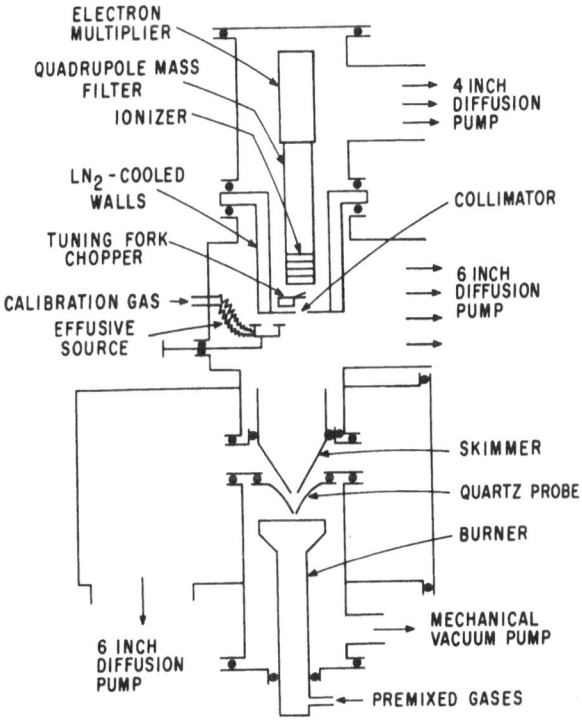

Figure 1.   Molecular beam mass spectrometer system for the study
of flat low-pressure premixed laminar flames.

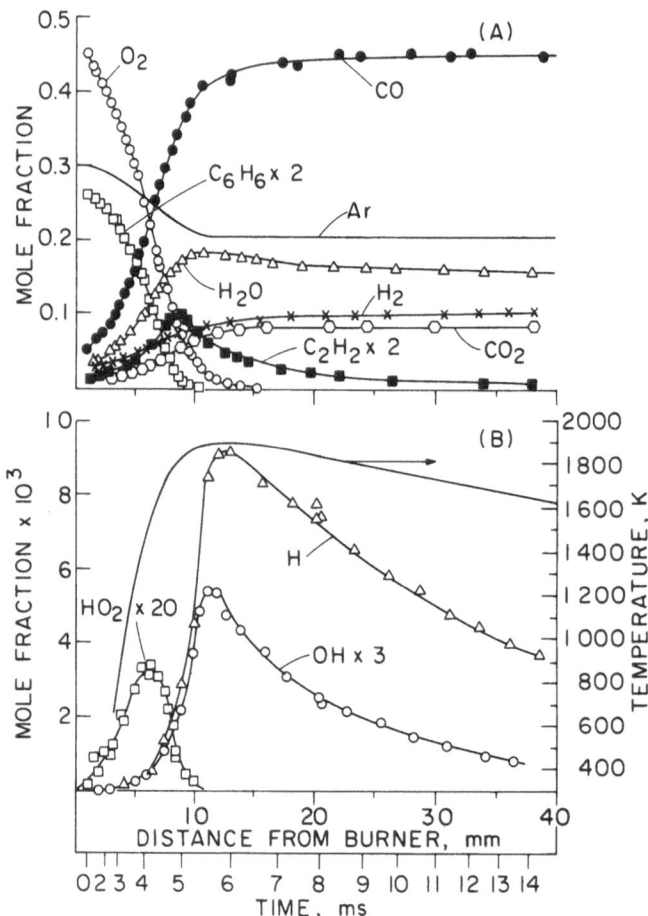

Figure 2. Mole fractions and temperature vs. distance from burner
in a near-sooting (∅ = 1.8) benzene (13.5% m%) - oxygen
(56.5 m%) - argon (30.0 m%) flame. Cold gas velocity =
0.5 m/s. Pressure = 2.67 kPa (20 torr). (A) Major stable
species. (B) H, OH, HO$_2$ and temperature.

where X_i is the mole fraction of species i, T is the temperature, A is the area expansion ratio, v is the bulk gas velocity, $D_{i,mix}$ is the diffusion coefficient of species i in the mixture, z is the distance from the burner, and R is the ideal gas constant[22].

The diffusion coefficients, $D_{i,mix}$, were calculated from the binary coefficients, D_{ij}, using the relationship of Fairbanks and Wilke[23] (Eq. 2).

$$D_{i,mix} = \frac{1 - X_i}{\sum\limits_{\substack{j=1 \\ j \neq 1}}^{n} \frac{X_i}{D_{ij}}}$$ (Eq.2)

For all non-polar pairs the binary diffusion coefficients were calculated from Eqs. 3 to 5[24]

$$D_{ij} = \frac{1.66 \times 10^{-3} \ [(M_i + M_j)/M_i M_j]^{1/2} T^{1.67}}{P\sigma_{ij}^2 (\varepsilon_{ij}/k)^{0.17}}$$ (Eq.3)

$$\sigma_{ij} = 1/2 \ (\sigma_i + \sigma_j)$$ (Eq.4)

$$\varepsilon_{ij}/k = [(\varepsilon_i/k) \ (\varepsilon_j/k)]^{1/2}$$ (Eq.5)

where D_{ij} is in cm^2/s, P is pressure in atmospheres, M is molecular weight, T is temperature in K, and σ and ε/k are the Lennard/Jones (12-6) potential parameters in A and K tabulated by Svehla[25] and Hirschfelder et al.[26]. When not available in references 25 or 26, the Lennard-Jones parameters were estimated from the critical pressure, critical temperature and acentric factor using the methods of Tee et al.[27] as described by Reid et al.[28]. For the only polar-polar pair considered, H_2O and C_6H_6O (assumed to be phenol), the method of Brokaw[29] (also described by Reid et al.[28]) has been used to estimate the binary diffusion coefficient. For the polar-non-polar interactions of H_2O and C_6H_6O with other species, the binary diffusion coefficients were calculated from Eqs. 3 to 5 using the Lennard-Jones parameters of H_2O and C_6H_6O estimated by the method of Brokaw[29].

Smooth curves were drawn through the 30 to 50 experimental data points (ratios of signal of species i relative to the argon signal) taken for each profile. These pencil-smoothed curves were entered into a computer program in the form of 151 evenly spaced points

(0.1mm) between 1 and 16mm above the burner. The argon and other
mole fraction profiles were calculated from the signal ratio profiles,
calibration factors, and an argon balance. These mole fraction profiles were further smoothed and differentiated by a simplified least
squares procedure[30,31].

    The importance of diffusion in determining the shapes of the
mole fraction profiles in this flame is apparent from the comparison
of the mole fraction and flux profiles for the major stable species
(Figure 3). The inlet fluxes of O_2 and C_6H_6 are indicated by the
hash marks on the ordinate. The constant C_6H_6 flux between 2 and
6mm indicates that the 45% decrease in the C_6H_6 mole fraction in
this region is caused by diffusion and not chemical reaction. The
agreement between the C_6H_6 flux at the inlet and that between 2 and
6mm (difference of 6%) is good considering the uncertainties in
the alignment of the temperature and mole fraction profiles and the
uncertainty ($\pm$ 20%) in the diffusion coefficients. The shape of
the O_2 flux profile between 2 and 6mm suggests that the steep
gradient in the O_2 mole fraction is distorted by the 0.7mm diameter
orifice in the sampling probe.

    The onset of significant benzene consumption between 6 and 7mm
is supported by the flux profiles of the products, CO, CO_2, and H_2O,
and the major hydrocarbon intermediate, C_2H_2 (Figure 3B). The large
mole fractions of these species present between 2 and 6mm (Figure 3A)
are not produced by chemical reaction in this region; rather, they
diffuse from the primary reaction zone, which extends from about
6mm to 11mm. This is established by the flux profiles of C_6H_6
(Figure 3B) and nine other carbon-containing intermediate and
product species (Figures 4 and 5).

    These flux profiles provide some insight into the production
of lower molecular weight hydrocarbons from benzene. The C_2H_2 flux
(Figure 3B) exhibits an induction time between the onset of benzene
decay at about 6mm and the sharp increase in C_2H_2 flux at about 8mm.
On the other hand, the CO flux starts to increase at 6mm, just as
the benzene flux begins to decay. This behavior is not sensitive
to temperature profile alignment or variations of diffusivities up
to 20%. It indicates that the initial reactions of benzene produce
CO. Acetylene apparently is not formed in the early stages of the
process that converts benzene to CO and CO_2, but only after several
intermediate steps. This is consistent with calculations described
elsewhere[20,21], which show that unimolecular fragmentations either
of phenyl radicals (Rxn.1)

$$C_6H_5^{\cdot} \longrightarrow \begin{array}{c} C_4H_3 \\ + \\ C_2H_2 \end{array} \longrightarrow C_4H_2 + H \qquad\qquad (Rxn.1)$$

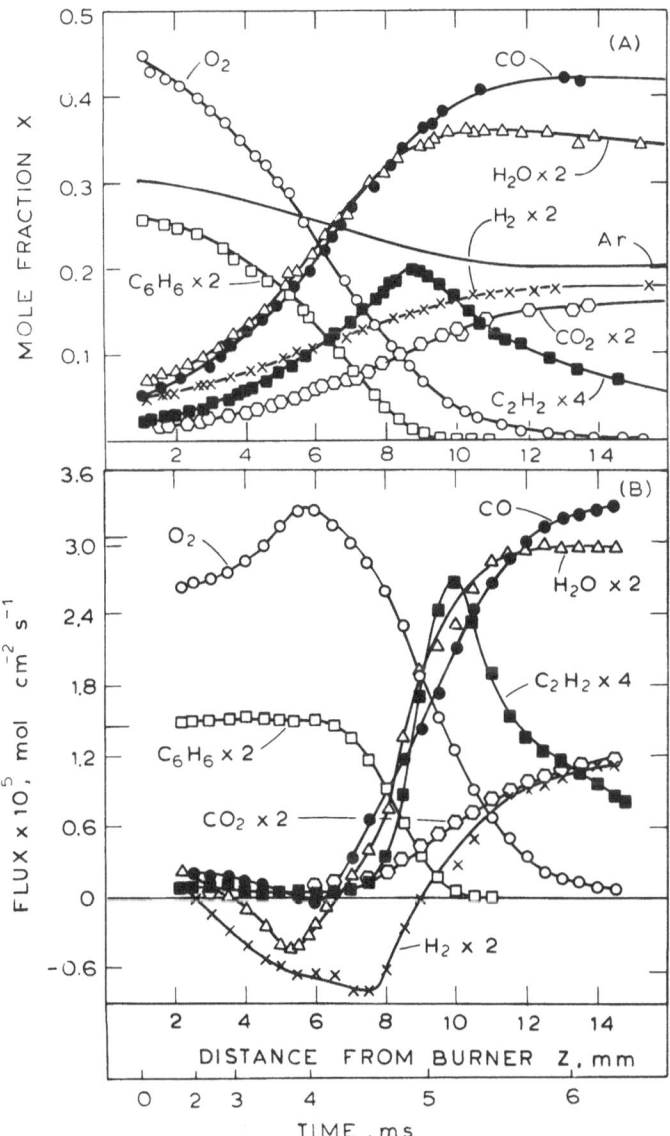

Figure 3. Mole fractions (A) and molar fluxes (B) of major stable
species vs. distance from burner in a near-sooting
(Ø = 1.8) benzene (13.5% m%) – oxygen (56.5 m%) – argon
(30.0 m%) flame. Cold gas velocity = 0.5 m/s.
Pressure = 2.67 kPa (20 torr).

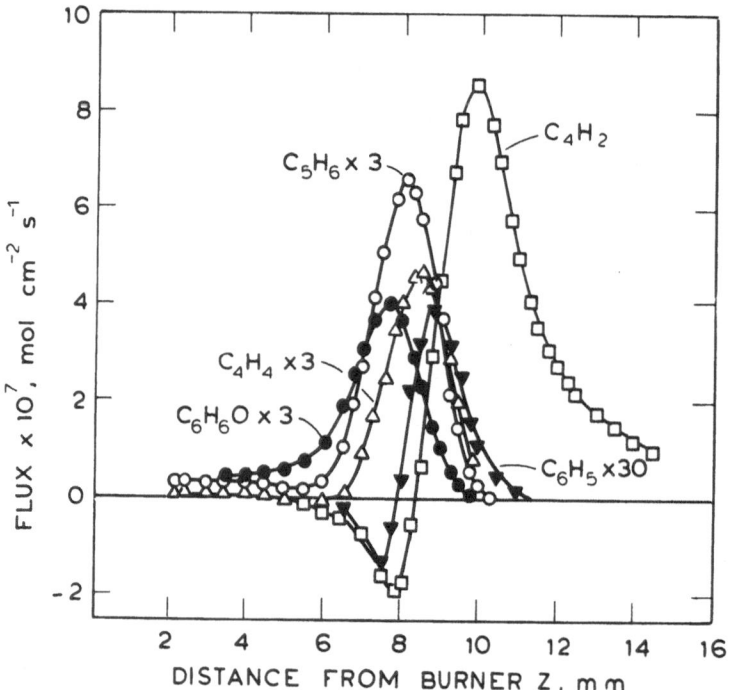

Figure 4.   Molar fluxes of C_6H_6O, C_6H_5, C_5H_6, C_4H_2 and
C_4H_4 vs. distance from burner in a near-
sooting ($\emptyset = 1.8$) benzene (13.5 m%) - oxygen
(56.5 m%) - argon (30.0 m%) flame. Cold gas
velocity = 0.5 m/s. Pressure = 2.67 kPa (20 torr).

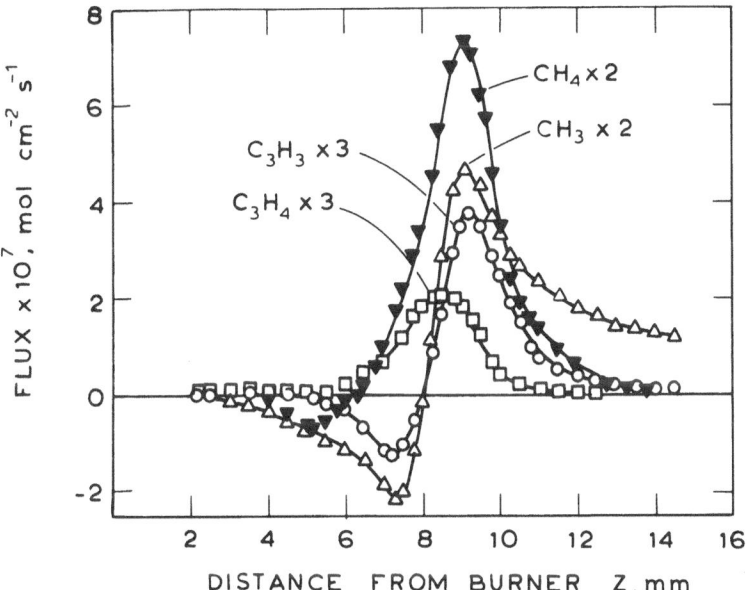

Figure 5. Molar fluxes of C_3H_4, C_3H_3, CH_4 and CH_3 vs. distance from burner in a near-sooting ($\emptyset$ = 1.8) benzene (13.5 m%) - oxygen (56.5 m%) - argon (30.0 m%) flame. Cold gas velocity = 0.5 m/s. Pressure = 2.67 kPa (20 torr).

or of cyclohexadienyl radicals (Rxn.2)

$$\text{(H, H structure)} \quad ----\rightarrow \quad \begin{array}{c} C_4H_5 \\ + \\ C_2H_2 \end{array} \quad \longrightarrow \quad \begin{array}{c} C_2H_3 \\ + \\ C_2H_2 \end{array} \quad \longrightarrow \quad C_2H_2 + H \qquad (Rxn.2)$$

which could produce C_2H_2 prior to CO, cannot account for the observed rate of benzene decay.

An alternative interpretation might be that in the early stages of benzene disappearance C_2H_2 is consumed as fast as it is produced. This seems improbable. It would imply that benzene decay is rate-limiting and that the C_2H_2 produced is consumed very rapidly to give a near-zero net C_2H_2 flux until about 8mm. Since there is considerable C_2H_2 (0.035 mole fraction) at 7mm (Figure 3A), ultra-fast conversion of C_2H_2 to CO should also consume the C_2H_2 that has diffused upstream from the zone of net C_2H_2 production. It would be remarkable if C_2H_2 production should so nearly equal its rate of destruction in this zone.

Although the mode of initial attack on the benzene ring is not entirely certain[21], the following reaction scheme, which is based on many studies of single reactions[32-46], is consistent with the early production of CO, the delayed production of C_2H_2, and the following sequence of positions (mm) of the flux profile maxima (Figures 3 to 6).

$C_6H_6O,7.7$; $C_5H_6,8.1$; $C_4H_4,8.4$; $C_3H_4,8.5$; $CH_4,9.1$;

$CH_3,9.1$; $C_3H_3,9.2$; $C_2H_2,9.8$; $C_4H_2,9.9$

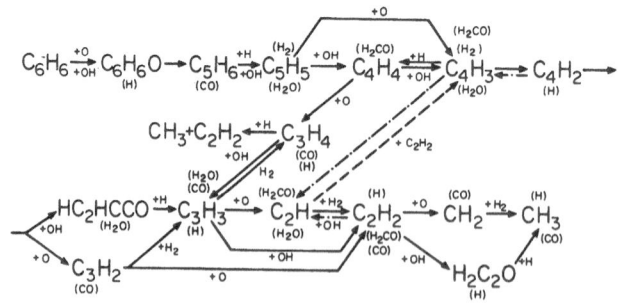

Analysis of the mole fraction data for this flame led to the initial conjecture[21] that benzene destruction was initiated by reaction with O atoms. Because the O-atom signal was obscured by CH_4 in this rich flame, it was not possible to confirm this hypothesis. However, the flux profiles for C_6H_6 and C_2H_2 indicate that the reaction of OH (rather than O) with benzene may be the primary source of C_6H_6O. If it is accepted, as postulated above, that the near-zero C_2H_2 flux between 6 and 7.5 mm is due to a very low C_2H_2 production rate and a near-zero consumption rate (rather than a delicate balance between high production and consumption rates), then a considerable fraction (22 mol %) of the benzene is consumed in a region (6 - 7.5mm) where little or no C_2H_2 is being consumed. Thus, $k_{OH+C_6H_6} \cdot X_{C_6H_6} \gg k_{OH+C_2H_2} \cdot X_{C_2H_2}$ or $k_{O+C_6H_6} \cdot X_{C_6H_6} \gg k_{O+C_2H_2} \cdot X_{C_2H_2}$. $k_{O+C_2H_2} = 5.2 \times 10^{13} \exp(-1860/T)$ cm^3 mol^{-1} s^{-1} [47]. $k_{OH+C_2H_2} = 1.3 + 0.3 \times 10^{12}$ cm^3 mol^{-1} s^{-1} [48]. Low temperature studies (298 - 462K)[49,50] of the O + C_6H_6 reaction are in excellent agreement and give $k_{O+C_6H_6} = 1.1 \times 10^{13} \exp(-2060/T)$ cm^3 mol^{-1} s^{-1}. Studies of the OH + C_6H_6 reaction[51,52] indicate that below 300K OH addition to the ring is the major channel. Between 300 and 500K both addition and abstraction occur simultaneously. A kinetic isotope effect indicated that abstraction is the major channel above 500K with $k_{OH+C_6H_6} = 1.4 \times 10^{13} \exp(-2260/T)$ cm^3 mol^{-1} s^{-1} (500<T<1000K)[52]. Extrapolating the benzene rate coefficients to flame temperatures, $k_{OH+C_6H_6} \cdot X_{C_6H_6}/k_{OH+C_2H_2} \cdot X_{C_2H_2} = 6.3$ and 2.1, and $k_{O+C_6H_6} \cdot X_{C_6H_6}/k_{O+C_2H_2} \cdot X_{C_2H_2} = 0.47$ and 0.14, at 6 and 7.5mm, respectively. Thus one might conclude that OH abstraction for form phenyl radical is the main channel for benzene consumption.

However, two other observations do not support this picture. Firstly, if OH + C_6H_6 is assumed to be the primary reaction for benzene consumption, the present flame data do not yield reasonable Arrhenius parameters. A mean value of $k_{OH+C_6H_6} = 6 \times 10^{13}$ cm^3 mol^{-1} s^{-1} could account for benzene decay between 7 and 9mm. This is over 15 times greater than the extrapolation of the value of Tully et al.[52] to flame temperatures. Secondly, if most of the benzene reacts via H-abstraction, reactions destroying phenyl radical must be rapid enough to explain its low mole fraction (maximum = 8×10^{-5} at 8.2mm). As indicated above and discussed in detail elsewhere[20], the unimolecular decomposition of phenyl radical is too slow to account for its low mole fraction. For the reaction of phenyl with O_2, OH or O to account for the observations, rate coefficients equal to or greater than the collision frequency would be required. Thus, H-atom abstraction by OH is only a minor path for benzene destruction.

A mechanism of benzene destruction that is consistent with both the flux profile analysis and the low phenyl radical mole fraction is the decomposition of the adduct formed by OH + C_6H_6 to phenol or other products that would yield CO and a C_5 species in

subsequent reactions. Although Sloane[53] has observed phenol in
single collision studies at 300K, Tully's study[52] provides no support
for adduct decomposition channels other than back to reactants for
temperatures up to 1150K.

Although we are unable to identify with certainty the main
mechanism for benzene destruction, the following statements can be
made. (1) Abstraction of hydrogen is only a minor channel. (2)
Without O-atom measurements an O-atom addition channel cannot be
confirmed. Since the only evidence that is in conflict with this
channel is based on the qualitative features of the flux profiles of
benzene and acetylene and extrapolation of low temperature rate coefficients to flame temperatures this channel cannot be eliminated.

The early appearance of a mass 66 species (C_5H_6) along with
the mass 94 species (C_6H_6O) (Figure 4) at 6mm where benzene decay
and CO production begin (Figure 3B) suggests that the initial attack
by O or OH is followed by production of CO and a C_5H_6 hydrocarbon.
Exclusion of a possible contribution to the mass 66 signal by a
fragment of mass 94 is difficult since the appearance potential of
the cyclo-$C_5H_6^+$ fragment ion from ground state phenol is low
(estimated to be 10.2 eV from heats of formation available in the
literature[54]). Fragmentation at lower electron energies of the
energy-rich adduct formed by reaction of benzene with O or OH cannot
be excluded even though the experimental ionization potential (IP)
of mass 66, referenced to the experimental IP of mass 94, was 8.4 eV;
However, two independent observations support the argument that the
mass 66 signal is from a neutral species generated in the flame.
First, two C_5H_6 hydrocarbons have been detected by GC-MS analysis of
batch samples collected from the flame via conventional microprobes[55].
Second, a natural flame ion unique to benzene flames has been observed
at mass 67. This ion presumably is formed, as many hydrocarbon flame
ions are, by protonation of an unsaturated natural species at mass
66[56].

An interesting feature of the proposed reaction scheme is the
production of C_2H_2 from benzene by removal of one carbon atom at a
time in the form of CO, H_2CO, or CH_3, if the routes depicted by the
dashed and dotted lines are neglected. The observation that the mass
flux of C_2H_2 (1.78 x 10^{-4}g cm^{-2} s^{-1} maximum) approaches but never
exceeds one-third the initial mass flux of benzene (5.67 x 10^{-4}g cm^{-2}
s^{-1}) is consistent with this mechanism. The closeness of the approach
of the C_2H_2 mass flux to one-third of the initial benzene mass flux
suggests a two-stage mechanism whereby each mole of benzene produces
a mole of C_2H_2 that is refractory relative to benzene and other
hydrocarbon intermediates. This is consistent with the aforementioned
hypothesis that OH is the major consumer of benzene and that
$k_{OH+C_6H_6} \cdot x_{C_6H_6} >> k_{OH+C_2H_2} \cdot x_{C_2H_2}$. Hence, prior to about 10mm, where
benzene is almost depleted and the OH mole fraction rises dramatically

C_2H_2 is essentially only <u>produced</u>, i.e., its destruction is slow. Even after 10mm, where OH mole fractions are high, the C_2H_2 decay is slow enough that C_2H_2 persists beyond 40mm (Figure 2).

Another contributor to the close approach of the C_2H_2 mass flux to one-third of the initial C_6H_6 mass flux may be that a significant fraction of the C_2H_2 is produced from C_4 species via the routes depicted by the dotted and dashed lines. The observation, at 9 - 10mm where the H-atom mole fraction rises rapidly, that H_2 and C_2H_2 become equilibrated with C_4H_2 through the overall reaction,

$$2 \ C_2H_2 \ \rightleftharpoons \ C_4H_2 + H_2 \hspace{4cm} \text{(Rxn.3)}$$

indicates that a fast route for production of two moles of C_2H_2 from the C_4 hydrocarbons exists. The fact that mole fractions of C_2H consistent with equilibration of the route shown by the dotted and dashed lines were not observed in this flame, although they should be well within the limits of detectability of the apparatus[20,21,57], together with the observation in rather similar flames of equilibria among hydrocarbon ions, C_2H_2, C_4H_2, and H_2[58], suggests that ion-molecule reactions may be responsible for the observed equilibrium (Rxn.3). This equilibrium is also responsible for the C_4H_2 flux maximizing near the C_2H_2 flux maxima rather than prior to that of C_3H_3.

The flux profiles of the C_1, C_3, and C_4 hydrocarbon inter-mediates (Figures 4 and 5) demonstrate the importance of diffusion of even relatively large molecules in the main reaction zone (6 - 10mm). The C_4H_4 flux (Figure 4) increases from zero at about 6.5mm. Between 6.5 and 8mm where the net flux of C_4H_4 is increasing, the C_4H_2 flux and its derivative with respect to distance, z, are negative. Thus, C_4H_2 formed at z > 7.8mm is diffusing towards the burner and being consumed at distances less than 7.8mm. Some of this C_4H_2 may form C_4H_4 via the C_4H_3 radical as suggested in the reaction scheme dis-cussed above. The sum of the C_4H_2 and C_4H_4 fluxes shown in Figure 6 indicates that C_4H_4 formation almost accounts for the net con-sumption of C_4H_2 between 6 and 8mm. Although this discrepancy may be within the experimental accuracy, the C_4H_6 and C_4H_3 mole fraction profiles (Figure 7B) indicate that formation of C_4H_6 and C_4H_3 may account for some of the consumption of C_4H_2 and C_4H_4 between 6 and 8mm.

Similar relationships exist for the fluxes of C_3H_3 and C_3H_4, and CH_3 and CH_4. The C_3H_4 flux (Figure 5) remains positive and maximizes closer to the burner than the C_3H_3 flux, which indicates net con-sumption of C_3H_3 at distances less than 7.2mm. The sum of the C_3H_3 and C_3H_4 fluxes (Figure 6) shows that the negative C_3H_3 flux is nearly offset by the positive C_3H_4 flux prior to 7.5mm. This suggests that C_3H_4 is formed from C_3H_3 that has diffused from the

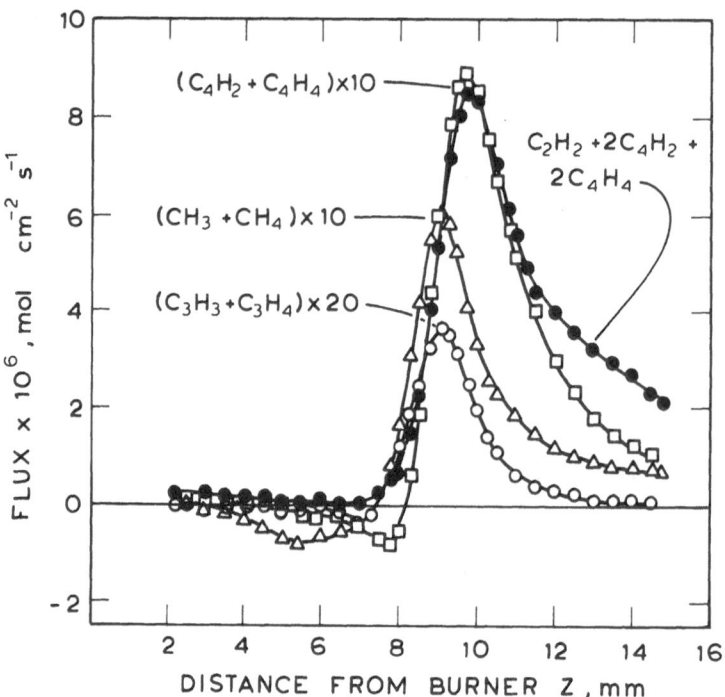

Figure 6.   Sums of fluxes of C_4H_2 and C_4H_4, CH_3 and CH_4,
            C_3H_3 and C_3H_4, and C_2H_2, 2 x C_4H_2 and 2 x C_4H_4 vs.
            distance from burner in a near-sooting ($\emptyset = 1.8$)
            benzene (13.5 m%) - oxygen (56.5 m%) - argon
            (30.0 m%) flame. Cold gas velocity = 0.5 m/s.
            Pressure = 2.67 kPa (20 torr).

region of net C_3H_3 production ($z > 7.2$mm). The CH_3 and CH_4 flux
profiles (Figure 5) also indicate that the stable CH_4 is formed from
the CH_3 radical as it diffuses towards the burner. The net consumption of CH_3 and CH_4 between 3 and 6mm (Figure 6), indicated by
the negative flux with a negative slope, may be due to reaction of
CH_3 to form many of the methyl substituted aromatics observed in
the flame or it may be caused by uncertainties in diffusion coefficients and alignment of temperature and mole fraction profiles.

The flux relationships just described suggest that species of the
same carbon number exchange hydrogen rapidly via reactions with H
and H_2. The more hydrogenated species are formed in the lower temperature region closer to the burner where their stability is more
favorable thermodynamically. They appear to be formed from less
hydrogenated species of the same carbon number that diffuse towards
the burner from the later stages of the primary reaction zone. As
discussed below, the presence of these more hydrogenated species may
contribute to the formation of polycyclic aromatic material in
benzene flames.

FORMATION OF PAH AND SOOT IN BENZENE FLAMES

Mole fraction profiles of several PAH observed in the $\emptyset = 1.8$
near-sooting benzene flame are shown in Figure 8. Possible structures
for these species and those at other masses at which signals were
observed are shown in Table I. The mole fractions of all PAH
maximize between 8 and 10mm above the burner. These relatively low-
mass PAH (≤ 228 amu) all disappear, as does benzene, by 10 or 11mm.
Much of the decrease in the PAH mole fractions is probably due to
oxidation. However, some of these species continue to grow, as
shown in Figure 9 by the high-mass signals for the $\emptyset = 1.8$ flame.
These signals were obtained by operating the quadrupole mass spectrometer as a high-pass filter that transmits all ions larger than a
specified mass. The signal from species of mass 200 and larger,
$I_{M > 200}$, maximizes at about 8.5mm, while $I_{M > 700}$ is much lower at
8.5mm and maximizes at 10.5mm. This progression from PAH in the
mass range 100-200 amu that maximize between 8 and 9mm to material
with mass greater than 700 amu that maximizes at 10.5mm indicates
that the PAH material continues to add carbon. Further indication
of this growth is a shift of the mass distribution of the 200-700 amu
material to higher masses from 8 to 10mm[59]. Since the mass number
at which signals are observable between 200 and 300 amu correspond
to aromatic structures it is believed that this high-mass material
is aromatic rather than polyacetylenic.

The correlation of the growth of this polycyclic material with
the onset of soot formation was demonstrated by measurement of
$I_{M > 700}$ as $\emptyset$ was increased at constant cold gas velocity, pressure,
and mole percent argon. The orange luminosity characteristic of

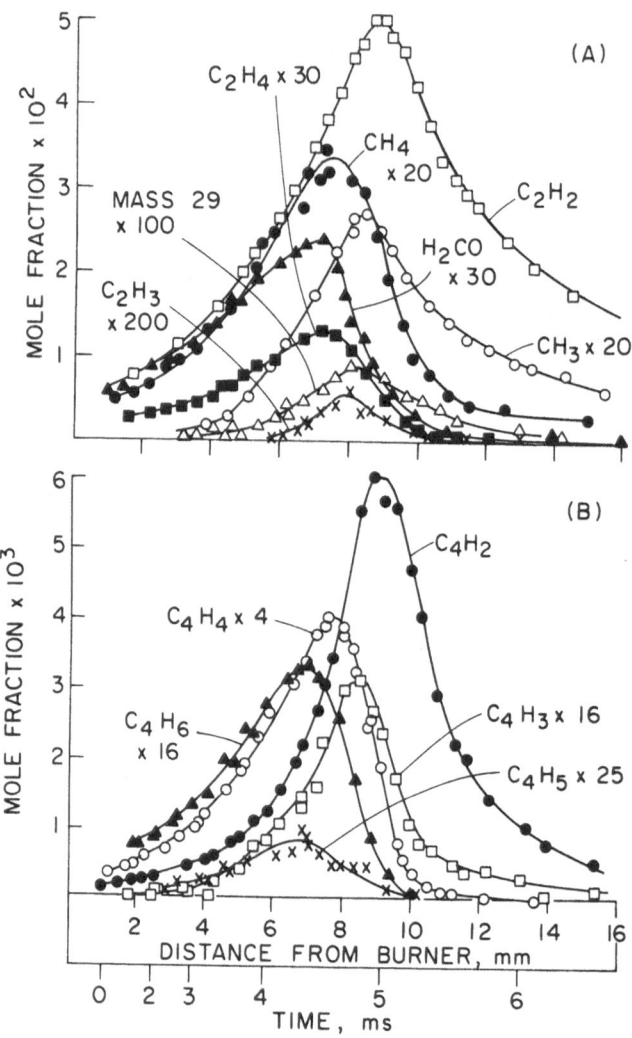

Figure 7.   Mole fractions of C_1 and C_2 (A) and C_4 (B) species vs. distance from burner in a benzene (13.5 m%) – oxygen (56.5 m%) – argon (30.0 m%) flame.   Cold gas velocity = 0.5 m/s.   Pressure = 2.67 kPa (20 torr).

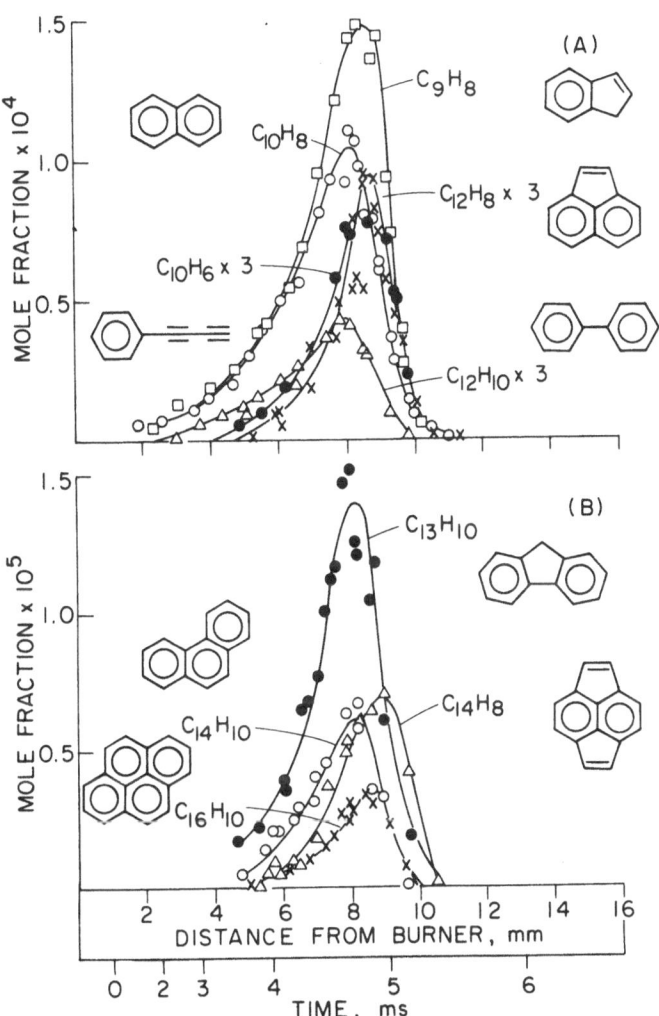

Figure 8.   Mole fractions of polycyclic aromatic hydrocarbons vs.
distance from burner in a near-sooting (∅ = 1.8) benzene
(13.5 m%) - oxygen (56.5 m%) - argon (30.0 m%) flame.
Cold gas velocity - 0.5 m/s.   Pressure = 2.67 kPa (20 torr).

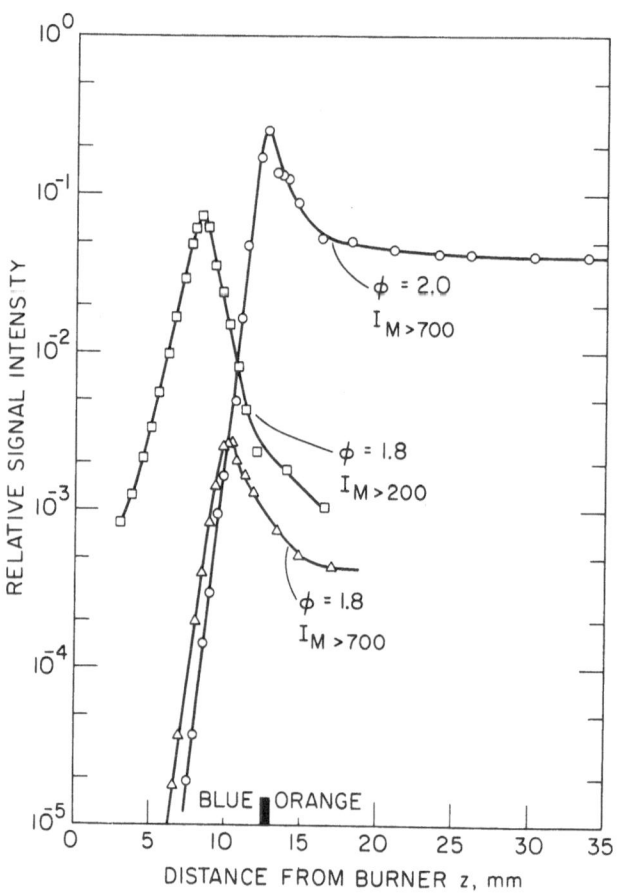

Figure 9. Relative intensities of signals from high mass mode
operation (see text) vs. distance from burner. (□) species
larger than mass 200, $I_{M>200}$; and (△) species larger than
mass 700, $I_{M>700}$, in near sooting (∅ = 1.8) benzene
(13.5 m%) - oxygen (56.5 m%) - argon (30.0 m%) flame. (o)
species larger than mass 700, $I_{M>700}$, in sooting (∅ = 2.0)
benzene (14.7 m%) - oxygen (55.3 m%) - argon (30.0 m%)
flame. Pressure = 2.67 kPa (20 torr), cold gas velocity
= 0.5 m/s for both ∅ = 1.8 and ∅ = 2.0 flames. Shaded
region at 13 mm designates blue-orange boundary in sooting
(∅ = 2.0) flame.

soot becomes visible at $\emptyset = 1.9$. The $I_{M>700}$ signal at $\emptyset = 2.0$,
shown in Figure 9 on the same basis as that at $\emptyset = 1.8$, rises
sharply to a maximum and then declines rapidly at 13mm where the
orange luminosity becomes visible. The $I_{M>700}$ signal maximum at
$\emptyset = 2.0$ is about one hundred times greater than that of the non-
sooting $\emptyset = 1.8$ flame. This signal should be proportional to the
number density of the high-mass species. Fluorescence signals
measured in atmospheric pressure C_6H_6-air flames show a similar
behavior[14].

Comparison of the mole fraction profiles of C_6H_6, C_2H_2, and
O_2 for the $\emptyset = 1.8$ and $\emptyset = 2.0$ flames (Figure 10) with the $I_{M>700}$
profiles provides some insight into the possible role of aromatics
in this rapid growth process. For both flames, the C_6H_6 mole
fraction decreases very rapidly where the $I_{M>700}$ signals of Figure
8 maximize [as indicated in Figure 10 by the dashed (---) and
dotted (-•-•-•) lines for $\emptyset = 1.8$ and $\emptyset = 2.0$, respectively]. It
is interesting that in the $\emptyset = 1.8$ flame, this point (10mm) also
corresponds to the region of the sharp increase in the mole fractions
of H and OH (Figure 2). Assuming that the high-mass aromatic
material is formed by a radical mechanism, the region of maximum
growth will be where both aromatic species and radicals are in high
concentration. The region of rapid increase in the number concentration of species with masses greater than 700 amu in the $\emptyset = 1.8$
flame (7-10mm), as indicated by $I_{M>700}$ in Figure 9, also is a region
of relatively high mole fractions of benzene, other monocyclic
aromatics, small unsaturated aliphatic species, and low-molecular-
weight PAH. The sudden decrease in net production of M>700 species,
as indicated by the sharp maximum and almost equally rapid decline
beyond 10mm, probably is due to the very rapid disappearance of
benzene and other low-molecular-weight aromatics near 10mm (Figure
10). As $\emptyset$ is increased to 2.0 the benzene mole fraction beyond 10mm
increases more than an order or magnitude, and $I_{M>700}$ continues to
increase two orders of magnitude beyond its maximum in the $\emptyset = 1.8$
flame. Again, the point at which $I_{M>700}$ maximizes corresponds to
the region of rapid benzene decay (12.7mm). Thus, both the benzene
mole fraction and $I_{M>700}$ between 10 and 13mm are very sensitive to
changes in $\emptyset$, while the C_2H_2 and O_2 mole fractions are relatively
insensitive to $\emptyset$. The mole fractions of the low-molecular-weight
aromatics may control the number concentration of rapidly growing
high-mass species. Most of the mass addition to the high-mass
material is probably from non-aromatics like C_2H_2, C_4H_4, and C_4H_2.

From the behavior of $I_{M>700}$ and the mole fraction profiles of
C_2H_2, O_2, and C_6H_6, no recognizable change in the flame structure
occurs at the soot limit with regard to the aromatics, such as C_6H_6
surviving in significant quantities into the post-flame zone or the
O_2 mole fraction declining to negligible values prior to the disappearance of C_6H_6. The appearance of soot in the $\emptyset = 2.0$ flame but

Table I. Possible Structures of Several Hydrocarbons
Observed Above Mass 90.

MASS	FORMULA	NAME	STRUCTURE	MASS	FORMULA	NAME	STRUCTURE
92	C_7H_8	Toluene		170	$C_{14}H_2$	Tetradecaheptayne	
98	C_8H_2	Octatetrayne		176	$C_{14}H_8$	Cyclopent[f,g]acenaphthylene	
102	C_8H_6	Phenylacetylene				Cyclopent[b,c]acenaphthylene	
104	C_8H_8	Styrene		178	$C_{14}H_{10}$	Anthracene	
116	C_9H_8	Indene				Phenanthrene	
		Ethynyltoluene		190	$C_{15}H_{10}$	4H-Cyclopenta[def]phenanthrene	
122	$C_{10}H_2$	Decapentayne		192	$C_{15}H_{12}$	Methylanthracene	
126	$C_{10}H_6$	1-Phenylbutadiyne				Methylphenanthrene	
128	$C_{10}H_8$	Naphthalene		202	$C_{16}H_{10}$	Fluoranthene	
140	$C_{11}H_8$	Butadiynyltoluene				Pyrene	
142	$C_{11}H_{10}$	Methylnaphthalene		216	$C_{17}H_{12}$	Benzofluorene	
146	$C_{12}H_2$	Dodecahexayne				Methylfluoranthene	
152	$C_{12}H_8$	Biphenylene				Methylpyrene	
		Acenaphthalene		226	$C_{18}H_{10}$	Benzo[ghi]fluoranthene	
154	$C_{12}H_{10}$	Acenaphthene				Cyclopenta[cd]pyrene	
		Biphenyl		228	$C_{18}H_{12}$	Chrysene	
166	$C_{13}H_{10}$	Fluorene				Benz[a]anthracene	
		Methylacenaphthalene					
		Phenalene					
168	$C_{13}H_{12}$	Methylbiphenyl					

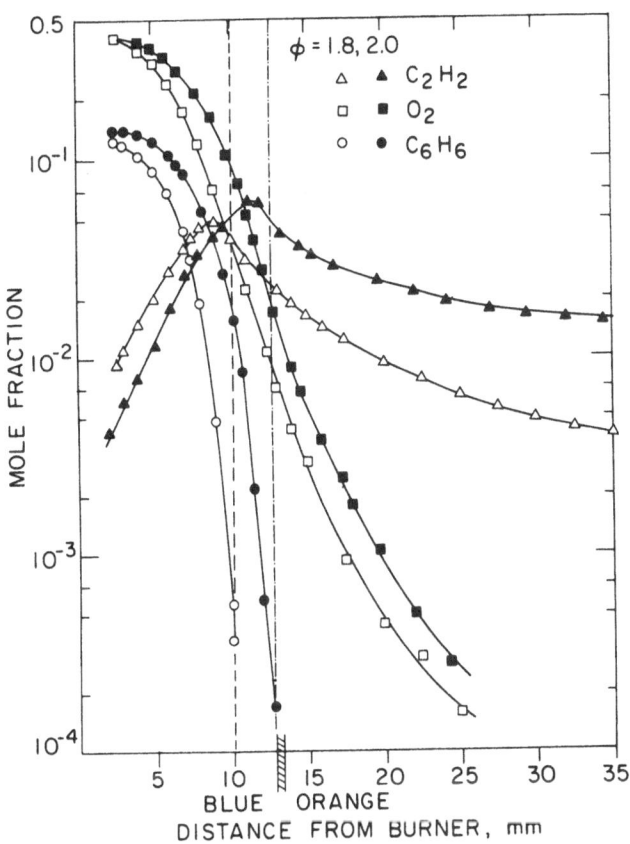

Figure 10.   Log of mole fraction of C_2H_2 (triangles), O_2 (squares)
and benzene (circles) vs. distance from burner for near-
sooting (Ø = 1.8, open symbols) and sooting (Ø = 2.0,
shaded symbols) flames.   For Ø = 1.8: benzene = 13.5 m%,
oxygen = 56.5 m%, argon = 30.0 m%.   For Ø = 2.0: benzene =
14.7 m%, oxygen = 53.3 m%, argon = 30.0 m%.   Pressure
= 2.67 kPa (20 torr), v = 0.5 m/s for both flames.
Dashed (---) and dotted (-.-.-) lines designate position
of maxima of $I_{M>700}$ in Ø = 1.8 and Ø = 2.0 flames,
respectively.   Shaded region at 13mm designates blue-
orange boundary in the sooting (Ø = 2.0) flame.

not the $\emptyset = 1.8$ flame apparently is due to a combination of the
high-mass species growing to larger sizes in the richer flame where
they emit in the orange continuum and their concentration increasing
to where the emission is detectable by the eye.

The structures of the lower-molecular-weight PAH formed in the
early stages of this rapid growth process (Figure 8, Table I) indicate
that they are products of reactions of aromatic species with C_2, C_3,
and C_4 species. Furthermore, the region of rapid increase in the PAH
mole fractions (7-8mm) corresponds to the region where the mole
fractions of more hydrogenated species, i.e., C_2H_4, C_4H_4, and C_4H_6
(Figure 7), and phenyl and benzyl radicals[21] maximize. These observations are consistent with those of Smith[60,61] in toluene pyrolysis
between 1173 and 2173K. He found the concentration of higher-molecular-
weight hydrocarbons (up to $C_{14}H_{10}$) to maximize between 1573 and 1673K
where the mole fractions of phenyl radical, C_4H_4, C_3H_3, C_3H_4 and other
more hydrogenated hydrocarbons were also near their maximum concentrations. Beyond 1673K the concentration of these lower-molecular-
weight species and the higher-molecular-weight hydrocarbons declined
as C_2H_2 and C_4H_2 continued to increase, similar to the behavior
observed in the percent $\emptyset = 1.8$ flame. Thus, from the observations
made in the present $\emptyset = 1.8$ benzene flame and Smith's toluene pyrolysis
experiments, it appears that a proper mix of monocyclic aromatic
species, e.g., benzene, phenyl radical, toluene, and benzyl radical,
and non-aromatics such as C_2H_2, C_4H_2, C_4H_4, C_4H_3, C_2H_4, etc. results
in rapid production of higher-molecular-weight PAH. The role of the
initial ring in this process may be to provide a structure that is
capable of stabilizing the radical adduct formed by addition of non-
aromatic species.

For rapid growth of aromatic structures by a free radical
mechanism, the stabilization of the energy-rich adduct formed by
addition of radicals to stable hydrocarbons may be a crucial step.
Removal of the energy of chemical excitation through collisional
processes may not be sufficient to prevent unimolecular fragmentation
at temperatures above 1500K where PAH and soot are formed in pyrolysis
systems[60,61], diffusion flames[62] and the present $\emptyset = 1.8$ and $\emptyset = 2.0$
benzene flames. The unimolecular decomposition of the thermalized
radical adducts can be very fast at 1500K and above. For example,
from transition state theory the half-life, $t_{1/2}$, of a radical that
decomposes unimolecularly through a process with an activation
energy of 209 kJ mol^{-1} (50 kcal mol^{-1}) is only about 10^{-7}s at 1500K
if it is assumed that the entropy of activation is zero. This is
of the same order as the time between collisions of a molecule having
the collision diameter of toluene with other species of possible
importance in the growth process (e.g., H, $2x10^{-6}$s; H_2, $5x10^{-8}$s;
C_2H_2, $2x10^{-7}$s; C_4H_2, $3x10^{-6}$s; C_4H_4, 10^{-5}s, at 7.5mm in the present
$\emptyset = 1.8$ flame). Thus, unimolecular reactions with activation energies
less than about 210 kJ mol^{-1} are much faster than bimolecular reactions.

Therefore, the channel of decomposition of the radical adduct which generally has one or more bonds with dissociation energies less than 210 kJ mol^{-1} is important in determining whether growth into fused-ring aromatic structures or fragmentation of the carbon framework occurs.

The exact role of the aromatic ring in stabilizing the radical adducts is uncertain. To gain insight into the rapid growth of aromatic structures, such as the form of the aromatic reactant (radical or stable molecule) and the nature of the "growth" species that are responsible for the major fraction of the mass that ends up as high-molecular-weight PAH and/or soot, several mechanisms have been considered on the basis of the enthalpies of reactions of the initial addition steps, the subsequent steps required to stabilize the adduct by formation of six-membered rings, and other processes that may stabilize the adduct without forming six-membered rings. The principal mechanisms considered have been (A) the addition of non-aromatic unsaturated radicals to stable aromatic structures (aromatic substitution reactions), and (B) the addition of aromatic radicals to non-aromatic stable species.[20]

For this analysis, thermochemical data for the stable structures considered were estimated from the group additivity method of Benson[63] and Stein et al.[64] The enthalpies of formation, ΔH°_{f298}, of radicals were estimated from ΔH°_{f298} of stable species and known bond dissociation energies [65] when available. The ΔH°_{f298} of C_4H_3 and C_4H_5 radicals (two isomers each) and those of similar structure were estimated from the methods of Cowperthwaite and Bauer,[66] heats of hydrogenation from Skinner and colleagues[67-69] and updated thermochemical data,[65] and are reported in ref. 20. When dissociation energies were not available for bonds which break to give π-bonding resonance-stabilized radicals, the resonance stabilization energies for alternant radicals were estimated using the methods of Stein and Bolden.[70]

(A) To summarize the results of the preliminary analysis of the addition of the non-aromatic unsaturated radicals C_2H_3, C_3H_3, and C_4H_5 to benzene: fused-ring compounds such as naphthalene and indene are not likely to be formed by rapid unimolecular rearrangement of the cyclohexadienyl-type radical adducts. In some case, e.g. addition of the 1,3-butadien-1-yl (1-C_4H_5) radical to benzene (Rxn.4), the favored channel is H-atom elimination from the cyclohexadienyl-type adduct (Rxn.5) to form a stable structure with an unsaturated side chain (D) rather than formation of the fused-ring compound (E) via Rxn.6.

ΔH°_{f298} 83 kJ mol^{-1} 344   320

$\Delta H^{\circ}_{Rxn.298}$ = −107 kJ mol^{-1}

(Rxn.4)

$$\rightarrow \quad 208 \text{ (D)} \quad + \quad \text{H} \atop 218 \qquad 106 \qquad \text{(Rxn. 5)}$$

$$\rightleftharpoons \quad 301 \quad \rightarrow \quad 265 \text{ (E)} \quad + \quad \text{H} \quad -19,182 \quad \text{(Rxn. 6)}$$

In other cases, e.g., C_3H_3 addition (Rxn.7), decomposition of the adduct back to reactants (Rxn.-7) is favored over H-atom elimination (Rxn.8).

$$+ \quad \bullet\!\!-\!\!\equiv \quad \rightleftharpoons$$

$$\Delta H^\circ_{f298} \quad 83 \text{ kJ mol}^{-1} \quad 362 \pm 5 \quad 405 \pm 21 \qquad \Delta H^\circ \text{ Rxn. 298} \atop -39 \text{ kJ mol}^{-1} \qquad \text{(Rxn. 7)}$$

$$\rightarrow \quad 302 \quad + \quad \text{H} \atop 218 \quad 114 \qquad \text{(Rxn. 8)}$$

Some of the substitution products (formed by H-atom elimination from the cyclohexadienyl-type radical) with unsaturated side chains may participate in subsequent bimolecular reactions that yield radicals which may then unimolecularly form fused aromatic ring structures. For example, the 1-phenyl-cis 1,3-butadien-4-yl radical formed by H-atom abstraction from the 4-position of 1-phenyl-cis 1,3-butadiene (D) formed in Rxn.5 may cyclize via an internal aromatic substitution (Rxn9).

$$\rightleftharpoons \quad 326 \quad \rightarrow \quad 150 \quad + \quad \text{H} \atop 218 \qquad \text{H}^\circ \text{ Rxn. 298} \atop -112,42 \quad \text{(Rxn. 9)}$$

$$\Delta H^\circ_{f298} \quad 438 \text{ kJ mol}^{-1}$$

(B) The analysis of the addition of aromatic radicals to non-aromatic unsaturated molecular species included addition of phenyl radical (C_6H_5) and benzyl radical (C_7H_7) to acetylene, butadiyne (C_4H_2), 1-buten-3-yne (C_4H_4), methylacetylene (C_3H_4) and allene (also C_3H_4). Of those considered, there are only two mechanisms, C_6H_5 plus C_4H_4 and C_7H_7 plus C_3H_4 (allene), whereby unimolecular rearrangement of the initial adduct to form fusing six-membered ring systems is favored over other decomposition channels. In general the initial adducts formed by C_7H_7 addition (e.g., to C_2H_2, Rxn.10) are unlikely to stabilize by H-atom elimination (Rxn.11) to form species with unsaturated side chains. The carbon framework is stabilized by

H-atom elimination only after the initial adduct has undergone subsequent internal addition reactions such as

$$\Delta H^\circ_{Rxn.298} \quad -45 \text{ kJ mol}^{-1} \quad \text{(Rxn. 10)}$$

ΔH_{f298} 188 kJ mol^{-1} 227   370

    302   +   H   150   (Rxn. 11)
            218

283   173   + H   -87, 108   (Rxn. 12)

cyclization to a five-membered ring (Rxn.12). On the other hand, initial adducts formed by C_6H_5 addition (e.g., Rxn.13) are likely to eliminate H-atoms directly and form species with unsaturated side chains (Rxn.14) rather than form fused-ring species.

$$\Delta H^\circ_{Rxn.298} \quad -226 \text{ kJ mol}^{-1}$$

ΔH°_{f298} 328   456   558   (Rxn.13)

(F)   558   +   H $_{218}$ (Rxn.14)
          218

However, the vinyl-type radicals (e.g., F) formed from either phenyl radical addition to a triple bond (Rxn.13) or H-atom addition to a triple bond of a side chain (Rxn.-14) may react with C_2H_2 (Rxn.15). The resulting vinyl-type radical may then cyclize to form a fused ring structure through

$$\Delta H^\circ_{Rxn298} \quad -118 \text{ kJ mol}^{-1}$$

ΔH°_{f298} 558 kJ mol^{-1} 227   667   (Rxn.15)

$-132,77$
(Rxn.16)

535   394   218

an internal aromatic substitution reaction (Rxn.16).

This brings the discussion again to the competition between unimolecular and bimolecular reaction steps. The half-life of vinyl-type radicals formed either by C_6H_5 addition to triple bonds (e.g., Rxn.13) or H-atom addition to the α-position of a triple bond conjugated with an aromatic ring (e.g., Rxn.-14) will be determined by the endothermicity of the H-elimination step (e.g., Rxn.14). The four vinyl-type adducts formed by C_6H_5 addition to C_2H_2, C_3H_4 (methylacetylene), C_4H_2, and C_4H_4 have enthalpies of reaction for H-atom elimination ΔH_{H-e} of 164, 157, 218, and 182 kJ mol^{-1}, respectively. Only the vinyl-type radical formed from C_4H_2 (F) has a ΔH_{H-e} larger than 210 kJ mol^{-1} and a half-life greater than 10^{-7}s at 1500K, which is of the same order as the time for a collision with C_2H_2 in the present $\emptyset = 1.8$ flame. It is the interaction of the unpaired electron with the triple bond of the vinyl-type radical (F) that is responsible for its added stability. Thus, aromatic structures with polyacetylenic side chains (C_4H and higher) may be more effective in forming fused-ring structures in an environment of high C_2H_2, H-atom, and C_4H_2 concentrations than species without the conjugated triple bonds.

The preceding analysis indicates that there are few mechanisms in which the favored channel for unimolecular stabilization of the radical formed by addition of aromatic and non-aromatic species might result in formation of fused aromatic ring systems. Many reactions may result in stable species with unsaturated side chains. The stability of radicals formed by addition of H-atoms to unsaturated side chains is important in determining whether bimolecular addition of more carbon, probably in the form of C_2H_2, can compete with unimolecular decomposition of the radical. Vinyl-type radicals formed from polyacetylenic side chains are more stable with respect to H-atom elimination than radicals from other types of side chains. Although phenyl radical addition to unsaturated non-aromatics is probably the dominant route for formation of aromatic species with side chains, the essential feature that the initial aromatic structure contributes to the PAH and soot formation process is the ability to undergo substitution reactions[60]. Internal aromatic substitution reactions (e.g., Rxn.16) are important for stabilizing the growing carbon network. Thus, the initial aromatic structure provides a foundation upon which a larger aromatic structure is built by addition

of non-aromatics such as C_2H_2, C_4H_4, and C_4H_2. Reactions such as
13, 15, and 16 appear to provide an especially favorable route to
ring-building. This is interesting in that it suggests that acetylene
and polyacetylenic species are essential to the formation of PAH in
aromatic flames, and that it is the combined presence of aromatics
and polyacetylenes in these flames that promotes the rapid growth
of PAH and soot.

Other mechanisms for growth of fused-ring aromatic structures
that require methyl substitution and rely on internal aromatic substitution reactions for adduct stabilization have been discussed in
a previous paper[21]. Our present thinking is that methyl substitution
reactions may well contribute to PAH and soot, but that mechanisms
of the type just discussed are more significant. Other mechanisms
considered[20] have included Diels-Alder reactions, 4-center reactions
(e.g. 2 $C_6H_6 \rightarrow C_{12}H_{10} + H_2$), and ionic mechanisms. The first two
are too slow to play significant roles and the second has the further
disadvantage of being unable to build fused-ring systems. Ionic
mechanisms, which often are appealing because of high rate constants,
may well contribute but information on ions in the present study is
insufficient to permit elucidation of their roles.

SUMMARY

Species and reactions involved in the early stages of soot
formation prior to the formation of particles were studied in flat
low-pressure flames of benzene using a molecular beam mass spectrometer system. In a near-sooting (equivalence ratio $\emptyset = 1.8$) benzene
oxygen-argon flame, the profiles of molar flux versus distance from
the burner for several reactant, intermediate, and product species
were calculated from measured mole fraction profiles and transport
considerations. The flux profiles indicate that CO, C_6H_6O and C_5H_6
are very early intermediates in the ring destruction process but
that C_2H_2 is not. Therefore, unimolecular fragmentation of phenyl
and cyclohexadienyl radicals to produce C_2H_2 is not significant in
this flame. Although it is not possible to identify with certainty
the main mechanism of benzene destruction, formation of phenyl radical
by hydrogen abstraction is only a minor route. Formation of C_6H_6O
may occur by reaction of benzene with either OH or O. The qualitative
features of the benzene and acetylene flux profiles suggest that
OH may be the major route. However, this is in conflict with studies
of the OH + C_6H_6 reaction[51,52].

A mechanism has been outlined for production of C_1 to C_5 species
from benzene that is consistent with the flux profiles of intermediate
species. The production of these smaller species appears to take
place through a sequence of steps each of which involves the loss of
a single carbon atom. Relationships between the fluxes of C_1, C_3
and C_4 species suggest that species of the same carbon number exchange

hydrogen rapidly via reactions with H and H_2. The more hydrogenated
species present closer to the burner may be formed from less hydrogenated species as they diffuse from near the end of the primary
reaction towards the burner.

Relative measurements of high molecular weight material (>200 amu)
as the fuel equivalence ratio is increased past the sooting limit
($\emptyset$ = 1.9) in benzene flames suggest a sequential growth process from
polycyclic aromatic hydrocarbons (PAH) in the mass range 120-210 amu
to higher molecular weight hydrocarbons and then to soot. An heuristic
analysis of reaction mechanisms suggests that the role of the aromatic
hydrocarbon is to provide a structure capable of stabilizing, by
internal aromatic substitution reactions, the radicals formed from
addition of aromatic radical species to non-aromatics. Specifically,
it appears that the combined presence of aromatic radicals, unsaturated aliphatics such as polyacetylenes and H atoms is particularly
favorable to growth of PAH and soot. Therefore, the production of
phenyl radical and the competition between its destruction, especially
by reactions with O_2, and addition to unsaturated aliphatics are
important factors in PAH formation.

ACKNOWLEDGEMENTS

J. D. Bittner and J. B. Howard are grateful to the U. S.
Environmental Protection Agency for support under Grant No. R803242
in the early part of this work. H. B. Palmer acknowledges support
from the Department of Energy under Contract DE-AC21-79MC12736.

REFERENCES

1. J. D. Bittner and J. B. Howard, Alternative Hydrocarbon Fuels:
   Combustion and Chemical Kinetics, (C.T. Bowman and J.
   Birkeland, Eds.) Prog. in Astro. and Aero., AIAA, Vol. 62,
   p.335 (1978)
2. H. F. Butze and R. C. Ehlers, "Effect of Fuel Properties on
   Performance of a Single Aircraft Turbojet Combustor", NASA
   TM-71789,(1975)
3. W. S. Blazowski, Sixteenth Symp. (Intern'l) on Comb., The Comb.
   Institute, Pittsburgh, p.1631 (1977)
4. N. J. Friswell, Comb. Sci. and Tech., 19:119 (1979)
5. S. T. Minchin, J. of Inst. Petrol. Tech., 17:102 (1931)
6. A. E. Clarke, T. G. Hunter and F. H. Garner, J. of Petrol. Tech.
   32:627 (1946)
7. R. A. Hunt, Jr., Ind. and Eng. Chem., 45:602 (1953)
8. R. L. Schalla and G. E. MacDonald, Ind. and Eng. Chem., 45:1497
   (1953)
9. K. P. Schug, Y. Manheimer-Timnat, P. Yoccarino and I. Glassman,
   Comb. Sci. and Tech., 22:235 (1980)

10. J. C. Street and A. Thomas, Fuel, 34:4 (1955)
11. J. J. Macfarlane, F. H. Holderness and F. S. E. Witcher,
    Comb. and Flame, 8:15 (1964)
12. F. J. Wright, Twelfth Symp.(Intern'l) on Comb., The Comb.
    Institute, Pittsburgh, p.867 (1969)
13. W. S. Blazowski, R. B. Edelman and P. T. Harsha, "Fundamental
    Characterization of Alternate Fuel Effects in Continuous
    Combustion Systems", Technical Progress Report prepared
    for DOE Contract EC-77-C-03-1543 (1978)
14. B. S. Haynes, H. Jander and H. Gg. Wagner, Ber. Bunsenges.
    Phys. Chem., 84:585 (1980)
15. G. Prado, J. Jagoda, K. Neoh and J. Lahaye, Eighteenth Symp.
    (Intern'l) on Comb., The Comb. Institute, Pittsburgh,
    p.1127, (1981)
16. S. C. Graham, J. B. Homer and J. L. J. Rosenfeld, Proc. Roy.
    Soc. Lond., 344:259 (1975)
17. D. B. Scully and R. A. Davies, Comb. and Flame, 9:185 (1965)
18. R. A. Davies and D. B. Scully, Comb. and Flame, 10:165 (1966)
19. C. P. Fenimore, G. W. Jones and G. E. Moore, Sixth Symp.
    (Intern'l) on Comb., The Comb. Institute, Pittsburgh,
    p.242 (1957)
20. J. D. Bittner, A Molecular Beam Mass Spectrometer Study of
    Fuel-Rich and Sooting Benzene-Oxygen Flames, Sc.D. Thesis,
    Department of Chemical Engineering, M.I.T., Cambridge, MA,
    (1981)
21. J. D. Bittner and J. B. Howard, Eighteenth Symp. (Intern'l)
    on Comb., The Comb. Institute, Pittsburgh, p.1105, (1981)
22. R. M. Fristrom and A. A. Westenberg,"Flame Structure",McGraw-
    Hill, N.Y., pp.74-80 (1965)
23. D. F. Fairbanks and C. R. Wilke, Ind. and Eng. Chem., 42:471
    (1950)
24. R. M. Fristrom and A. A. Westenberg, op. cit., p.276
25. R. A. Svchla, "Estimated Viscosities and Thermal Conductivities
    of Gases at High Temperatures", NASA Technical Report
    R-132 (1962)
26. J. O. Hirschfelder, C.F. Curtiss and R. B. Bird,"Molecular
    Theory of Gases and Liquids", J. Wiley and Sons, N.Y.,
    p.1112 (1964)
27. L. S. Tee, S. Goton and W. E. Stewart, Ind. and Eng. Chem.
    Fundam., 5:356 (1966)
28. R. C. Reid, J. M. Prausnitz and T. K. Sherwood,"The Properties
    of Gases and Liquids", 3rd ed., McGraw-Hill, N.Y., pp. 24
    and 551 (1977)
29. R. S. Brokaw, Ind. Eng. Chem. Process Des. Dev., 8:240 (1969)
30. A. Savitsky and M. J. E. Golay, Anal. Chem., 36:1627 (1964)
31. J. Steinier , Y. Termonia and J. Deltour, Anal. Chem.,
    44:1906 (1972)
32. R. Q. Bonanno, P. Kim, J. H. Lee and R. B. Timmons, J. Chem.
    Phys., 57:1377 (1972)

33. T. M. Sloane, J. Chem. Phys., 67:2267 (1977)
34. G. Boocock and R. J. Cvetanovic, Can. J. Chem., 39:2436 (1961)
35. R. Kim, J. H. Lee, R. J. Bonanno and R. B. Timmons, J. Chem. Phys., 59:4593 (1973)
36. M. C. Sauer, Jr., and B. Ward, J. Phys. Chem., 71:3971 (1967)
37. R. A. Perry, R. Atkinson and J. N. Pitts, Jr., J. Phys. Chem. 81:296 (1977)
38. I. Mani and M. C. Sauer, Jr., Adv. Chem. Ser., 82:142 (1968)
39. S. J. Sibener, R. J. Buss, P. Casavecchia, T. Hirooka and Y. T. Lee, J. Chem. Phys., 72:4341 (1980)
40. R. Cypres and B. Bettens, Tetrahedron, 30:1253 (1974)
41. F. Slemr and P. Warneck, Ber. Bunsenge. Phys. Chem., 79:152 (1975)
42. J. R. Kanofsky, D. Lucas, F. Pruss and D. Gutman, J. Phys. Chem., 78:311 (1974)
43. H. Gg. Wagner and R. Zellner, Ber. Bunseges. Phys. Chem., 76:518 (1972)
44. K. H. Homann, W. Schwanebeck and J. Warnatz, Ber. Bunseges. Phys. Chem., 79:536 (1975)
45. H. Niki and B. Weinstock, J. Chem. Phys., 45:3468 (1966)
46. K. H. Homann, J. Warnatz and C. Wellman, Sixteenth Symp. (Intern'l) on Comb., The Comb. Institute, Pittsburgh, p.853 (1977)
47. J. Peeters and G. Mahenen, Combustion Institute European Symposium, Academic Press, N.Y., p.53 (1973)
48. J. D. Bittner and J. B. Howard, Nineteenth Symp. (Intern'l) on Comb., The Comb. Institute, Pittsburgh, 1983, to be published
49. R. Atkinson and J. N. Pitts, Jr., J. Phys. Chem., 79:295 (1975)
50. A. J. Colussi, J. L. Singleton, R. S. Irwin and R. J. Cvetanovic, J. Phys. Chem., 79:1900 (1975)
51. R. A. Perry, R. Atkinson and J. N. Pitts, Jr., J. Phys. Chem., 81:296 (1977)
52. F. P. Tully, A. R. Ravishankara, R. L. Thompson, J. M. Niocovich, R. C. Shah, N. M. Kreuter and P. H. Wire, J. Phys. Chem., 85:2262 (1981)
53. T. M. Sloane, Chem. Phys. Letters, 54:269 (1978)
54. F. H. Field and J. L. Franklin,"Electron Impact Phenomena and the Properties of Gaseous Ions", Academic Press, N.Y., pp.239-522 (1970)
55. S. M. Faist, Analysis of Stable Species in a Benzene-Oxygen-Argon Laminar Premixed Flame by Chemical and Spectroscopic Techniques: Applications to Soot Formation and Combustion Chamber Deposits, M. S. Thesis, Department of Chemical Engineering, M.I.T. (1979)
56. D. B. Olson and H. F. Calcote, Eighteenth Symp. (Intern'l) on Comb., The Comb. Institute, Pittsburgh, p.453 (1981)

57. J. D. Bittner, Comment following paper by P. Michaud,
     J. L. Delfau and A. Barassin, Eighteenth Symp. (Intern'l)
     on Comb., The Comb. Institute, Pittsburgh, p.451 (1981)
58. P. Michaud, J. L. Delfau and A. Barassin, Eighteenth Symp.
     (Intern'l) on Comb., The Comb. Institute, Pittsburgh,
     p.443 (1981)
59. J. B. Howard and J. D. Bittner, Structure of Sooting Flames,
     this volume
60. R. D. Smith, Comb. and Flame, 35:179 (1979)
61. R. D. Smith, J. Phys. Chem., 83:1553 (1979)
62. J. H. Kent, H. Jander and H. Gg. Wagner, Eighteenth Symp.
     (Intern'l) on Comb., The Comb. Institute, Pittsburgh,
     p.1117 (1981)
63. S. W. Benson, "Thermochemical Kinetics", J. Wiley and Sons, N.Y.,
     (1976)
64. S. E. Stein, D. M. Golden and S. W. Benson, J. Phys. Chem.,
     81:314 (1977)
65. "Handbook of Chemistry and Physics", 60th ed., (R. C. Weast,
     Ed.), CRC Press, Boca Raton, FL, p.F-231 (1979)
66. M. Cowperthwaite and S. H. Bauer, J. Chem. Phys., 36:1743
     (1962)
67. T. Flitcroft, H. A. Skinner and M. C. Whiting, Trans. Faraday
     Soc., 53:784 (1957)
68. T. Flitcroft and H. A. Skinner, Trans. Faraday Soc., 54:47
     (1958)
69. H. A. Skinner and A. Snelson, Trans. Faraday Soc., 55:404
     (1959)
70. S. E. Stein and D. M. Golden, J. Org. Chem., 42:839 (1977)
71. M. J. Perkins, in "Free Radicals", (J. Kochi, Ed.), J. Wiley
     and Sons, N.Y., p.231 (1973)

DISCUSSION

P. Cadman (The University College of Wales)

Is there any information known about the structure of the
very important species C_6H_6O ? Is it for e.g. phenol, species
with O across the ring or across the double bond of benzene ?

H. Gg. Wagner

In experiments about the reaction of O-atoms with C_6H_6 under
molecular beam conditions the complex C_6H_6O has been found and
three isomeres could be identified by appearance potential
measurements.

RADICAL CHEMISTRY IN SOOTING FLAMES

Jürgen Warnatz

Institut für Physikalische Chemie
Technische Hochschule Darmstadt
6100 Darmstadt, West Germany

INTRODUCTION

The purpose of this paper is (1) to review our knowledge on radical chemistry leading to soot precursors, and (2) to use this knowledge to give a quantitative explanation of measured concentration profiles of some higher hydrocarbons in sooting acetylene flames.

The paper shall be confined to the radical chemistry in premixed flames of aliphatic fuels. Flames of aromatic fuels are discussed elsewhere (see contribution by Bittner and Palmer).

Today, there is agreement that the formation of intermediate acetylene is strongly connected with the process of sooting in flames of aliphatic fuels, leading to special interest in acetylene combustion. Therefore, two topics of this paper will be (1) the formation of acetylene, and (2) its reactions leading to higher hydrocarbons which are potential soot precursors.

FORMATION OF ACETYLENE IN FLAMES OF ALIPHATIC FUELS

Experimental Results on Acetylene Formation

Figures 1a and 1b show examples of the formation of intermediate acetylene in methane-oxygen flames.[1,2] These measurements use molecular beam sampling and mass spectrometric analysis of low pressure burner-stabilized flat flames.

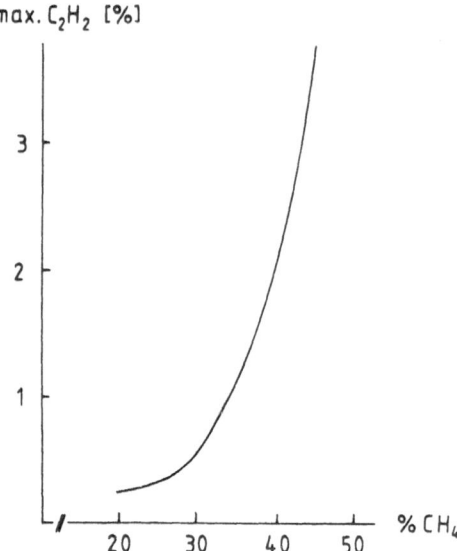

Fig. 1(a).   Maximum C_2H_2 mole fraction in burner-stabilized CH_4-O_2
flames at P = 46 mbar. Experiments by Wagner and co-
workers[1]

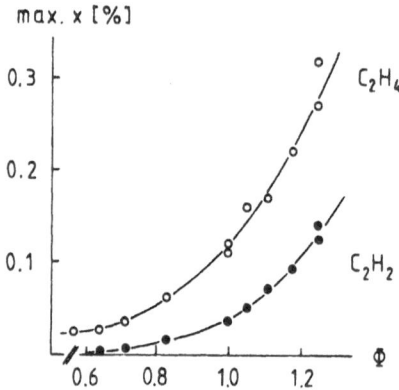

Fig. 1(b).   Maximum C_2H_2 and C_2H_4 mole fractions in burner-stabilized CH_4-O_2-Ar flames (43 % Ar) at P = 26 mbar.
Experiments by Harvey and McColl [2]

Both figures demonstrate a drastic increase of the maximum
C_2H_2 concentration with increasing equivalence ratio Φ. This increase can be described by a law

$$[C_2H_2]_{max} \sim \Phi^n$$

with n = 3 in Fig. 1a and with n = 4 in Fig. 1b.

Mechanism of Acetylene Formation

A quantitative description of this behavior of intermediate C_2H_2 can be given by use of a recently developed mechanism of the high temperature oxidation of aliphatic fuels.[3,4,5]   The validity of this mechanism has been demonstrated by its ability to explain quantitatively flame velocities and structures of flames of alkanes and alkenes up to C_4-compounds.

Alkanes are initially attacked (see Fig. 2a) by H, O or OH, which are formed by the chain-branching reactions of the H_2-O_2 system. The alkyl radicals formed in this way always decompose to smaller alkyl radicals by fast thermal elimination of alkenes. Only the relatively slow thermal decomposition of the smallest alkyl radicals CH_3 and C_2H_5 competes with recombination or disproportionation and with oxidation reactions by O atoms and O_2.

This part of the mechanism (see Fig. 2b) is rate-controlling in the combustion of alkanes and alkenes, and is therefore the reason for the similarity of all alkane and alkene flames (for details see [3,4]). CH_3 is converted to CO via CH_2O and CHO, whereas C_2H_5 is decomposed via C_2H_4, C_2H_3, C_2H_2, or CH_2CO. A further characteristic of this mechanism is the fast CH_3 recombination leading to considerable C_2-hydrocarbon formation in methane combustion.

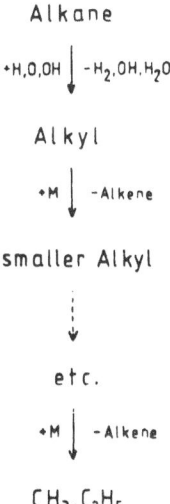

Alkane

$+H,O,OH$ | $-H_2,OH,H_2O$

Alkyl

$+M$ | $-Alkene$

smaller Alkyl

etc.

$+M$ | $-Alkene$

CH_3, C_2H_5

Fig. 2a.   Conversion of alkanes to CH_3 and C_2H_5.

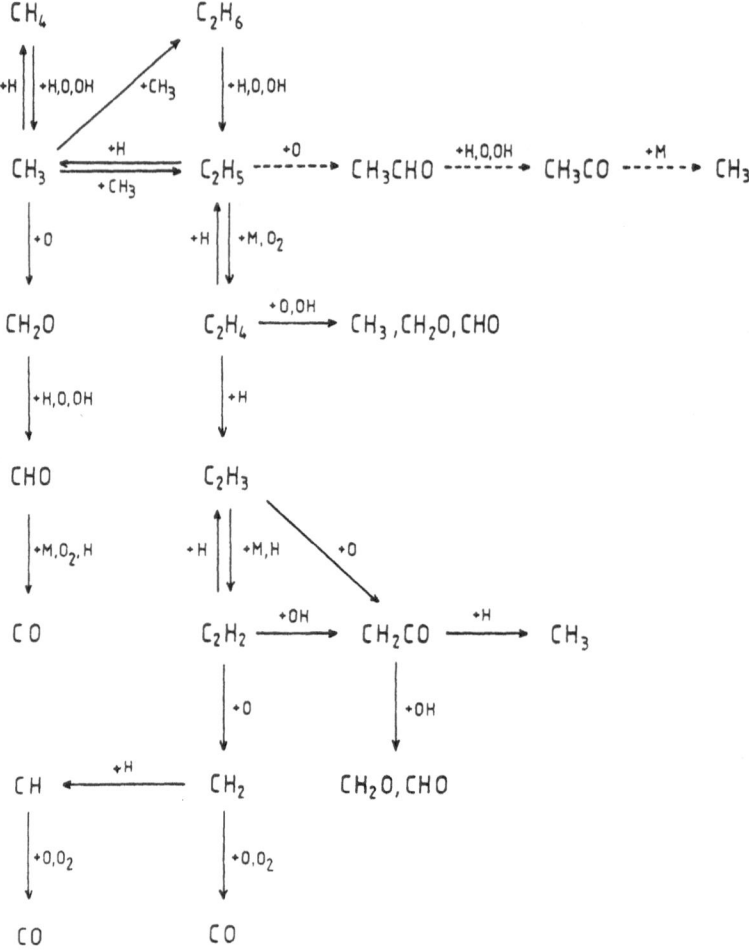

Fig. 2(b). Mechanism of the high temperature oxidation of CH_3 and C_2H_5

Results of simulations of one-dimensional freely propagating C_2H_4-air flames using the mechanism described above are given in Fig. 3 in a log-log presentation (simulations of the burner-stabilized CH_4-O_2-Ar flames studied by Harvey and McColl are in preparation). The calculated result

$$[C_2H_2]_{max} \sim \phi^{2.9}$$

shows satisfactory agreement with the experimental correlations. Nevertheless, there is need of further experimental and computational studies of the influence of temperature, pressure, and different fuels on this correlation.

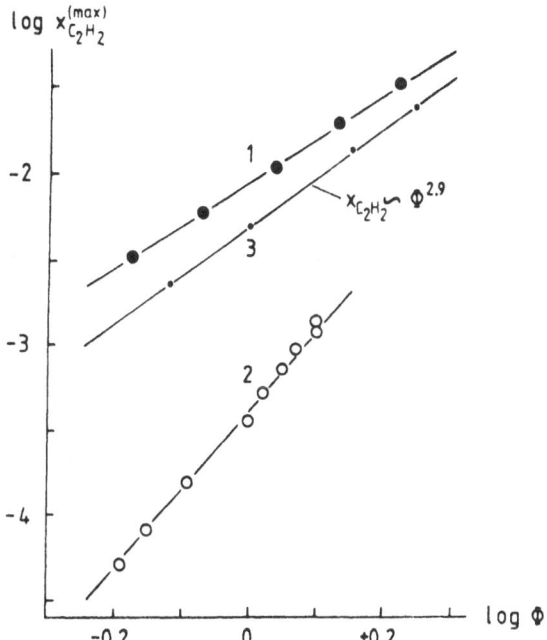

Fig. 3. Log-log presentation of measured and calculated maximum
         C_2H_2 mole fractions 1 and 2: see Fig. 1; 3: calculated re-
         sult for freely propagating C_2H_4-air flames at P = 1 bar,
         T_u = 298 K.

FORMATION OF HIGHER HYDROCARBONS IN THE OXIDATION OF ACETYLENE

Concentration profiles of some typical higher hydrocarbons
(C_3H_4, C_4H_2, C_4H_4, C_6H_2, C_8H_2) formed in sooting flames are shown
in Fig. 4. Furthermore, this illustration shows the reactants and
major products CO, CO_2, H_2, and H_2O in a weakly sooting burner-
stabilized C_2H_2-O_2 low pressure flame.[6]

Since the formation of the higher polyacetylenes C_6H_2, C_8H_2 etc.
will be analogeous to that of C_4H_2, and since C_4H_4 is a product of
the hydrogenation of C_4H_2 (see below), the following considerations
shall be confined to the explanation of C_3H_4 and C_4H_2.

Primary Reactions of Acetylene

The primary reactions of acetylene with O, OH, and H atoms
are subject of discussion up to the present.

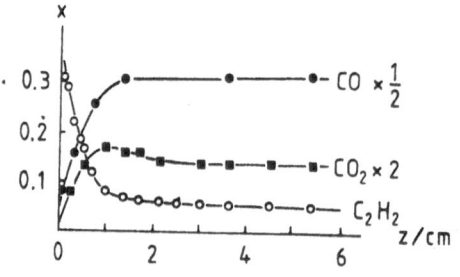

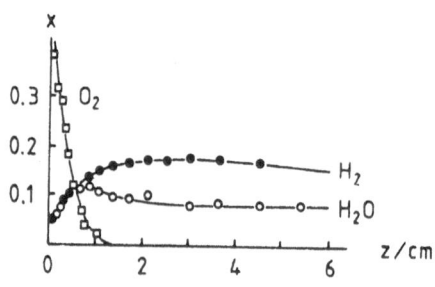

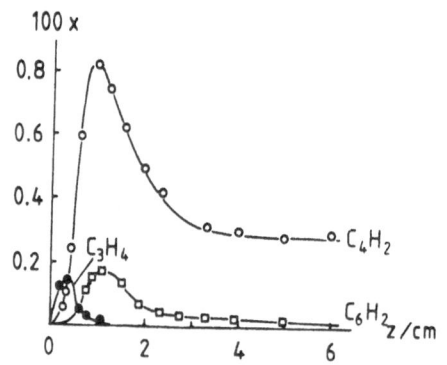

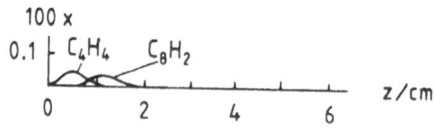

Fig. 4. Mole fraction profiles in a weakly sooting burner-stabilized
C_2H_2-O_2 flame at P = 26 mbar, C/O= 0.95, v_u =50 cm/s .[6]

$$O + C_2H_2 \longrightarrow CO + CH_2$$

$$\text{(above 1600 K)} \longrightarrow CHCO + H$$

$$OH + C_2H_2 \longrightarrow CH_2CO + H$$

$$\text{(above 1600 K)} \longrightarrow C_2H + H_2O$$

$$H + C_2H_2 \longrightarrow C_2H_3$$

$$\text{(above 1600 K)} \longrightarrow C_2H + H_2$$

The reaction of O atoms with C_2H_2 mainly leads to CO and CH_2[7], but there is evidence of a second channel[8], leading to CHCO and H atoms ($\sim$25% at 2000 K). Since there is complete lack of knowledge on subsequent reactions of CHCO, this minor channel shall not be taken into account in the flame simulation presented below.

The reaction of OH with acetylene leads to CH_2CO and H atoms[9]; besides at high temperature a channel with large activation energy seems to become important[10], forming C_2H and H_2O.

The reaction of H atoms with acetylene leads at low temperature to the addition product C_2H_3; again, at high temperature there is evidence of a high activation energy channel with the products C_2H and H_2 (see below).

Conversion of Acetylene to C_3H_4

The conversion of C_2H_2 to C_3H_4 (propyne or allene) has been studied by Vinckier and Debruyn at low temperature using a fast flow system with mass spectrometric analysis

$$O + C_2H_2 \longrightarrow CO + \boxed{CH_2}$$

$$CH_2 + C_2H_2 \longrightarrow \boxed{C_3H_3} + H$$

$$C_3H_3 + H \xrightarrow{(M)} \boxed{C_3H_4}$$

In contrary to former work postulating direct formation of C_3H_4 by reaction of CH_2 with C_2H_2, this study leads to the conclusion that C_3H_4 is formed via intermediate C_3H_3. This result is confirmed by work of Homann and Schweinfurth[12] using a similar experimental set-up.

$$O + C_2H_2 \longrightarrow CO + \boxed{CH_2}$$

$$CH_2 + C_2H_2 \longrightarrow \boxed{C_3H_3} + H$$

$$CH_2 + H \longrightarrow \boxed{CH} + H_2$$

$$CH + C_2H_2 \longrightarrow \boxed{C_3H_3}$$

$$C_3H_3 + H \xrightarrow[(M)]{} \boxed{C_3H_4}$$

In addition, this study leads to the formulation of a second path leading to C_3H_3 via CH.

Consumption of C_3H_4

The consumption of C_3H_4 (propyne or allene) in flames is due to reactions with H, O, and OH, which have been isolatedly investigated by use of fast flow systems at low temperature.

$$H + CH_3-CCH \longrightarrow C_3H_5$$

$$H + CH_2=C=CH_2 \longrightarrow C_3H_5$$

$$O + CH_3-CCH \longrightarrow CO + CH_3CH$$

$$OH + CH_3-CCH \longrightarrow products$$

The reactions of propyne and allene with H atoms have been studied by Wagner and Zellner[13,14] and by Whytock and coworkers.[15] The primary reaction leads to the addition product C_3H_5, which then can decompose to form CH_3 and C_2H_2, or add a further H atom to form C_3H_6.

The reaction of propyne with O atoms has been investigated by Herbrechtsmeier and Wagner[16] and by Arrington and Cox.[17] It leads to CO and CH_3CH, which then can form ethylene by a pressure-dependent isomerization reaction.

The products of the reaction of OH radicals with C_3H_4 are unknown up to the present. The only information on this reaction is a rate constant at room temperature given by Bradley et al.[18]

Conversion of Acetylene to C_4H_2

The conversion of C_2H_2 to C_4H_2 and higher polyacetylenes has been studied by Homann et al.[12,19] in a fast flow reactor at low temperature using mass spectrometric analysis. However, simulations of C_4H_2 concentrations in sooting C_2H_2-O_2 flames (see below) show this mechanism only to explain about one percent of the experimental C_3H_2 formation.

High temperature measurements in a shock tube by Gardiner and coworkers[20] lead to the conclusion that polyacetylene formation is started mainly by the high activation energy reaction of H atoms with C_2H_2 (activation energy 88 kJ/mol):

$$C_2H_2 + M \longrightarrow \boxed{C_2H} + H + M \qquad 448\,kJ/mol$$

$$C_2H_2 + H \longrightarrow \boxed{C_2H} + H_2 \qquad 88\,kJ/mol$$

$$C_2H + C_2H_2 \longrightarrow \boxed{C_4H_2} + H$$

$$C_2H + C_4H_2 \longrightarrow \boxed{C_6H_2} + H$$

$$C_2H + C_6H_2 \longrightarrow \boxed{C_8H_2} + H$$

$$C_2H_2 + C_2H_2 \longrightarrow \boxed{C_4H_3} + H \qquad 192\,kJ/mol$$

$$C_4H_3 + M \longrightarrow \boxed{C_4H_2} + H + M$$

This strong temperature depencence of polyacetylene formation explains the absence of this reaction path at low temperature.

Consumption of C_4H_2

Again, the consumption of C_4H_2 by O and H atoms has been investigated in a fast flow reactor at low temperature using mass spectrometric analysis. Up to the present, there are no data on the reaction of OH radicals with C_4H_2 existing in the literature.The reaction of H atoms with C_4H_2 has been studied by Schwanebeck and Warnatz .[21]

$$H + C_4H_2 \longrightarrow C_4H_3$$

$$H + C_4H_3 \longrightarrow C_4H_4^*$$

$$C_4H_4^* + M \longrightarrow C_4H_4 + M$$

$$C_4H_4^* \longrightarrow C_2H_2 + C_2H_2$$

C_4H_2 adds an H atom to form C_4H_3; this reaction is in its high pressure region at room temperature in the mbar range. Addition of a second H atom yields an excited $C_4H_4^*$ which can be desactivated to form C_4H_4 (probably vinyl acetylene), or can decompose to form two acetylene molecules. This competition of desactivation and decomposition is in its transition region at room temperature below 10 mbar.

The reaction of O atoms with C_4H_2 has been studied by Homann et al.[22] The primary products of this reaction could not be identified definitely, but the main reaction path seems to be:

$$O \quad + \quad C_4H_2 \quad \longrightarrow \quad CO \quad + \quad C_3H_2$$

SIMULATION OF A SOOTING C_2H_2-O_2 FLAME

The knowledge summarized in the preceding sections now shall be used to try the simulation of the weakly sooting C_2H_2-O_2 low pressure flame (C/O = 0.95) illustrated in Fig. 4.

This simulation is done by solution of the corresponding one-dimensional conservation equations taking into account chemical reaction, transport, and flow:

$$\varrho \frac{\partial w_i}{\partial t} = -\varrho v \frac{\partial w_i}{\partial z} - \frac{1}{A} \frac{\partial(A \cdot j_i)}{\partial z} + r_i \quad \text{with} \quad j_i = -D_{iM} \varrho \frac{\partial w_i}{\partial z}$$

Here temperature T and mass fractions w_i of species i are the dependent variables, time t and the cartesian space coordinate z are the independent variables. These equations contain a convection term including the total mass flux ϱv, a diffusion term including the diffusion flux j_i, and a chemical reaction term including the rate of formation r_i. The correction factor A takes into account the increasing area ratio in a stabilized flat flame.[23] Details are given elsewhere.[4,23,24] The temperature profile has been estimated from temperature measurements in similar flames.[6,25]

The reaction mechanism consists of a mechanism developed for C_2H_2 oxidation in lean and moderately rich mixtures[4,5], and furthermore of the reaction listed in Table 1, which describe the formation and consumption of C_3H_4 and C_4H_2 discussed before.

Due to the lack of corresponding data the following assumptions have been made: (1) The reaction OH + C_3H_4 is assumed to have the same activation energy as the reaction OH + $C_2H_2 \rightarrow CH_2$ + H. (2) The reaction OH + C_4H_2 is assumed to have the same activation energy as the reaction OH + $C_2H_2 \rightarrow CH_2CO$ + H; its room temperature rate constant is assumed to be that of O + C_4H_2 and H + C_4H_2, namely $(1.3 \cdot 10^{12}$ cm^3/mol.s).

Table 1. Reactions Describing C_3H_4 and C_4H_2 Formation and Consumption in a Rich C_2H_2-O_2 Premixed Flame;
$k = A \exp (-E/RT)$

Reaction	A cm^3/mol·s	E kJ/mol
$OH + C_2H_2 \rightarrow H_2O + C_2H$	$1.0 \cdot 10^{13}$	29
$H + C_2H_2 \rightarrow H_2 + C_2H$	$3.0 \cdot 10^{13}$	88
$C_2H + C_2H_2 \rightarrow C_4H_2 + H$	$3.0 \cdot 10^{13}$	0
$H + C_4H_2 \rightarrow C_4H_3$	$1.6 \cdot 10^{13}$	6.3
$OH + C_4H_2 \rightarrow products$	$8.0 \cdot 10^{12}$	4.6**)
$H + C_4H_3 \rightarrow 2 C_2H_2$	$-$	$-$ *)
$CH_2 + C_2H_2 \rightarrow C_3H_3 + H$	$3.0 \cdot 10^{11}$	0
$CH + C_2H_2 \rightarrow C_3H_3$	$4.5 \cdot 10^{13}$	0
$C_3H_3 + H \rightarrow C_3H_4$	$2.0 \cdot 10^{13}$	0
$H + C_3H_4 \rightarrow CH_3 + C_2H_2$	$1.2 \cdot 10^{13}$	9.6
$OH + C_3H_4 \rightarrow products$	$3.6 \cdot 10^{12}$	4.6**)

*) reaction assumed to be very fast (quasi-steady state
   concentration of C_4H_3)
**)products assumed to react very fast to form CO and H_2
   (quasi-steady state concentration of these products)

The results of this flame simulation are given in Fig. 5. For comparison the experimental values are given once more. Fig. 5a shows
the profiles of the reactants and major products. There is agreement
of experiments and calculations with one exception: The H_2 and H_2O
profiles show discrepancies which cannot be definitely explained.
These discrepancies may result partly from the fact that exact mass
spectrometric calibration of these species is extremely difficult.
Furthermore, they may be partly due to uncertainties of the temperature profile which influences the water gas equilibrium including
H_2 and H_2O. A further explanation of this deviation may be that the
rate constant used for $OH + C_4H_2$ is too small. This is supported by
the discrepancy of measured OH mole fraction ($5 \cdot 10^{-4}$ [25]) and its
calculated value ($1.8 \cdot 10^{-3}$).

Fig. 5b shows the profiles of C_3H_4 and C_4H_2. Taking into
account that the calculation depends on a number of assumptions,
there is excellent agreement of measured and simulated profiles.

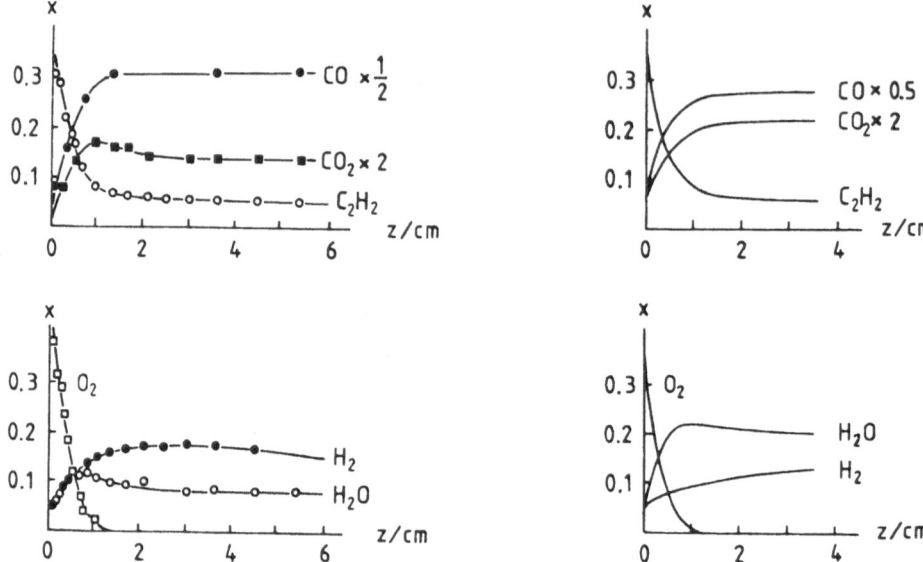

Fig. 5a.   Experimental and calculated mole fraction profiles of
reactants and major products in a weakly sooting burner-
stabilized $C_2H_2-O_2$ flame (see Fig. 4.).

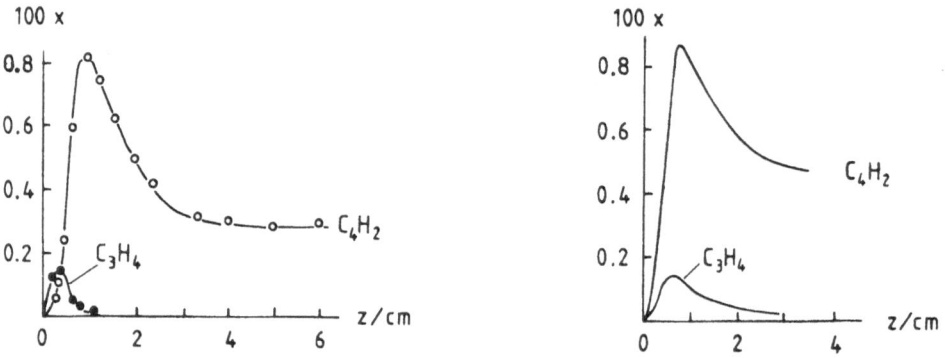

Fig. 5b.   Experimental and calculated mole fraction profiles of
C_3H_4 and C_4H_2 in a weakly sooting burner-stabilized
$C_2H_2-O_2$ flame (see Fig. 4.).

A very important result of the calculations are rates of
chemical formation and consumption for each species in each reaction at each position in the flame. One result is, that reactions
of O atoms are unimportant for the formation (e.g. via
$O + C_2H_2 \rightarrow OH + C_2H$) and for the consumption of C_4H_2; both effects
are due to the relatively small O atom concentration.

A further important result is that about 55 percent of the C_2H_2
consumed are decomposed via C_4H_2. In contrast, for instance only
2.5 percent of the C_2H_2 are decomposed via C_3H_4 (the other higher
hydrocarbon considered here), and only 0.6 percent via CH (which
has been discussed to be precursor of ion formation in sooting
flames [26]). This is strong support for the hypothesis to assume C_4H_2
to be an important precursor in the process of formation of soot.
It should be emphasized that this latter result is independent of
details of the (relatively uncertain) mechanism of C_4H_2 consumption,
as can be shown by use of modified mechanisms.

CONCLUSIONS

     Some global statements on potential reaction routes leading to
soot are possible from consideration of an illustration (Fig. 6.)
given by Homann.[27] It is shown that the polyacetylene theory leads
with increasing molar mass to particles containing too much carbon
atoms, and that even precursors with a C/H ratio corresponding to
polycyclic aromatic hydrocarbons do not lead to a correct correlation of molar mass and C/H ratio.

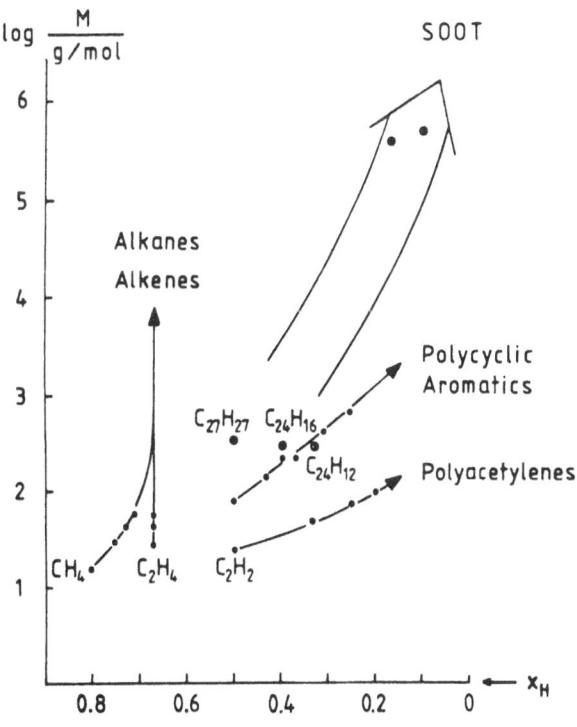

Fig. 6.   Correlation of molar mass and C/H ratio in hydrocarbons
          and soot.

As shown in the preceding section, diacetylene seems to be starting point of formation of higher hydrocarbons. If soot formation via higher polyacetylenes is excluded, it is obvious that in the next step diacetylene must be converted to radical species with a smaller C/H atomic ratio. Two ways seem to be possible: (1) decrease of the number of C atoms (e.g. by reaction with O atoms probably leading to CO and C_3H_2 radicals[22]) or (2) increase of the number of H atoms (e.g. by reaction with H atoms leading to C_4H_3 radicals[21]).

Therefore, it seems to be necessary to collect more data on details of reactions of diacetylene (primary products and subsequent reactions of the O atom attack, rate constant and mechanism of the reaction with OH radicals etc.). Furthermore, more detailed studies of the reactions leading to C_2H and of its reactions with radicals present in flames (H, O, OH etc.) are desirable.

ACKNOWLEDGEMENT

The author is grateful to Prof. Dr. K.H. Homann and Prof. Dr. H.Gg. Wagner for their sustained interest in this work and many helpful discussions. The financial support of the Deutsche Forschungsgemeinschaft and the Fonds der Chemischen Industrie is gratefully acknowledged.

REFERENCES

1. H. Gg. Wagner, 9th Symp. (Intern'l) on Combustion, p.572, Academic Press, New York (1963)
2. R. Harvey and A. McColl, 17th Symp. (Intern'l) on Combustion, p.857, The Combustion Institute, Pittsburgh (1979)
3. J. Warnatz, Progress in Astronautics and Aeronautics, (1981), in press
4. J. Warnatz, 18th Symp. (Intern'l) on Combustion, p.369, The Combustion Institute, Pittsburgh (1981)
5. J. Warnatz, in: W.C. Gardiner, Jr., Combustion Chemistry, Springer, New York, (1982), in press
6. K. H. Homann and H. Gg. Wagner, Ber. Bunsenges. Phys. Chem. 69:20 (1965)
7. B. Blumenberg, K. Hoyermann and R. Sievert, 16th Symp. (Intern'l) on Combustion, p.841, The Combustion Institute, Pittsburgh (1977)
8. P. Roth and R. Löhr, Ber. Bunsenges. Phys. Chem. 85, in press
9. K. Hoyermann, Habilitationsschrift, Göttingen (1979)
10. W. G. Browne, R. P. Porter, J. D. Verlin and A. H. Clark, 12th Symp. (Intern'l) on Combustion, p.1035, The Combustion Institute, Pittsburgh, (1969)

11. C. Vinckier and W. Debruyn, J. Chem. Phys. 83:2057 (1979)

12. K. H. Homann and H. Schweinfurth, Ber. Bunsenges. Phys. Chem. 85:569 (1981)

13. H. Gg. Wagner and R. Zellner, Ber. Bunsenges. Phys. Chem. 76:518 (1972)

14. H. Gg. Wagner and R. Zellner, Ber. Bunsenges. Phys. Chem. 76:667 (1972)

15. D. A. Whytock, W. A. Payne and L. J. Stief, J. Chem. Phys. 65:191 (1976)

16. P. Herbrechtsmeier and H. Gg. Wagner, Z. Physik. Chem. NF 93:143 (1974)

17. C. A. Arrington and D. J. Cox, J. Phys. Chem. 79:2684 (1975)

18. J. N. Bradley, W. Hack, K. Hoyermann and H. Gg. Wagner, J. Chem. Soc. Faraday I 69:1889 (1973)

19. K. H. Homann, J. Warnatz and C. Wellmann, 16th Symp. (Intern'l) on Combustion, p.853, The Combustion Institute, Pittsburgh, (1977)

20. T. Tanzawa and W. C. Gardiner, Jr., 17th Symp. (Intern'l) on Combustion, p.563, The Combustion Institute, Pittsburgh (1979)

21. W. Schwanebeck and J. Warnatz, Ber. Bunsenges. Phys. Chem. 79:530 (1975)

22. K. H. Homann, W. Schwanebeck and J. Warnatz, Ber. Bunsenges. Phys. Chem. 79:536 (1975)

23. J. Warnatz, Ber. Bunsenges. Phys. Chem. 83:950 (1979)

24. J. Warnatz, Ber. Bunsenges. Phys. Chem. 82:193 (1978)

25. U. Bonne and H. Gg. Wagner, Ber. Bunsenges. Phys. Chem. 69:35 (1965)

26. D. B. Olson and H. F. Calcote, 18th Symp. (Intern'l) on Combustion, p.453, The Combustion Institute, Pittsburgh (1981)

27. K. H. Homann, FVM Frankfurt 327:137 (1978)

DISCUSSION

J. B. Howard (Massachusetts Institute of Technology)

What mole fraction of C_2H is predicted in your study? The study from which the data you employed were taken reported C_2H at a mole fraction of 10^{-3}. Our measurments under the same conditions were unable to reproduce this result. (Bittner, J.D. and Howard, J.B., Eighteenth Symposium(Intern'l) on Combustion, The Combustion Institute, Pittsburgh, 1981).

Warnatz

The predicted maximum mole fraction of C_2H in the acetylene-oxygen flame presented here is 9×10^{-5}. Discrepancies between

measured and predicted values can result from incorrect mass
spectrometric calibration of C_2H, especially due to the unknown
temperature dependence of C_2H^+ formation in ionization of C_2H_2.

Dr. Feugier (Institut Français du Pétrole)

   Was radical C_2 identified in acetylene flame? What is its
mechanism of formation?

Warnatz

   C_2 can be identified by laser diagnostic methods (McDonald
and Pasternak at NRL, Washington D.C,). At present we do not
have valid information on the mechanism of its formation.

P. Cadman (The University College of Wales)

   Can the speaker comment on the discrepancies for the production of H_2O as only one route has been discussed and this was
for higher temperature only?

Warnatz

1) The mass spectrometric calibration of H_2O is very difficult
and can be in error.
2) The temperature profile used for the calculations has been
estimated from temperature profiles of similar flames. This
can lead to incorrect predictions, since the water gas equilibrium is strongly temperature dependent.
3) H_2O is mainly formed by OH + RH $\longrightarrow$ H_2O + R, OH + OH $\longrightarrow$
H_2O + O and OH + H_2 $\longrightarrow$ H_2O + H. These reactions are well
known, and should not lead to incorrect predictions.

J. B. Howard (Massachusetts Institute of Technology)

   Regarding the disagreement between measured and computed H_2O
mole fraction, the reaction OH + H_2 = H_2O + H has been found to
be equilibrated in the post flame zone of an almost sooting
(ϕ= 1.8) low-pressure $C_6H_6/O_2/Ar$ flame (Bittner, J.D. and Howard,
J.B., Eighteenth Symposium (Intern'l) on Combustion, The Combustion
Institute, Pittsburgh, 1981). Is it possible that this condition
might apply also to the post flame zone of the flame you studied
and, if so, would it shed light on the H_2O question?

Warnatz

   At temperatures above 1800 K, the reaction OH + H_2 = H_2O + H
always should be equilibrated. Potential reasons for the disagreement of measured and calculated H_2O profiles have been discussed
before (see question by P. Cadman).

P. Cadman (University College of Wales)

The slide showing the molar mass versus mole fraction of hydrogen suggested that polyacetylenes or even aromatics did not correlate with the molar mass vs C/H ratio in soot. We know, however, that acetylenes or aromatics do give soot and species like C_2, C_3 etc... and may also be involved. So, in practice, the situation may be not as simple as this slide suggests. As soot consists of more than one species, i.e. a mixture of carbon containing species together with many PAHs, it is also arguable that one cannot really talk about a molar mass for soot.

Warnatz

This slide shows only a general tendency of the route of soot formation, and excludes some death ways (polyacetylene theory, formation via saturated polycyclic systems, etc...). It cannot lead to a constructive idea of the main path leading to soot.

H. F. Calcote (AeroChem Research Laboratories, Inc.)

Would you care to comment on the source of CH^* in flames?

Warnatz

To our present knowledge, CH^* is produced by the reaction $C_2H + O \longrightarrow CO + CH$, which can correctly predict chemiluminescence experiments in a fast flow system (Homann et al., to be published).

S. Galant (Société Bertin)

Do you take account of diffusion fluxes in your computer simulations?

Warnatz

Yes, this is a complete one-dimensional flame model, including multicomponent diffusion and thermal conduction, which has been tested extensively in hydrogen-oxygen, carbon monoxide-oxygen and hydrocarbon flames.

SOOT PARTICLE NUCLEATION AND AGGLOMERATION

Gilles Prado, Jacques Lahaye[*] and Brian S. Haynes[**]

C.R.P.C.S.S., Mulhouse, France
[*]Université de Haute-Alsace, Mulhouse, France
[**]Institute of Earth Resources, North Ryde, Australia

INTRODUCTION

In considering the overall phenomenon of soot formation, we are dealing with the evolution of a carbonaceous aerosol in hot reactive gases. If we assume that this aerosol is well described as a monodispersion of spherical particles, its characterization can be made with a knowledge of the particle number density N, the diameter of the particles d, and the particulate volume fraction f_v. Obviously only two of these parameters are independent ($f_v = \pi/6 \, Nd^3$) and any two are sufficient.

Phenomenologically, it is convenient to consider two stages occurring in the formation of the aerosol: (i) the inception of the first small particles (nucleation) and (ii) their subsequent growth. Particle growth itself occurs as a result of two distinct processes (coagulation and surface growth). In the first process two particles collide and stick to form one larger particle, thus reducing the particle number density without changing the particulate loading. This process will be considered in detail in this paper. Apart from the morphological effects arising from whether the collisions are coalescent or agglomerate-forming, coagulation is now fairly well understood in terms of fundamental particle kinetics.

The second particle-growth mechanism is surface growth. It corresponds to the attachment to, and incorporation in, existing particles of species smaller than the first observable soot particles (usually 1.5 nm). These species can be gaseous hydrocarbons or small nuclei.

As will be discussed in another contribution to this Workshop,[1] there is much yet to be learned on the details of surface growth kinetics. It suffices here to note that surface growth does not alter the particle number density but does lead to an increase in soot volume fraction.

Clearly now, the ideal choices for the independent parameters in our characterization of the soot aerosol are f_v and N. Particle growth can be expressed then in terms of the changes in these quantities through surface growth and coagulation respectively:

$$d \ln d = \frac{1}{3} d \ln f_v - \frac{1}{3} d \ln N$$

In the strictest sense, the transition from particle inception to particle growth is probably not a sharp one. Moreover the limit between inception and growth is depending upon the resolution of the method used to observe nuclei and to measure their site. The growth of very large molecules by addition of smaller hydrocarbons is obviously not far removed from the addition of similar species to "particles". Similarly, large molecules can react among themselves in the same way as particles coagulate. However, in both premixed [2-4] and laminar diffusion flames,[4-7] the zones of particle nucleation and of particle growth do appear to be spatially distinct. New particle formation is restricted to a narrow region very near the main reaction zone where temperatures and radical and primary ion concentrations are highest. Although very many (N $\gtrsim 10^{18}$ m^{-3}) are formed initially (N $\sim 10^{21}$ m^{-3} sec^{-1}), they are very small (d ~ 1.5 nm) at their inception and represent a very light soot loading ($f_v \sim 10^{-9}$). Except under weakly sooting conditions (e.g. a premixed flame operating near the sooting limit), this initial volume fraction is negligible compared with that contributed by surface growth which occurs over extended periods as the gases move away from the primary reaction zone, even into relatively cool, unreactive regions. Therefore, we shall not attempt to define the limit between molecule and particle any more precisely than to accept, for example, the appearance of continuous yellow emission as signalling the occurrence of particles.

In some systems, oxidation of the soot particles after their formation may also occur. In the simplest sense, this process is the reverse of surface growth in that soot-phase material is gasified by oxidants such as OH and O_2. In addition, however, there is recent evidence that some break-up of soot agglomerates may occur on extensive oxidation so that oxidation may also act as a weak source of particles.[5,8]

The phenomenology of soot formation can now be summarized as in Table I.

In the subsequent discussion, we shall be dealing with the

Table I. Phenomenology of soot formation

	N	f_v
Source	Nucleation	Nucleation Surface growth
Sink	Coagulation	Oxidation

processes of nucleation and coagulation. As a preface to this
discussion, and those of oxidation and surface growth, we pose the
following question : is it really necessary to understand nucleation in order to be able to describe the evolution of the soot
aerosol? We shall see that, as a result of coagulation, N rapidly
becomes independent of the initial number of nuclei and given that
particle inception does not contribute significantly to the total
mass of soot in most systems, it may be enough to treat nucleation
as no more than a boundary condition in that it either does or does
not occur.

Two approaches have been used to predict the influence of
experimental parameters on particle number density: the classical
nucleation theory, and free-molecule coagulation. The first approach
appears restricted to formation of soot particles at low temperatures
($\leq$ 1500 K), whereas the free-molecule coagulation may govern at
higher temperatures (1500-2000 K).

CLASSICAL NUCLEATION THEORY

A relatively high concentration of gaseous molecules of several
hundred atomic mass units are formed very early in soot forming
systems. Physical condensation of these molecules has been proposed
to explain the appearance of the first condensed spherical particles
observable by electron microscopy.[9,10] The number density N_o of
"critical nuclei" can be derived from the equilibrium constant of
the nucleation reaction:

aM $\rightleftharpoons$ Ma

$$N_o = [M] \exp(-\Delta G^\circ/kT) \qquad (1)$$

where [M] = concentration of gaseous precursors
      ΔG° = Gibbs free energy of formation of the critical nucleus
      T   = temperature.

Before nucleation, the hydrocarbon precursors are in a super-
saturation state (above equilibrium concentration). The formation
of nuclei drastically reduces the precursor concentration such that

the formation of additional nuclei becomes impossible. Intense
surface growth through hydrocarbon deposition accounts for the
rapid increase of particle size, the number density remaining
essentially constant. The droplets are pyrolysed into a solid
material, with some chain forming agglomeration occurring
simultaneously.

This mechanism explains well the main features of soot formation in thermal systems at relatively low temperature. As an example,
let us review briefly experimental results obtained when pyrolising
benzene and methane in inert atmosphere.

These results were obtained in a flow system, where hydrocarbons were pyrolysed diluted in nitrogen.[10,11,12] The volume
fraction, number density and particle size were measured as a
function of the following parameters:

- Residence time (30 to 500 ms)

- Initial hydrocarbon volume fraction (0.25 to 100%)

- Temperature (1323 to 1723 K)

- Nature of initial hydrocarbon (benzene and methane)

According to the classical nucleation theory, Eq. (1), the
particle number density should be independent of residence time
and initial hydrocarbon volume fraction, and should depend only
on the temperature and the nature of the condensing species. This
is indeed observed experimentally.

Influence of residence time

As shown in previous papers[11,13,14] the influence of residence
time on the mass balance of carbonaceous material in the system,
as well as observations of intermediate species by electron microscopy, lead to the conclusion that, at relatively low temperature
(1383 K), the precursors of soot particles are viscous droplets
of tars.

Influence of initial hydrocarbon volume fraction

The effect of increasing the initial hydrocarbon volume fraction at constant temperature and residence time is summarized in
Table 2. When increasing the proportion of benzene from 0.25 to
10%, the soot volume fraction increases by more than a factor of
200. This increase is totally accounted for by the increase in
size of the spherical particles, their number remaining constant
within the experimental uncertainty, in agreement with Eq.(1).

Table 2. Benzene pyrolysis in nitrogen. T = 1383 K. Residence
          time = 0.5 s.

Benzene Volume Fraction	.25	.5	1	2	4	6	8	10
Soot Volume Fraction ($10^7 \times f_v$)	.94	3.02	12.2	31.4	70.0	147	172	222
D (nm)	25.8	37.8	52.4	71.7	91.4	100.5	132.3	138.5
$10^{-10} \times N$ (cm^{-3})	.48	.51	.78	.79	.83	.87	.72	.72

Influence of temperature

The influence of temperature, at constant benzene volume fraction and residence time, on soot volume fraction, diameter and
number density is summarized on Table 3. For the higher temperatures,
a different oven which did not permit quantitative recovery of soot
was used, and only the particle size could be measured. As apparent
on Table 3, both the volume fraction and number density increase
significantly with temperature, the particle size being drastically
reduced. At the highest temperature, the particle diameter is close
to particle size of soot formed in flames.

Influence of initial hydrocarbons

The results reported above are referring to benzene.
Methane can be used as a representative of aliphatic hydrocarbons.
Higher temperatures and/or molar fractions of hydrocarbons are
necessary to form soot.[14] For a given temperature, more individual
particles are formed with benzene than with methane, the difference
being larger at lower temperature. It is worth noting that in the
pyrolysis of mixtures of methane and benzene, the particle size
distribution was unimodal, indicating that the two hydrocarbons[13]
did not react independently with regard to physical mechanisms.

In summary the classical nucleation theory explains well the
results obtained during pyrolysis of hydrocarbons at low temperature.
This process is used in industry to produce a variety of carbon
black (thermal black). It may also be relevant to reactions occurring in the inner core of a diffusion flame, where only small amounts
of oxygenated compounds are present. A similar analysis has also
been recently applied to soot formation in diesel engine.[15,16,17]
However, its validity has been strongly questioned for processes
occurring at flame temperature.[18,19] Equilibrated physical condensa-

Table 3. Benzene pyrolysis. 6% benzene in nitrogen.
          Residence time: 0.5 s.

Temperature (K)	1311	1383	1419	1473	1593	1700	1720
$10^6 \times f_v$		8.1	14.7	16.6	20.4		
D (nm)		97.3	100.5	93.4	76.9	47.3	34.4	28.6
10^{-10} N (cm^{-3})		.76	.87	1.42	2.75		

tion at these high temperatures requires indeed molecules with a
very high mass, typically 2000 to 3000 a.m.u. The mass of these
molecules is so close to the mass of the first observed particles
that they correspond probably to the same entity. In other words,
it appears more reasonable to postulate that the first particulate
phase is formed through a long sequence of chemical reactions.

FREE-MOLECULE COAGULATION

     During a sticking collision process, the rate of change of
particle number density N with time t can be expressed as:

$$\frac{dN}{dt} = \dot{N}_u - \dot{N}_c$$

where
     $\dot{N}_u$ = the rate of nucleation, defined for practical purpose
          as the rate of appearance of the smallest observable
          particles (2 nm)
     $\dot{N}_c$ = the coagulation rate.

     $\dot{N}_u$ has an important contribution only very early in the process.
Very rapidly, the number density is controlled by the coagulation
rate, which can be expressed in the form of the Smoluchovski
equation:

$$\frac{dN}{dt} = - k(D) N^2 \tag{2}$$

where the rate constant k depends on the particle diameter D, or
equivalently, on the mean volume $\bar{v}$. For particles small compared
with gas mean free path (Knudsen number > 10), Brownian motion,
and k, are described in terms of free molecule theory; for soot
particles in a typical flame environment this requires that
$d \lesssim 60$ nm. For such a system in which uncharged spherical particles
coalesce on every collision, this leads to a coagulation rate of :

$$\frac{dN}{dt} = - \frac{6}{5} k_{th} f_v^{1/6} N^{11/6} \tag{3}$$

where
$$k_{th} = \frac{5}{12} (\frac{3}{4\pi})^{1/6} (\frac{6kT}{\rho})^{1/2} G \cdot \alpha \tag{4}$$

f_v = particulate volume fraction
ρ = density of the particles
G = a factor which takes into account the inter-particle dispersion forces and can be expected to have a value of about 2 for spherical particles
α = a weak function of the particle size distribution, reflecting the variation in collision rate with different particle sizes.

Upon integration at constant volume fraction and temperature, Eq. (3) yields (S.I. units):

$$N = N_0 [1 + 9.03 \times 10^{-13} N_0^{5/6} T^{1/2} f_v^{1/6} t]^{-6/5} \tag{5}$$

where

N_0 = the initial particle number density
N = the particle number density at time t

For large enough values of N_0 (typically $N_0 > 10^{18}/m^3$), Eq. (5) can be reduced to:

$$N = 2.84 \times 10^{14} [T^{1/2} f_v^{1/6} t]^{-6/5} \tag{6}$$

Coalescent coagulation is followed in any soot forming system by chain forming collision. In this process, the colliding particles stick together but do not fuse and so give rise to an aggregate in which the individual particles retain their identity. As a first approximation, the above equation can be applied to describe this process, the aggregate being assimilated to a sphere of equal volume.

Eq. (6) predicts that the number of particles, N, is independent of the initial number N_0, and only weakly dependent on soot volume fraction, f_v, and temperature, T. This has been verified on pre-mixed flames by different investigators, burning different fuels, including methane[20,21] ethylene and benzene.[2,22] We will illustrate this behaviour with experimental results obtained on a pre-mixed propane/oxygen flame, burning at atmospheric pressure.[3]

The quantities measured included the soot volume fraction, the size and number density of aggregates and spherules. The techniques used were electron microscopy of collected soot, and

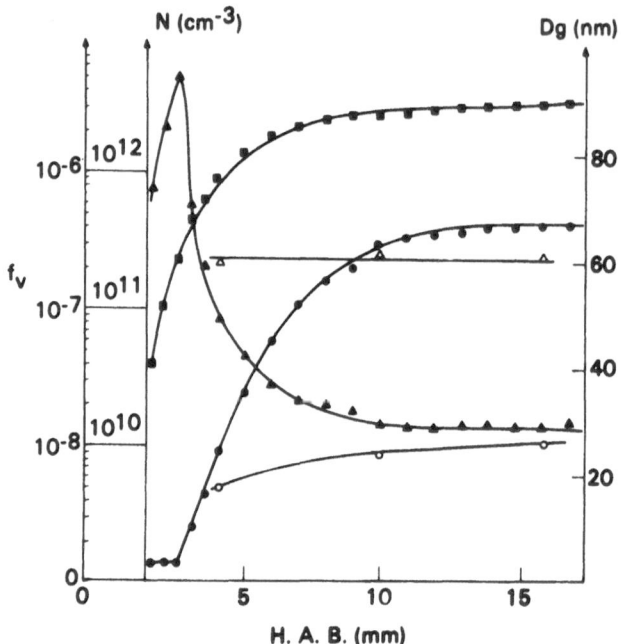

Fig. 1. Agglomerate volume fraction f_v (■). Number density N, cm^{-3} (▲). Diameter Dg, nm (●). Spherical units number density N, cm^{-3} (△). Diameter D, nm (○), as a function of height above burner (HAB, mm). Propane/oxygen flame. Fuel equivalence ratio = 2.9. Cold gas velocity = 5.5 cm/s.

laser scattering and extinction. Varying parameters were fuel equivalence ratio and temperature.

The informations obtained on one flame are plotted in Fig. 1. The conditions are ϕ = 2.9, v = 5.5 cm/s. The following trends, illustrated in Fig. 1, are representative of nearly all the flames studied: The agglomerate number density reaches a maximum very early in the process, whereas the number density of spherical units, when measurable, remains constant. Number densities are in-situ values, measured at flame temperature. The agglomerate mean geometric diameter increases strongly and the sphere diameter increases slightly, depending on the increase of soot volume fraction. As with pyrolysis experiments, particles collected early in the process could not be measured by electron microscopy, due probably to intensive coalescence and/or surface deposition during sampling. In Fig. 1, 1 mm corresponds approximately to 1.2 ms.

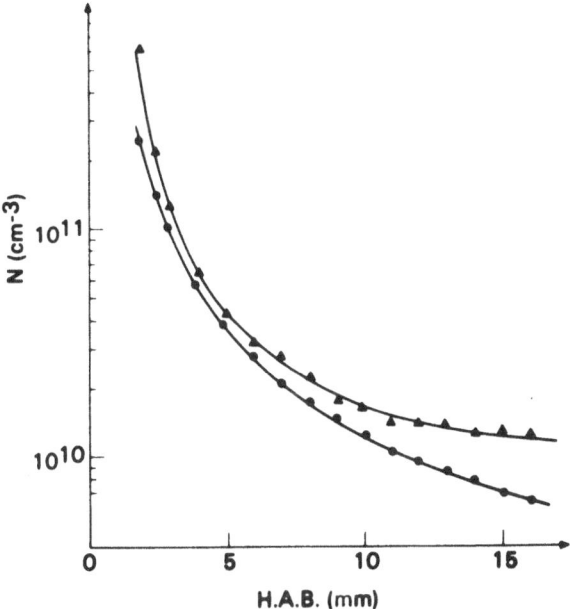

Fig. 2.   Experimental (▲) and computed (●) number density of
          agglomerates (N, cm^{-3}). Propane/oxygen flame. Fuel
          equivalence ratio = 2.5. Cold gas velocity = 5.5 cm/s.

When comparing the particle number density curve with predictions of Eq. (6), (Fig. 2), a good agreement is obtained, except
at the end of the process, where the number of particles measured
optically might be inaccurate due to the non-sphericity of the
agglomerates. Other reasons for the divergence occurring at larger
residence time might also include: (i) a sticking coefficient for
the collisions smaller than one, and (ii) aggregates becoming too
large for collisions to occur in the free-molecule regime.[23]

Effect of fuel-equivalence ratio

     The fuel equivalence ratio (ϕ) was varied from 2.1 to 3, by
increasing the percentage of propane in the propane/oxygen mixture,
at a constant cold gas velocity of 5.5 cm/s. Below ϕ = 2.1, no
soot was formed (blue flame); above ϕ = 3.0, the flame was unstable,
due probably to molecular diffusion. The flame temperature, measured with the Kurlbaum method, decreases when ϕ increases, and with
increasing height above the burner. In the soot nucleation zone
(3 mm), the flame temperature was only slightly affected by changes

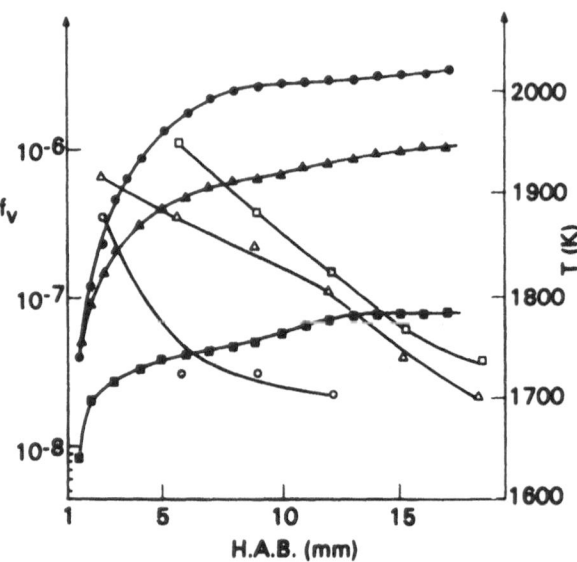

Fig. 3. Soot volume fraction (f_v, filled symbols) and temperature
(T, open symbols) as a function of height above burner
(H.A.B., mm). Propane/oxygen flame. Cold gas velocity =
5.5 cm/s. Fuel equivalence ratio = 2.2 (■), 2.5 (▲),
2.9 (●).

in ϕ (70 K difference from ϕ = 2.2 to 2.9). The effect is stronger
at larger distance above the burner (160 K difference at 9 mm),
and is attributed to increased radiative transfer with increased
soot volume fraction.

For clarity, only three fuel equivalence ratios are plotted
in Figs. 3 and 4. The soot volume fraction increases considerably
with ϕ (Fig. 3), whereas the agglomerate number density changes
only slightly (Fig. 4), with the exception of ϕ = 2.2 which is
at the very beginning of soot formation. For other values of ϕ
(here 2.5 and 2.9), all the values of N fit on one unique curve.
Furthermore, at the end of the reaction zone (18 mm), all the
curves, including ϕ = 2.2, converge towards a common limit, close
to 10^{10} agglomerates/cm^3. The agglomerate mean geometric diameter
increases considerably (Fig. 4) with increasing fuel-equivalence
ratio and height above burner.

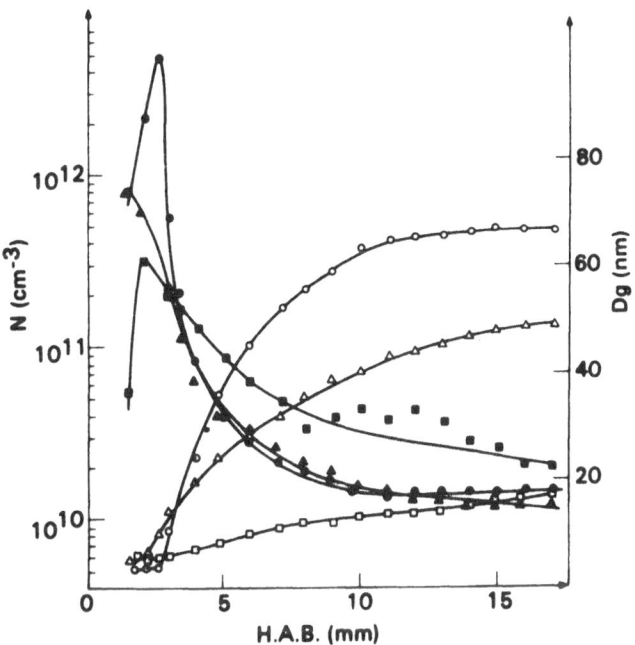

Fig. 4.   Agglomerate number density (N, cm^{-3}, filled symbols)
and diameter (Dg, nm, open symbols) as a function of
height above burner (H.A.B., mm). Propane/oxygen
flame. Cold gas velocity = 5.5 cm/s. Fuel equivalence
ratio = 2.2 (■), 2.5 (▲), 2.9 (●).

On examining the electron microscopy results, the picture
is somewhat different (Table 4).   The number of spherical units,
when measurable, remains constant at φ = 2.2 and 2.5, and increases
by about 50% between 4 and 16 mm for φ = 2.9.   At 4 mm, φ = 2.2
and 2.5, the material collected was poorly defined (no spherical
units), and no measurements could be done.   In Table 6, one can
also compare results from electron microscopy and from the optical
technique.   Only at φ = 2.2, chain forming agglomeration is not
dominant.   At 10 mm, φ = 2.2, the discrepancy in size is probably
due to some particle surface growth during collection.   At 16 mm,
the diameters compare well.

Table 4. Geometric mean diameter (D, nm) and number density (N, $10^{10}\,\mathrm{cm}^{-3}$) of soot particles measured by electron microscopy and by in-situ optical techniques. Propane/oxygen flame. Cold gas velocity = 5.5 cm/sec.

Fuel equivalence ratio		2.2	2.2	2.2	2.5	2.5	2.5	2.9	2.9	2.9
Height above burner (mm)		4	10	16	4	10	16	4	10	16
Electron microscopy	D	—	16.2	15.4	—	17.3	18.5	18.2	24.4	26.4
	N	—	2.2	3.6	—	20.0	23.9	22.9	25.3	24.2
In-situ optical techniques	D	7.7	12.9	17.2	20.0	40.1	48.8	24.7	63.1	67.1
	N	10.2	4.7	2.2	6.5	1.6	1.3	8.6	1.5	1.5

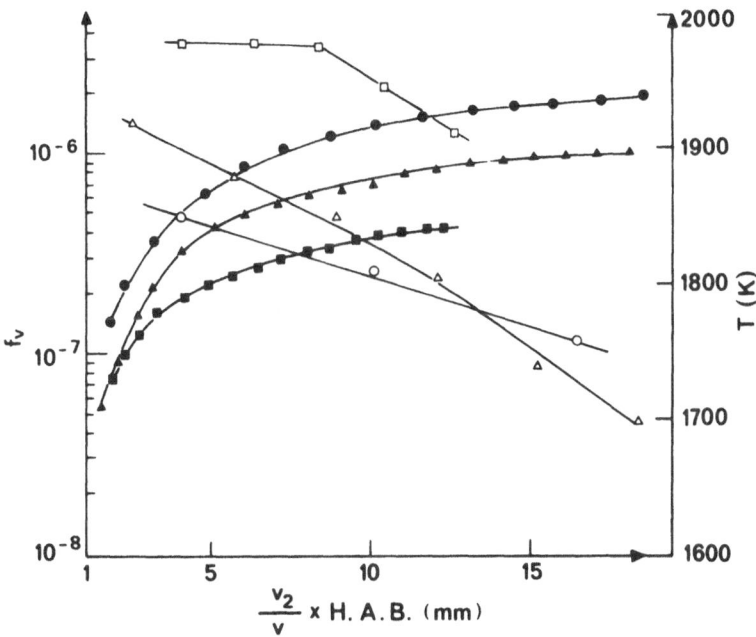

Fig. 5.   Soot volume fraction (f_V, filled symbols) and temperature
(T, open symbols) as a function of height above burner
(H.A.B., mm). Propane/oxygen flame. Fuel equivalence
ratio = 2.5. Cold gas velocity v_1 = 3.92 cm/s ($\bullet$),
v_2 = 5.5 cm/s ($\blacktriangle$), v_3 = 6.76 cm/s ($\blacksquare$).

Effect of temperature

        The flame temperature was varied at a given fuel-equivalence
ratio (2.5) by changing the cold gas velocity. Three velocities
were used: 3.92, 5.50 and 6.76 cm/s; the maximum temperature
differences in the nucleation zone (3 mm) was about 250 K. For
these three velocities the values of T, f_V, N and D as a function
of height above burner are plotted in Figs. 5 and 6. The height
above burner is corrected by a factor proportional to the velocity
ratio (this ratio was eroneously inverted in previous publications
[3,14]), in order that the abscissa be proportional to the reaction
time (about 1.2 ms par mm). This does not take into account the
small differences in reaction time due to temperature differences.

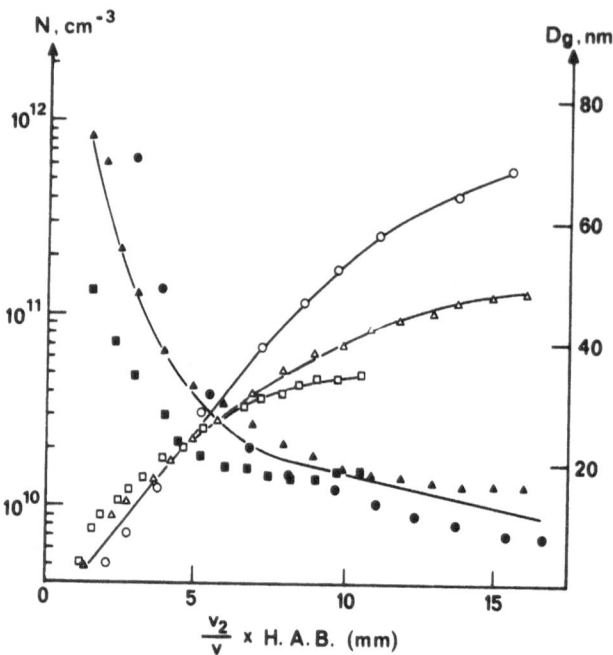

Fig. 6.   Agglomerate number density (N, cm^{-3}, filled symbols)
          and diameter (Dg, open symbols) as a function of height above
          burner (H.A.B., mm). Propane/oxygen flame. Fuel equiva-
          lence ratio = 2.5. Cold gas velocity v_1 = 3.92 cm/s (●),
          v_2 = 5.5 cm/s (▲), v_3 = 6.76 cm/s (■).

Table 5.   Geometric mean diameter (D, nm) and number density
           (N, 10^{10} cm^{-3}) of soot particles measured by
           electron microscopy and by in-situ optical techniques.
           Propane/oxygen flame. Fuel equivalence ratio =
           2.5.   Height above burner = 10 mm.

Cold gas velocity (cm/s)		3.92	5.5	6.26
Electron microscopy	D	21.0	14.3	16.4
	N	18.0	19.0	7.2
In-situ optical techniques	D	64.5	40.1	31.9
	N	0.9	1.6	1.5

The soot volume fraction decreases strongly, by more than one order of magnitude, when the temperature increases by about 250 K (Fig. 5). This indicates a stronger effect of temperature on oxidation than on pyrolysis rates. The number density of aggregates is not affected by temperature, with the exception of the lowest temperature (Fig. 6). The aggregate size (Fig. 6) as well as the sphere diameters (Table 5) decreases when the temperature increases.

REFERENCES

1. H. Gg. Wagner, Mass Growth of Soot, in this volume
2. B. S. Haynes, H. Jander and H. Gg. Wagner, Ber. Bunsenges. Phys. Chem. 84:585 (1980)
3. G. Prado, I.J. Jagoda, K. Neoh and J. Lahaye, Eighteenth Symp. (Intern'l) on Comb., The Comb. Institute, Pittsburgh, p. 1127 (1981)
4. B. S. Haynes and H. Gg. Wagner, Progr. Energy Comb. Sci., 7:229 (1981)
5. I. J. Jagoda, G. Prado and J. Lahaye, Comb. Flame 37:261 (1980)
6. B. S. Haynes and H. Gg. Wagner, Ber. Bunsenges. Phys. Chem. 84:499 (1980)
7. J. Kent, H. Jander and H. Gg. Wagner, Eighteenth Symp. (Intern'l) on Comb., The Comb. Institute, Pittsburgh, p. 1117 (1981)
8. K. Neoh, J. B. Howard and A.F. Sarofim,"Particulate Carbon Formation During Combustion",D.C. Siegla and G.W. Smith, Eds., Plenum Press, New York-London, p. 261 (1981)
9. C. M. Sweitzer and G. L. Heller, Rubber World 134:855 (1956)
10. J. Lahaye and G. Prado, Carbon 12:24 (1974)
11. G. Prado and J. Lahaye, J. Chem. Phys. 11:1678 (1973)
12. G. Prado and J. Lahaye, J. Chem. Phys. 4:683 (1975)
13. J. Lahaye and G. Prado, C.R. Acad. Sci., Paris, Ser. C, 283:425 (1976)
14. G. Prado and J. Lahaye, "Particulate Carbon Formation During Combustion", D.C. Siegla and G.W. Smith, Eds. Plenum Press, New York-London, p. 143 (1981)
15. G. W. Smith, Kinetic Aspects of Diesel Soot Coagulation, in this volume.
16. G. W. Smith, Research Publication GMR-3502, General Motors Research Laboratories, Warren, Michigan, U.S.A.
17. D. M. Roessler, F. R. Faxvog, R. Stevenson and G. W. Smith, "Particulate Carbon Formation During Combustion, D.L. Siegla and G.W. Smith, Eds., Plenum Press, New York-London, p. 57 (1981)
18. S. C. Graham, J. B. Homer and J.L.J. Rosenfeld, Proc. Roy. Soc. Lond. A 344:259 (1975)

19. K. H. Homann, Comment, Sixteenth Symp. (Intern'l) on
     Comb., The Comb. Institute, Pittsburgh, p.717 (1977)
20. A. D'Alessio, A. DiLorenzo, A.F. Sarofim, F. Beretta,
     S. Masi and C. Venitozzi, Fifteenth Symp. (Intern'l)
     on Comb., The Comb. Institute, Pittsburgh, p.1427
     (1975)
21. A. D'Alessio, A. DiLorenzo, A. Borghese, F. Beretta
     and S. Masi, Sixteenth Symp. (Intern'l) on Comb.,
     the Comb. Institute, Pittsburgh, p.695 (1977)
22. B. S. Haynes, H. Jander and H. Gg. Wagner, Seventeenth
     Symp. (Intern'l) on Comb., The Comb. Institute,
     Pittsburgh, p.1365 (1979)
23. G. D. Ulrich, Comment, Eighteenth Symp. (Intern'l) on
     Comb., The Comb. Institute, Pittsburgh, p. 1135 (1981)

DISCUSSION

D. Rivin (Cabot corporation)

The carbonaceous products obtained form pyrolysis of benzene
have morphologies similar to those in environmental soots. Carbonaceous microgel is prevalent at the early stages of pyrolysis
whereas aciniform carbon is the major product at longer time.

Is the large amount of tar formed early in the pyrolysis
eventually converted to aciniform carbon by liquid phase accretion
and dehydrogenation or by pyrolysis to small fragments which then
add to the growing particles from the vapor phase (i.e. surface
growth)?

Prado

Tars formed early in the pyrolysis consist of molecules with
masses in the range 200 a.m.u. - 800 a.m.u. (higher mass observed
when introducing directly tars in a mass spectrometer). They can
be considered as fragments small relative to growing soot particles
and contribute to increase soot mass through direct deposition on
particle surface.

G. W. Smith (General Motors Research Laboratories)

Does not the fact that pyrolytic systems form soot lend support
to "nucleation" rather than to competition of formation and oxidation
processes? Thus, I would support the view that there is a barrier.

Prado

No. In pyrolytic system there is no competition between formation and destruction of soot precursors, as there are no oxidation

processes. In flames, there is ample evidence that soot appearance is the result of shift of relative rates of formation and destruction of soot precursors. (cf. Paper of Howard and Bittner in this volume)

H. F. Calcote (Aerochem. Lab., Princeton)

You point out that the nucleation step can be ignored or simply treated as a boundary condition because most of the soot mass results from surface growth. Is nucleation also unimportant in determining the critical concentration at which soot occurs?

Prado

We pointed that, in practice, to describe evolution of soot particle mass and number density,nucleation may be restricted to a boundary condition. Understanding of nucleation mechanisms is certainly very important with regard to the control of first soot appearance.

M..E. Weill (C.N.R.S., Rouen)

Is the type of size distribution associated with a given process: for instance, a Gaussian distribution with the nucleation or/and a log normal with the coagulation ?

Prado

In our experiments of soot formation during pyrolysis of benzene, we carefully measured the size distribution of the particles. We observed that at the early stages (nucleation), the distributions were Gaussian, and progressively evolved toward log normal distribution during particle growth. We attributed this evolution to limited coagulation.[1]

1) G. Prado and J. Lahaye, J. Chem. Phys. 4:683 (1975)

KINETIC ASPECTS OF DIESEL SOOT COAGULATION

George W. Smith

Physics Department
General Motors Research Laboratories
Warren, MI 48090-9055

INTRODUCTION

Recently we developed a simple physical model[1] for the formation of the fundamental building blocks (spherules) which are the basis of particulate carbon agglomerates such as diesel soot (Fig. 1). This model explained the limited size range of the spherules over a wide range of formation conditions by invoking depletion of soot precursor species to limit growth and incorporating this mechanism into a homogeneous nucleation theory of particle formation. The model was able to explain the spherule size reasonably well. However, it did not include the effects of coagulation on the spherule size,[2] nor did it deal with kinetic aspects of diesel soot formation. In this communication we shall address both questions, assessing the influence of coalescent coagulation on spherule radius and constructing a semiquantitative view of the sequence of events for diesel soot formation. The framework around which we shall build our temporal model will be the well-developed theory of coagulation kinetics. We shall also briefly consider several explanations for the transition from coalescent to chain-forming coagulation.[3] A more extensive discussion will be submitted for publication elsewhere.[4]

INFLUENCE OF COALESCENT COAGULATION ON SPHERULE SIZE

Previously we derived the following expression for r_{sph}, the mean radius of a spherule:

$$r_{sph} = \left[\frac{3M}{4\pi A\rho} \exp(4\pi r_c^2 \sigma/3kT) \right]^{1/3} \tag{1}$$

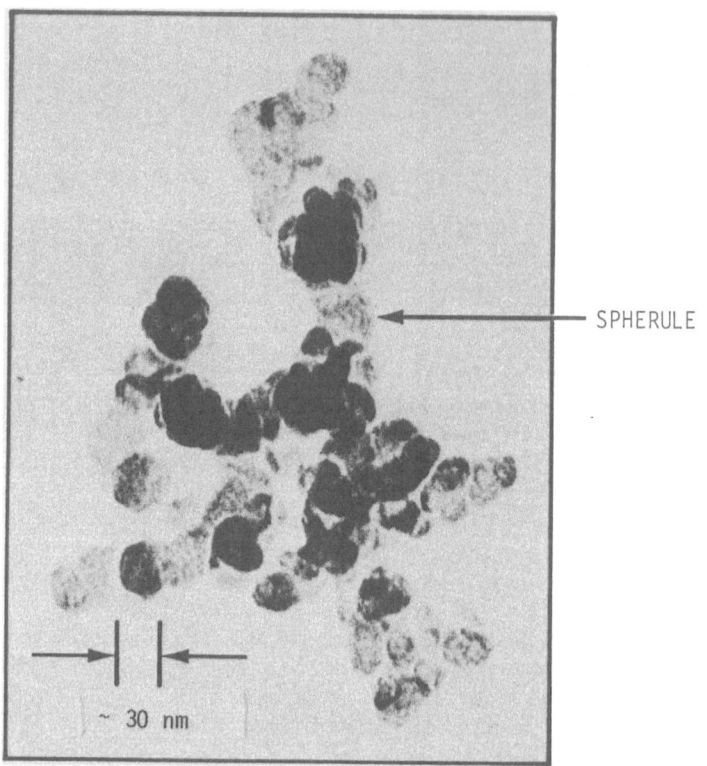

Fig. 1 Transmission electron micrograph of a particle formed
        during diesel combustion. The particle is an agglomerate
        of fundamental spherules which are approximately 30 nm in
        diameter. (TEM courtesy of R. Stevenson).

where M is the molecular weight of the condensible precursor
species, A Avogadro's number, ρ the density of the condensed phase
of precursor, r_c the radius of a critical sized nucleus, σ the
surface energy of the condensed precursor, k Boltzmann's constant,
and T the absolute temperature. Equation (1) is not a function of
pressure or concentration of precursor species, in agreement with
the observed insensitivity of spherule size to those parameters; it
was derived on the assumption that each critical nucleus grows to
form a single spherule. However, it has been suggested[2,3,5] that
coalescent coagulation in the early stages of particle formation
would cause each spherule to contain anywhere from a few to many
"nuclei". As a result, the value of r_{sph} calculated from equation
(1) would be too small. To take account of coalescent coagulation
we must multiply equation (1) by a factor $(N_c)^{1/3}$ where N_c is the

number of coalescing nuclei per spherule. Results of several experiments[5-7] suggest that N_c is in the range of a few to 30, so that equation (1) underestimates r_{sph} by no more than a factor of 2 or 3 ($10^{1/3}$ to $30^{1/3}$). This is within the order-of-magnitude range of validity claimed for equation (1). (At any rate, the nucleation/depletion model was originally proposed as a basis from which to evaluate precursor depletion as a mechanism to explain spherule size universality,[1] not as a quantitative description of soot formation.)

TEMPORAL ASPECTS OF DIESEL SOOT FORMATION

We shall concern ourselves with the soot formation process in a small volume element V_{local} of the diesel combustion chamber, an increment of volume whose size is determined by the distance over which material can be transported and mixed by turbulent processes within a time τ_{mix}. The soot formation events in a given V_{local} will be regarded as isolated from those in other volume elements if the formation processes are fast compared to τ_{mix}. (Various estimates[8-10] suggest that $\tau_{mix} \gtrsim 1$ ms.) We shall assume that the formation of critical soot nuclei takes place essentially instantaneously after local formation of condensible precursor species. Bauer[11] has pointed out that the nucleation process should take only a few microseconds, consistent with estimates of Tesner, et al.[12] The kinetic processes governing initial coalescent coagulation and subsequent chain-forming coagulation are identical. What differs is the resulting structural form of the product agglomerate. Coalescent coagulation and surface growth yields a quasi-spherical structure (the "spherule") and chain-forming the ultimate clustered ("aciniform") appearance of Fig. 1.

The groundwork for the kinetics of coagulation was laid by Smoluchowski[13] with his well-known equation:[3,14]

$$\frac{dn}{dt} = -kn^2 \tag{2}$$

where n is the number density of coagulating particles and k is a rate constant whose functional form depends on the particle size and the pressure, temperature and molecular size of the gaseous medium in which coagulation is occurring[14] (i.e. on the Knudsen number K_n). For particles large compared to the mean free path of the gaseous medium (small K_n) coagulation occurs in the "continuum regime". For small particles (large K_n) "free molecule" coagulation occurs. The details of the theory have been worked out be several groups of workers.[15-17] For continuum coagulation ($K_n < 1$) the solution to equation (2) is[15]

$$n = \frac{n_o}{1 + 4kT\,n_o\,t/3\mu} \; , \tag{3}$$

where n_o is the initial number density of coagulating particles (equal to the number density of nuclei), t the time, and μ the viscosity of the gaseous medium. In the free molecule regime[3,7,14-17] ($K_n > 10$) the time dependence of n is

$$n = \frac{n_o}{[1 + \frac{5}{12} \left(\frac{3}{4\pi}\right)^{1/6} G\alpha \, n_o^{5/6} \left(\frac{6kT}{\rho}\right)^{1/2} f_v^{1/6} t]^{6/5}} \qquad (4)$$

where G is a parameter which takes into consideration the enhancement of the collision cross section by dispersion forces, α is a factor which accounts for the particle size distribution, and ρ is the density of the particle material. For the case of diesel soot formation it is easy to show that K_n is on the order of unity so that we expect the kinetic behavior of diesel soot coagulation to be bracketed by equations (3) and (4).

DIESEL SOOT COAGULATION CALCULATION

In order to calculate the kinetics of coagulation — coalescent or chain-forming — we must determine n_o, f_v, μ, ρ, and T. We have estimated n_o and f_v for diesel pressures by several methods (extrapolation from low pressure results[7,18-21] and calculation from in-situ soot concentration measurements[22] and emission index[23]), obtaining $n_o \approx 2.5 \times 10^{-3}$ and $f_v \approx 10^{-6}$ to 10^{-5}. Estimates of the other parameters were obtained from the literature [T $\approx$ 1500K; μ = 5.4 $\times$ 10^{-5} kg/m-s (air at 1500K); G = 2.04; α $\approx$ 6.67; ρ $\approx$ 2 $\times$ 10^3 kg/m^3].

The results of the calculations are plotted in Fig. 2 which shows how n decreases with time due to coagulation. A striking feature of Fig. 2 is that the two coagulation curves calculated for the continuum and free-molecule regimes do not differ significantly from one another (less than one order of magnitude) over the time region of interest for diesel combustion (t $\lesssim$ a few ms), probably due to the fact that the Knudsen number for diesel particle coagulation lies intermediate to the values for continuum and free-molecule coagulation.

It should be mentioned that Khan, et al.,[9] and Dolan, et al.,[10] have each calculated coagulation behavior for diesel soot. However, both computations consider initial particles to be spherule size (25 nm diameter) and hence do not treat the early stages of coagulation where coalescence and surface growth are dominant. Furthermore, Khan's value of the coagulation constant (5.93 $\times$ 10^{-5} cm^2/s) is some 10^5 higher than the value found experimentally for other aerosols or that calculable from equation (3). He ascribes the difference to high air velocities and turbulence in the engine. However, as we have seen, ordinary continuum and free molecule equations both give fast enough coagulation without invoking turbulence. Also, it seems unlikely that turbulence

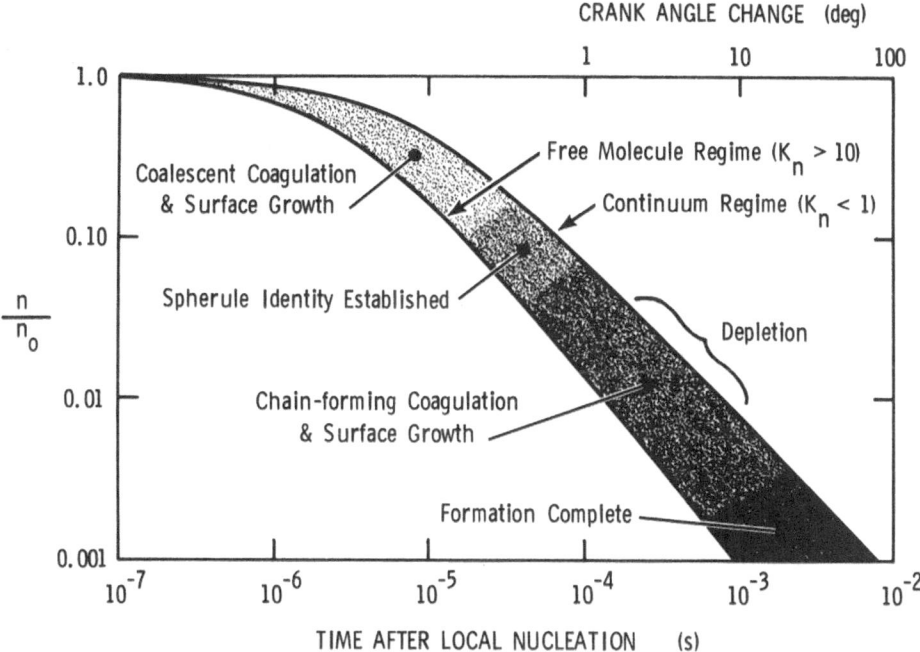

CRANK ANGLE CHANGE (deg)

Fig. 2   Plot of calculated particle number density versus time for
         diesel soot formation in a local volume element.   Soot
         chain formation is essentially complete within 1-2 milli-
         seconds after local nucleation, corresponding to a change
         in crank angle of about 10 degrees.

microscales would be sufficiently small for turbulence to be
important in the coagulation process.

DISCUSSION

     Let us now summarize semiquantitatively the diesel soot forma-
tion events of Fig. 2. We emphasize again that the processes are
assumed to occur in a local volume element which can be regarded as
more-or-less isolated from neighboring volume elements.   Furthermore, time zero (time of nucleation) for one volume element may be
shifted relative to that in another volume element by as much as
τ_{mix} (~1 ms).   After local formation of precursors (Tesner's
"radical nuclei"), nucleation occurs with a time constant on the
order of a few μs.[11]   The number density of critical nuclei
(Tesner's "smallest soot particles") is taken to be equal to
n_o(~2.5 x 10^{20} m^{-3}).   Immediately following nucleation, collision
is coalescent and remains so until the particle number density
decreases from n_o to ~n_o/10 (or even ~n_o/30); from Fig. 2 this

decrease takes only about 0.05 ms. At the end of this time spherule identity has become established, and further coagulation is chain-forming. A simple kinetic calculation suggests that gas-phase precursor molecules will not become depleted until a few tenths of a ms after nucleation, so that surface growth by condensation of precursors proceeds during coalescent coagulation and continues well into the chain-forming stage. There is some evidence[3,7] that non-sticking collisions occur toward the end of soot formation in lower pressure systems, and this may take place for diesel soot as well. At any rate, 1000-fold coagulation occurs for diesel soot within 1-2 ms after local nucleation. At the end of this time soot formation is essentially complete, and cooling occurs due to cylinder expansion.

An extensive discussion of dehydrogenation and oxidation reactions (which occur simultaneously with these processes) is beyond the scope of this paper. It has been suggested[3,7,16] that an increase of particle viscosity due to dehydrogenation is responsible for the transition from coalescent to chain-forming coagulation. The Frenkel relation[24] (that the time for fusion of two identical viscous spheres is proportional to the ratio of the product of sphere viscosity and radius to the surface energy) provides a possible quantitative basis for evaluation of this mechanism. However, dehydrogenation may proceed too slowly to govern the coalescence/chain-forming transition. An alternative view[5] is that "coalescence" is the result of aggregation of individually-nucleated crystallites while surface mobility is still high or surface growth still rapid. A third possibility[25] is that an energy accommodation mechanism involving atom-atom forces may be responsible for coalescence when particles are small and that such a process may be inoperative for larger particles.

It appears that although many aspects of diesel soot formation are incompletely understood (e.g. precursor formation, nucleation, dehydrogenation, oxidation, and the coalescence/chain-forming transition mechanism), the kinetic behavior of diesel soot coagulation is adequately described by well-tested coagulation models.

REFERENCES

1. G. W. Smith, Extended Abstracts, 15th Biennial Conference on Carbon, American Carbon Society and University of Pennsylvania, Philadelphia, (1981); D. M. Roessler, F. R. Faxvog, R. Stevenson, and G. W. Smith in "Particulate Carbon: Formation During Combustion," D. C. Siegla and G. W. Smith, eds., Plenum Press, New York (1981).

2. B. S. Haynes in "Particulate Carbon: Formation During Combustion," D. C. Siegla and G. W. Smith, eds., Plenum Press, New York (1981).

3. G. Prado and J. Lahaye in "Particulate Carbon: Formation During Combustion," D. C. Siegla and G. W. Smith, eds., Plenum Press, New York (1981).

4. G. W. Smith, to be submitted to Society of Automotive Engineers.

5. J. B. Howard, B. L. Wersborg, and G. C. Williams, in "Faraday Symposium, No. 7, Fogs and Smokes," Faraday Division, Chemical Society London (1973).

6. B. S. Haynes, H. Jander, and H. Gg. Wagner, in "Seventeenth Symposium (International) on Combustion," The Combustion Institute, Pittsburgh, PA (1978).

7. G. Prado, J. Jagoda, K. Neoh, and J. Lahaye, in "Eighteenth Symposium (International) on Combustion," The Combustion Institute, Pittsburgh, PA (1981).

8. A. Vranos, Combust. and Flame 22: (1974); A. M. Mellor, Prog. Energy Combust. Sci. 1: 111 (1976).

9. I. M. Khan and G. Greeves, in "Heat Transfer in Flames," J. M. Beer and N. H. Afgan, eds., Scripta Book Co., Washington (1974); I. M. Kahn, C. H. T. Wang, and B. E. Langridge, Combust. and Flame 17: 409 (1971); I. M. Khan, Proc. Inst. Mech. Engineers 184, Pt. 3J: 36 (1969-70).

10. D. F. Dolan, Ph.D. Thesis, University of Minnesota, December 1977; D. F. Dolan and D. B. Kittelson, SAE paper 780110, Society of Automotive Engineers, Warrendale, PA (1978).

11. S. H. Bauer, in "Particulate Carbon: Formation During Combustion," D. C. Siegla and G. W. Smith, eds., Plenum Press, New York (1981).

12. P. A. Tesner, T. D. Snegiriova and V. G. Knorre, Combust. and Flame 17: 253 (1971).

13. M. v. Smoluchowski, Z. Physik Chem. 92: 129 (1917).

14. H. Gg. Wagner, in "Seventeenth Symposium (International) on Combustion," The Combustion Institute, Pittsburgh, PA (1978).

15. G. M. Hidy, J. Colloid Sci. 20: 123 (1965); G. M. Hidy and J. R. Brock, J. Colloid Sci. 20: 477 (1965); G. M. Hidy and D. K. Lilly, J. Colloid Sci. 20: 867 (1965); F. S. Lai, S. K. Friedlander, J. Pich, and G. M. Hidy, J. Colloid and Interface Sci. 39: 395 (1972).

16. G. D. Ulrich, Combust. Sci. and Tech. 4: 47 (1971); see also G. D. Ulrich, in "Twelfth Symposium (International) on Combustion," The Combustion Institute, Pittsburgh, PA (1969).

17. S. C. Graham and J. B. Homer, in "Faraday Symposium, No. 7, Fogs and Smokes," Faraday Division, Chemical Society, London (1973); S. C. Graham, J. B. Homer, and J. L. J. Rosenfeld, Proc. Roy. Soc. Lond. A344: 259 (1975).

18. B. L. Wersborg, A. C. Yeung, and J. B. Howard, in "Fifteenth Symposium (International) on Combustion," The Combustion Institute, Pittsburgh, PA (1974).

19. T. Kadota, N. A. Henein, and D. A. Lee, 5th International Automotive Propulsion Systems Symposium, April 1980; T. Kadota and N. A. Henein, in "Particulate Carbon: Formation During Combustion," D. C. Siegla and G. W. Smith, eds., Plenum Press, New York (1981).

20. B. S. Haynes, H. Jander, and H. Gg. Wagner, Ber. Bunsenges. Phys. Chem. 84: 585 (1980).

21. A. D'Alessio, A. DiLorenzo, A. F. Sarofim, F. Beretta, S. Masi, and C. Venitozzi, in "Fifteenth Symposium (International) on Combustion," The Combustion Institute, Pittsburgh, PA (1974); A. D'Alessio, F. Beretta, and C. Venitozzi, Combust. Sci. & Tech. 5: 263 (1972).

22. G. Greeves and J. D. Meehan, in "Combustion in Engines," Mechanical Engineering Publications, Ltd. (1973); V. K. Duggal, T. Priede, and I. M. Khan, SAE paper 780227, Society of Automotive Engineers, Warrendale, PA (1979).

23. S. L. Plee, T. Ahmad, J. P. Myers, and D. C. Siegla, in "Particulate Carbon: Formation During Combustion," D. C. Siegla and G. W. Smith, eds., Plenum Press, New York (1981).

24. J. Frenkel, J. of Physics (Moscow) 9: 385 (1945).

25. J. G. Gay, General Motors Research Laboratories (private communication).

MASS GROWTH OF SOOT

Heinz Georg Wagner

Institut für Physikalische Chemie der Universität
Tammannstr. 6
D-3400 Göttingen, West Germany

INTRODUCTION

How much soot is emitted in unit time by a combustion device? This is the question, an engineer will ask and his task usually is to reduce this quantity of soot emitted. His problem, therefore, is not so much one of quality but of quantity.

In premixed systems it is fairly easy to avoid soot formation by staying sufficiently far away from the (visible) limits of soot formation which correspond roughly to about 10^{-9}g soot per cm^3, a value which depends on temperature and to a certain extent on other circumstances. Diffusion flames can also be operated such that very little soot is emitted, because the soot particles are slowly oxidized.

For many combustion devices like furnaces, combustion chambers or diesel engines, there are good reasons not to operate completely under premixed conditions. The time available for the combustion of a certain amount of fuel is definitely restricted. In addition, temperature, pressure overall C/O ratio and flow conditions of the combustion processes vary within certain limits. When we discuss some of the material available on the growth rate of soot mass, we shall try to stay as close to these conditions as possible.

The Soot Mass Growth Rate

Investigations of the kinetics of soot formation[1] deal with

171

the variation of particle number and mean diameter. These two quantities are, at least in principle, well defined. They also have to deal with the total mass of soot and its growth. This quantity needs, as we will see, a definition and it will be much easier to give "rate constants" than to describe details of the growth process.

Before going into details, we can try to give an overall estimation of a mean value of the growth rate of soot.

Measurements of the amount of soot formed in well maintained technical combustion systems as diesel engines, burner flames or other devices often give values of about 10^{-7} g soot per cm^3 burned gas, contained in approximately 10^9 soot particles. The passage time of some volume element through the combustion zone often lies around 10 milliseconds. Combining these figures we obtain as a zeroth approximation a value of about 10^{-5} g s^{-1} cm^{-3} or 10^{-6} mol cm^{-3} s^{-1} (based on C atoms) for a rate of soot mass growth in these combustion systems. For different conditions, as during carbon black production, we will expect different values. We also do know that the amount of soot finally emitted by the combustion device need not necessarily represent the amount of soot formed in the combustion process. Soot particles may be oxidized or they can grow by the condensation of hydrocarbons from the gas phase on the surface when they are on their way out of the combustion system.

Studies of the rate of growth of the soot mass on surfaces have been performed by various authors[2-4], especially by Tesner[2] and co-workers. They measured the growth of soot layers at solid surfaces by pyrolysis of hydrocarbons. There, they give the growth rate as the velocity with which the soot layer grows perpendicular to the surface as V in cm s^{-1}, a presentation which has also been adopted by Howard[5] et al., or as V g cm^{-2} s^{-1} bar^{-1}. Some corresponding expressions are (activation energies in kcal/mol)

$$7.86 \cdot 10^2 \ \exp -65/RT \ \text{g} \ cm^{-2} \ s^{-1} \ bar^{-1} \ \text{for} \ CH_4 \ (970 - 1280K)$$

$$3.3 \cdot 10^2 \ \exp -55/RT \ \text{g} \ cm^{-2} \ s^{-1} \ bar^{-1} \ \text{for} \ C_6H_6 \ (1000 - 1250K)$$

They and others observed that the deposition of soot at the wall takes place already at lower temperatures than the formation of particles in the gas phase.

In order to obtain information about the rates of soot mass growth comparatively simple experimental conditions have to be chosen and we will look into three of them: a) one dimensional isothermal flow for pyrolysis experiments; b) premixed flat flames and c) laminar diffusion flames. At first, however, the meaning of soot mass shall be discussed.

The Soot Mass

The determination of the mass of soot is most easily performed in sooting flat flames by pumping the burned gas through a fine filter. This method may, however, give misleading results in the soot formation zone. A rather reliable and direct, but very laborious and in no way perfect technique to obtain information about the soot mass within the region of soot formation is to collect soot particles by molecular beam sampling[5,6]. On an electron micrograph one can count particles and thus determine their number density N per cm^3 gas. One can also measure the diameters d of the images of the particles. A distribution function results for these diameters. Assuming ball shape, the particle volume v can be evaluated. From the distribution function, whether it is Gaussian, log-normal or self-preserving cannot easily be decided[5], a mean value $\bar{v}$ of the particle volume can be calculated. It is important to realize here that the resolution of an electron microscope is limited. Therefore, information about particles below a certain diameter will be lost.

With these quantities N and $\bar{v}$ one can calculate a volume fraction of soot $f_s = \Sigma N_i v_i = N \bar{v}$ which can be considered as one direct measure for the mass m_s of soot per cm^3 gas, because $m_s = f_s \rho_s$ where ρ_s is the density of the soot particles[7-9]. Here any matter which cannot be seen on the electron micrograph is neglected. Therefore, f_s evaluated in this way depends, especially in the early phase of the formation process, on the resolution of the electron microscope.

Let us consider a simple example: The condensation of super-saturated vapour. After condensation, when the equilibrium vapour pressure is reached, the mass of substance contained in droplets is m_s. For an isothermal condensation this corresponds exactly to that amount of vapour which is condensed and which had caused the super-saturation. Therefore, m_s does not change during condensation (and f_s will also remain nearly constant during the transition from molecules to droplets). If, however, we can see only particles above a certain size, then that part of m_s (or f_s), we can see, will increase with time.

Besides the electron microscope, several other methods[2,10] have been applied to measure the growth rate of soot particles and soot mass which shall not be discussed here. In recent years the investigation of aerosols in general and especially of soot formation by laser light scattering and absorption proved to be very useful[11] and sufficiently fast to make systematic studies[12].

If there are N particles in a volume of one cm^3 with volume $v = \pi d^3/6$, then the light extinction coefficient k_{ext} (for $\pi d/\lambda < 0.3$) is

$$k_{ext} = - 6\pi\lambda^{-1} \ Im \ \{(m^2-1)/(m^2+2)\} \cdot N \ v$$

and the light scattering factor Q_{vv} is

$$Q_{vv} = (3\pi)^2 \ \lambda^{-4} \ |(m^2-1)/(m^2+2)|^2 \cdot N \ v^2$$

The relations become simple if the particles are spheres, if the complex refractive index m is constant and if the distribution function of the particles is known (for details see paper of d'Alessio[11]). Under these assumptions f_s can directly be determined from k_{ext}, the particle number from k_{ext}^2/Q_{vv} and the mean particle volume from Q_{vv}/k_{ext}.

As realized already in[10,13-15] light absorption within a main part of the soot formation zone is not only due to soot particles but also to other species. The contribution of these species to light absorption varies with wavelength and even small molecules already from C_2 up show more or less light absorption at high temperature between 200 nm and 5000 nm[17]. The determination of soot mass by optical measurements therefore requires great care, scattering measurements are influenced by these facts as well. Fluorescence measurements and investigation of depolarisation ratios can give some help[11,15].

The assumption of constant refractive index may also be questioned to a certain extent. Measurements by Meyer[18] indicate that the refractive index changes in a way pointing to higher electrical conductivity for older and larger soot particles, for those with less hydrogen content.

The assumption of spherical particles does definitely not hold in later stages, when soot particles agglomerate to form chains for which the free zone between the spheres is not filled by further growth of the soot mass. Here, it is important to state whether the chains or the spheres in the chains are counted as particles. In the early phase of the process, particles do look similar to spheres on an electron micrograph and coagulated particles seem to attain a compact form very rapidly again, probably by transport processes and by surface growth.

The total soot mass determined from optical (absorption) measurements in a region where df/dt becomes small does not always agree with that calculated from electron micrographs[19] or that obtained by direct sampling[20,21]. This holds already for flames not to far away from the limit of soot formation. In some cases reported, these discrepancies for normal pressure flames are rather substantial.[19]

The different methods look at the soot mass from different points of view and no one monitors exactly that f which is needed for a reaction kinetic evaluation of the data.

Relations between f_v, N and v

It is well known that during soot formation particle volume and particle number change at the same time and the shape of the particle distribution function may also vary. For reason of simplicity, let us first consider some simple relations. With the definition $f_v = N \bar{v}$ we need two quantities in order to determine the third one. Here, we are mainly interested in the variation of f_v for particles above a certain size. Therefore, we need not discuss details of the very beginning of the process. The change of f_v is given by

$$df_v = N \, d\bar{v} + \bar{v} \, dN \qquad\qquad\qquad 1)$$

with $N = (\partial f_v / \partial \bar{v})_N$ and $\bar{v} = (\partial f_v / \partial N)_{\bar{v}}$. Variations of f_v can be due to an increase of the particle volume v, to the generation of new particles or the removal of particles, e.g. by oxidation.

Coagulation of particles does not change f_v (It is assumed for simplicity that the density of the matter in the particles is the same for all particles and remains constant.) For that process $df_v = 0$ and $(\partial N / \partial \bar{v})_f = -N/\bar{v}$. Measurements of the variation of particle number with time in the later phase of soot formation in flames[5,10] and in shock tube pyrolysis experiments[9,22] do show that the particle number densities strongly decrease towards the "burned gas side"[5,10], that dN/dt can be quite well described by a coagulation model, where the particles collide and stick[9,12]. Therefore, $dN/dt = -k_{co} N^2$. (The formation of particles at the beginning of the process is neglected.) The rate coefficient k_{co} varies proportional to $\bar{v}^{1/6}$. Its value around 1700 to 1800 K is

$$k_{co} / \bar{v}^{1/6} \approx 2 \cdot 10^{-6} \ cm^{5/2} \ s^{-1}$$

for unseeded flames.

Using this expression one can write

$$df_v/dt = N(d\bar{v}/dt - f_v k_{co}) \qquad\qquad\qquad 2)$$

Here, the second term corrects for that part of particle volume growth which is due to coagulation.

For the growth of particle volume one can write

$$d\bar{v} = N^{-1} df_v - (f_v/N^{+2}) dN \qquad\qquad\qquad 3)$$

where $(\partial \bar{v}/\partial f) = N^{-1}$ and $(\partial \bar{v}/\partial N)_f = -f_v/N^{+2}$. With the same assumption about particle coagulation as above, the change of particle volume is given by

$$d\bar{v}/dt = \bar{v} \, d \ln f_v/dt + f_v k_{co} \qquad\qquad\qquad 4)$$

This expression shows nicely the two factors which contribute to the particle growth in the later phase of the soot formation process: coagulation and surface growth.

If the quantities f, N and $\bar{v}$ are known, one can also calculate approximately the total surface area F represented by the particles in cm^2/cm^3 and their total "radius" R (For an accurate calculation the particle distribution functions must be used). The relations are

$$F = \sum_i N_i F_i = 3f_v / \bar{r}$$

where $\bar{r}$ is the mean particle radius and

$$R = \sum_i r_i N_i = (3/4 \ \pi) \ f_v / \bar{r}^2 \ .$$

Upper Bounds for Surface Growth

The number of collisions on a surface of 1 cm^2 in one second in an ideal gas is approximately given by $n \cdot \bar{w}$ where n is the number of particles per cm^3 and $\bar{w}$ their mean velocity. For air at 1800 K this number is $3.5 \cdot 10^{23} \ cm^{-2} \ s^{-1}$.

When we assume that from a certain time on there are no new particles formed, then the total increase of soot mass can be attributed to the difference of the substance added to the particles minus that released from the particles. For a certain molecule hiding the particle surface there are various possibilities:
  - it can be reflected or released after some time
  - it can be incorporated and react either at the
    surface or inside. The reaction products stay
    on the particle or they are (partly) released
    into the gas phase, e.g. in form of molecules
    with higher hydrogen content or with high thermo-
    dynamic stability.
Both processes, addition and release of molecules should, in a first approximation, be proportional to the surface area of the soot particles (when the mean free path is large compared to the particle diameter)

$$df/dt = k_1 \ \phi(S) \ F - k_2 \ \psi \ (R_c) \ F \qquad\qquad 5)$$

S represents the concentration of species which add to the particles and R_c the (over)concentration of those species which can leave the particles. ϕ and ψ are functions. This simple expression contains the assumption that the activity of particles with respect to surface reaction does only depend on their surface area which seems to be incorrect in soot formation.

For continuum conditions (see e.g.[7] or [8]), when the mean free
path is smaller than the particle size , the processes are diffusion
controlled and instead of F the quantity R has to be used[7,8]. If not
the transport to and from the particles is rate determining, but the
surface reaction, then it is again the active surface which has to
be used instead of R.

With the above assumptions and $\psi(R_c) = R_c$; $\phi(S) = S$ one obtains

$$\bar{r}/3 \ d \ln f/dt = k_1 \cdot (S) - k_2 \cdot (R) \ . \qquad\qquad 6)$$

This expression shows that the accurate determination of soot mass
growth rates requires extremely accurate measurements of $\bar{r}$ and f_v.
At present, neither S nor R can be accurately identified. What we
do often know, is the total amount of substance added to the particles after some time t_e (in the order of milliseconds) when the
soot formation process already has slowed down and the particle
number has reached its "final" value. There df/dt ≈ 0 and dN/dt ≈ 0.
The f_v values obtained there shall be called f_{max}. The time t_e,when
f_{max} is reached,is generally determined by the combustion process
itself. If we would keep the system for much longer time at high
constant temperature, another value of f_{max} would result because then
the system would come closer and closer to equilibrium.

Soot Volume Fraction in Flat Flames

Soot profiles measured in flat flames exhibit some typical
properties: The amount of soot,f_v, at first rises rapidly and then
the increase of soot mass df/dt steadily decreases further downstream. Often, f_v seems to reach a maximum. It must be mentioned,
however, that the flame temperature decreases steadily behind the
main reaction zone where soot formation proceeds and the density
change may cause some change in f_v. The particle number where it can
be measured more reliably, decreases over two to three orders of
magnitude to values around $10^9 - 10^{10}$ particles per cm^3 after 20 to
40 ms. This value shows only little variation with pressure and
temperature. It can be changed by certain additives. The particle
radius determined by optical methods continues to increase steadily
also after df/dt has reached its "maximum value" in that range where
particles coagulate to form chain like aggregates.

For systems where the influence of the C/O ratio on the amount
of soot produced in a flat flame has been measured[6,10,23-25], the
value of f_m increased strongly with the surplus hydrocarbon
$f_m \sim (C/O - (C/O)_{limit})^n$. For C_2H_4 the n value is close to 3,for
C_6H_6 even larger[25]. This increase of soot mass with increasing C/O
ratio leads (mostly) to an increase in the mean particle diameter,
while the "final" N seems to vary very little.

Gas analysis in the burned gases shows that besides the water-gas components and the soot particles there are various hydrocarbons present. The total amount of these hydrocarbons is normally larger than that of soot. For the standard C_2H_2-O_2 low pressure flame the hydrocarbon mass in the burned gas (9 cm above the burner) increases more than 20 times stronger with increasing C/O ratio than the amount of soot. For low pressure benzene/oxygen flames (~ 2000 K) the amount of soot formed increases stronger with increasing C/O ratio than for C_2H_2 flames, but the increase of the mass of hydrocarbons is still much faster than that of soot[23]. Measurements in atmospheric C_2H_4-air flames show that 4 cm above the burner the mass of C_2 hydrocarbons is at least 10 times larger than the soot mass (near C/O_{crit} Data of D'Alessio also indicate relatively high hydrocarbon concentrations in the burned gas besides soot in methane-oxygen flames[26].

This relatively high hydrocarbon concentration where $df/dt \approx 0$ requires that the net df/dt due to the addition of these hydrocarbons in that domain must be much smaller than in the main soot forming region of the flame, even though the total surface area of the soot particles is quite similar. There are two reasons discussed for that effect: either the surface of the larger soot particles becomes less "active" or the substrate which adds easily to the particles is consumed. The fact, that the situation in the main soot forming region of a flame is distinctly different from that in the burned gas, is very convincingly demonstrated by the results of Long et al.[27].

Measurements of df_v/dt in Flat Flames

Values of df_v/dt taken from optical measurements in flat flames show the following tendency (see Fig. 1): at the beginning where optical measurements can be evaluated, df_v/dt is highest. It, then, decreases continuously into the the burned gas. Towards shorter times the evaluation of df_v/dt is no longer possible, however, df_v/dt must go to zero somewhere within the reaction zone. The area under the curves is given by $f_m = \int (df_v/dt) dt$. Figure 1 shows that df_v/dt at the beginning of the curves increases with increasing C/O ratio. Values reported for df_v/dt in the early parts of flames range from about $5 \cdot 10^{-5}$ s^{-1} for benzene-air flames (C/O $\approx$ 0.78); $1.3 \cdot 10^{-5}$ for a C_2H_4-air flame (C/O = 0.76) and about 10^{-6} s^{-1} for the low pressure standard C_2H_2-O_2 flame (C/O = 1.4).

The values of df_v/dt at $f_{1/2} = f_M/2$ seem to be proportional to f_M. Because f_M increases strongly with increasing C/O ratio, this is also true for the "initial rate" of soot mass growth. In order to be able to compare the rates of soot formation in different flames, it is necessary to find a rational representation of the data. For various C_2H_4-air and C_6H_6-air flames plots of df_v/dt as a function of f_v could be represented approximately by straight lines over a fairly wide range of f_v. Some examples for different

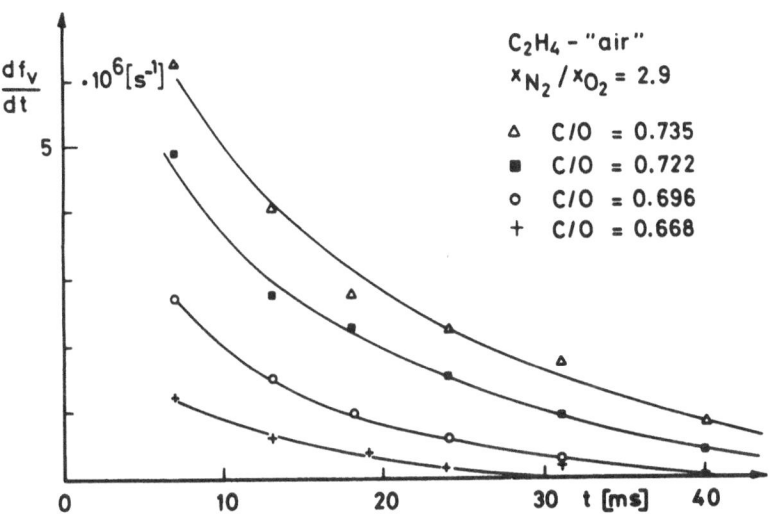

Fig. 1.   Rate of soot mass growth as a function of time (Haynes).

conditions are shown in Fig. 2. The scatter of the data is relatively large, it does however not seem to be a systematic one, their reproducibility is moderate. Values for the apparent rate constant
$k = -\Delta(df/dt)/\Delta f_v$ are in the order of magnitude of 100 s^{-1}. Data
taken from D'Alessio's papers are of the same order of magnitude.

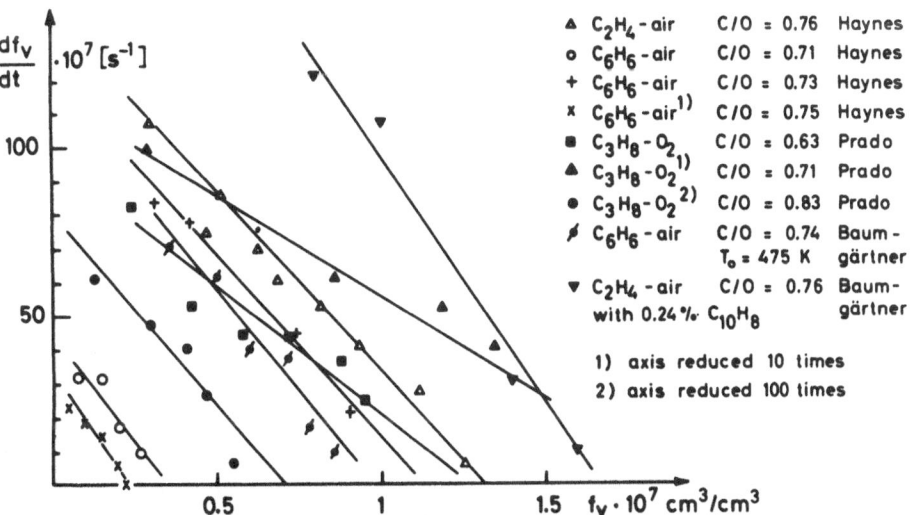

Fig. 2.   Values of df_v/dt over f_v for different flames.

The lines can be extrapolated to df/dt = 0. From this extrapolation a fictive value f_M results which can be used to set up an
equation for the straight part of these curves which reads

$$df_V/dt = k\, f_M\, (1 - f_V/f_M) \qquad\qquad 7)$$

It is obvious that this expression does not hold for very short and
for very long times but it represents that part of the process fairly well in which most of the soot mass appears in particles.

The fact that k values from different sources and for different
flames are rather similar may be partly due to the fact that flame
conditions, especially if optical methods shall be used for the
measurements can only be varied over a relatively small rang. Values
extracted from the measurements of Prado et al.[20] which cover a
rather large range of C/O seem to indicate that k increases slightly
with increasing C/O ratio.

Influence of Additives on the Rate of Soot Mass Growth

Certain additives can strongly influence the total amount of
soot formed in a premixed flame. For metal additives like Ba or Mo
this effect is very pronounced. It is known that this reduction of
the amount of soot in most cases can be mainly attributed to a decrease of the particle volume[12]. The absolute values of df_V/dt and
$d\bar{v}/dt$ in these flames decrease correspondingly.

If, for metal additives however, the rate of soot mass formation
df_V/dt is evaluated from the measured f_V curves in the way described
above, the variations in the k values obtained are small and lie
within the limit of experimental error: the change of soot mass is
represented by f_M. For gaseous additives like SO_2, SO_3, NH_3, H_2, etc.
the situation is very similar[25]. Changes in the amount of soot
finally formed do not appear as changes of the k values.

The situation seems to be slightly different if naphthalene is
used as an additive. The addition of 0.24% naphthalene to a C_2H_4-air
flame with an initial C/O ratio of 0.69 gave an increase in k of
about 50%. Naphthalene-air flames seem to have a higher k value than
those reported in Fig. 2.

The Influence of Temperature on the Soot Mass Growth

In flames, which do not form a polyhedral structure, a relation
between the maximum flame temperature and the amount of soot formed
can be given. For C_2H_4-air flames, at a given C/O ratio the amount
of soot formed decreases nearly exponentially with increasing temperature [28] (in the range from 1750 to 1950 K). (For the same soot mass
per cm^3 Floßdorf gives a linear relation between C/O and temperature). This is linked to the well-known fact that for a given system

the C/O value at the limit of soot formation increases with increasing temperature[28,29].

    If soot limits for different fuel-air systems are determined
at the same temperature in the burned gases the corresponding C/O
values are much closer together than those for the adiabatic flame
temperatures of the different systems. This means that the total
amount of soot formed in premixed laminar flames under near limit
conditions depends only little on the type of fuel, at least for
smaller hydrocarbon molecules. (For polyhedral flames, the situation
is more complicated.)

    In order to check the influence of temperature on the growth
rate of soot, the maximum flame temperature can be varied by changing
the gas flow, the fresh gas temperature, the oxygen content and the
inert gas component. (The effects of these variations are slightly
different.)

    H. Jander[30] performed experiments where gas flow and fresh gas
temperature have been varied. For a given fresh gas temperature the
amount of soot finally formed decreased with increasing gas flow
velocity and the maximum flame temperature increased. Flame stability
and the temperature dependence of flame velocity allowed both quantities to be varied only over relatively small ranges. As a result,
the maximum flame temperature increased with increasing flow velocity
(from 1660 K to 1890 K) while the amount of soot for a constant C/O
of about 0.87 decreased by a factor of about 25. The variations of
the df_v/dt values were correspondingly large, the k values, however,
changed less. There seems to be a tendency towards larger k for increasing temperature for C_2H_4 and C_6H_6-air flames. The measurements
do not allow to separate the influence of flow velocity and of
temperature quantitatively[30,31].

    In experiments of Bonne et al. increase of the flow velocity
for the C_2H_4-O_2 low pressure standard flame did also reduce the
amount of soot formed. In these experiments there was, however, no
pronounced variation of the maximum burned gas temperature. The
injection of cold inert gas into the burned flame gases, however,
increase soot formation[6].

<u>Time Scale</u>

    The amount of soot formed not to far away from the limits of
soot formation of the corresponding flame is about $10^{-7}g/cm^3$, this
means $5 \cdot 10^{15}$ C atoms $\cdot$ cm^{-3} are present as soot. A characteristic
time the soot formation process in premixed flames derived
from df/dt is around 10 ms(k^{-1}). The particle number variation can,
as mentioned already, be well described by a simple coagulation model
(except at the very beginning of the process). Both processes,
dN/dt and df_v/dt are closely interlinked in flames and proceed on

about the same time scale. The time necessary for the formation of
the first observable soot particles within and after the main oxidation zone of the flame is definitely shorter. The amount of soot
present at the place in the flame where the first particles can be
located by light scattering seems to vary somewhat. In C_2H_4-air
flames it is around 15% for N $\approx 10^{12}$cm^{-1}. Bockhorn[32] reports even
lower values for f_v of about 5%. With increasing temperature that
percentage seems to increase.

Surface Area of Soot

For the mass growth of soot particles the surface area or the
diameter of the soot particles might be an important quantity.

The soot particle area $F = 3f_v/\bar{r}$ per cm^3 is shown for several
systems in Fig. 3 as a function of time (C_2H_2-O_2 Bonne et al.,
C_3H_8-O_2 from Prado et al., C_2H_4 from Haynes, C_6H_6-O_2 from Baumgärtner
et al. ϕ is for a flame with 200 C fresh gas temperature). An
important point is that the variation of F as a function of time is
very small for times longer than 5 ms, that is for the time where
most of the soot mass is added to the particles. This property seems
to be independent of the absolute value of F (here from 0.01 to
5 cm^2/cm^3). With increasing C/O ratio the value of F increases, as
can be seen from the C_3H_8 and from the C_6H_6 data.

The values which represent the particle radius R are shown in
Fig. 4. In the time interval where most of the soot mass grows (t >
5 ms), these values do also not change strongly. It should also be
noted that their dependence on the C/O ratio is smaller than that
of the parameter F.

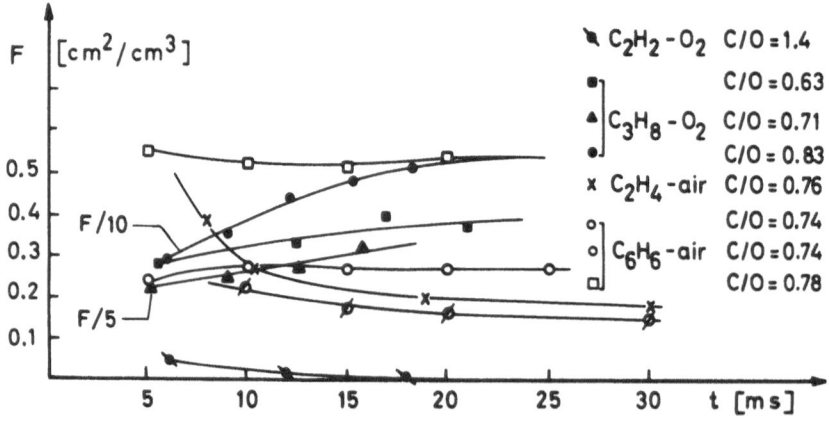

Fig. 3.   Particle area F for different flames.

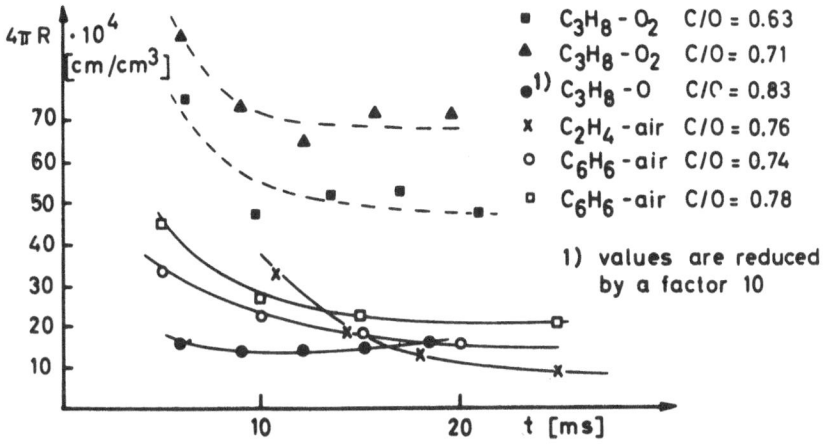

Fig. 4. 4πR as a function of time. Different flames

Growth of Soot in Pyrolysis

In premixed flames the preparation of matter for soot formation
is essentially governed by the oxidation reaction and its products,
in pyrolysis experiments the situation is different, the reaction is
initiated by thermal decomposition of the initial fuel molecules.

Soot formation by pyrolysis can be readily observed in shock
waves in a temperature range which is similar to that in combustion
processes and also to higher temperatures.

A typical picture which can be used to described the situation
is shown in Fig. 5. After passage of the shock some time nothing
seems to happen, then after a certain induction period, the absorption signal rises and approaches a constant value. The light scattering intensity (~ N $\bar{v}^2$) also rises indicating an increase of the
soot particle size. While the scattering signal is a good indication
for particles, this is not necessarily so with absorption signals.
Measurements by Fussey[33] et al. and by Graham[9] et al. did show that,
as in flames, during soot formation by pyrolysis absorption takes
place, which is wave-length dependent and which is usually stronger
at lower wave-length and weaker at longer wave-length. Survey
experiments in shock waves did also show that at high remperatures
small hydrocarbons show more or less absorption up to wave-length
of 5 μm. The absorption observed by Fussey et al. and Graham are
interpreted to be due to the absorption of larger than C_2 molecules.
They are considered to be intermediates on the way to soot formation.

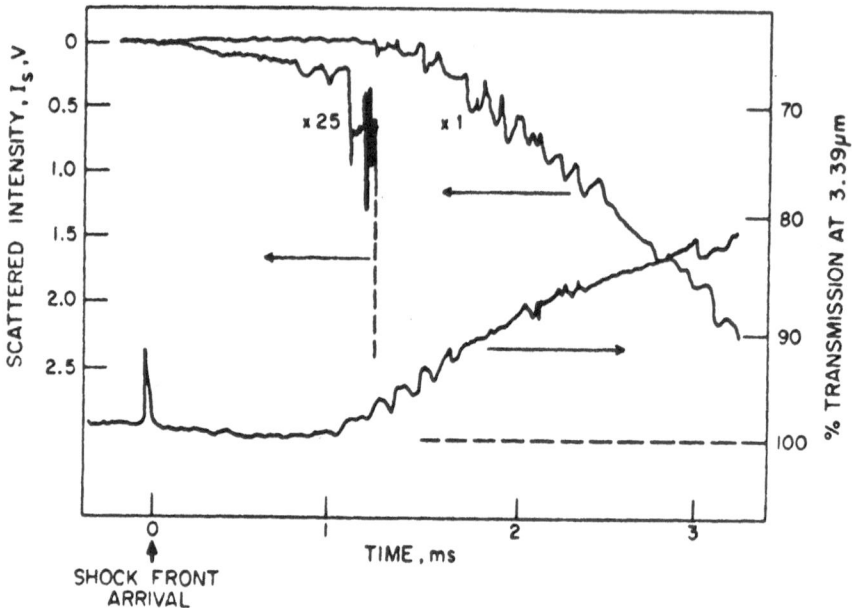

Fig. 5.   Scattering and absorption profile for ethylbenzene pyro-
          lysis in shock waves (Graham et al.).

        Another indicator for soot particles is the light emission:
after an induction period, the luminosity of the particles increases
and approaches a plateau. As long as the soot concentration remains
below $f_v = 5.10^{-7}$ (in a 8 cm diameter shock tube), the emission is a
direct measure of this soot concentration[28] ($\lambda \approx 600$ nm), of f_v. This
limitation in soot mass brings some uncertainty for these measure-
ments. In flames it has been observed for soot volume fractions up
to 10^{-6} (corresponding to about 5.10^{16} C atoms in soot per cm^3) reaction times up to 30 ms and longer and temperatures from 1600 to
2000 K that the mass of gaseous hydrocarbons exceeded that of soot
by a substantial factor. It has not been measured whether the situation in high temperature pyrolysis is the same or not, but there
are good reasons to assume that it should not be too different at
least in the temperature range considered for flames.

Induction Times for Soot Formation

        As mentioned above, measurements of light emission, light scat-
tering and light absorption do indicate induction periods between
the shock front and the first observable signals due to soot particles. The values of these induction times depend on the experimental
method used (especially for light absorption technique, where they
become wave-length dependent). In Fig. 6 induction periods for soot
formation for different hydrocarbons are plotted as a function of

temperature. For benzene, cyclohexadiene, cyclohexene, ethylbenzene indene, these data are from Graham et al., the concentration is always very close to $2 \cdot 10^{17}$ C atoms per cm^3, for the data for acetylene, ethylene and ethene, taken from Fussey, the concentrations are also close to $2 \cdot 10^{17}$ C atoms per cm^3 but they increase slightly towards lower temperatures.

The pressure range, a few bars, is the same for both sources. The time is laboratory time, for real time the values have to be multiplied by approximately 3.5.

Apparently the "activation energies" of these induction times, in the range of experimental conditions considered here, are not too different, ranging from 100 to 150 kJ/mol. (They change with pressure and concentrations.) The absolute values of the induction periods, however, show a definite trend, depending on the type of the substance.

Even for the C_2 hydrocarbons the induction times reported here are long enough to convert most of the initial hydrocarbon into products (based on available kinetic data for the pyrolysis of these molecules) before soot particles become visible. For the larger molecules that should be similar. The initial concentrations in the experiments discussed here are sufficiently high so that bimolecular reactions determine the rate of pyrolysis, including radical and condensation reactions. Because temperatures are high even those

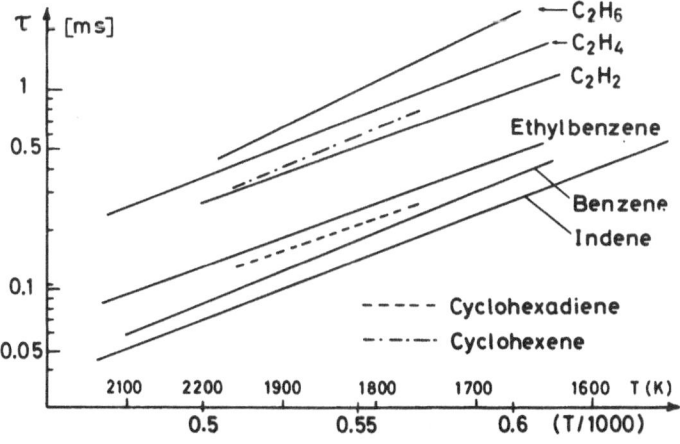

Fig. 6.   Induction times for onset of soot formation as function of temperature. C_2 data from Fussey, others from Graham et al.

reactions, which under normal conditions can only proceed as photo-reactions, like the plane by plane addition of anthracene, have a good chance (with bimolecular rate constants of around 10^{10} cm^3/mol·s).

Rates of Soot Mass Growth in Pyrolysis Experiments

The most complete measurements on soot formation in shock waves are those of Graham[9,22] et al. They could show that coagulation of particles can be well determined by a collide and stick model. Most of his data are represented using equation 4) where $k_{co} = k_{coag} \bar{v}^{1/6}$ and $k_{coag} = 1/2(3/4\pi)^{1/6} (6kT/\rho)^{1/2} \cdot \alpha \cdot G'$ and adjusting measured profiles with G' as an adjustable parameter (α is a constant). It is therefore not easy to extract the df/dt values from the published data. For ethylbenzene from data fiven in[36] f_v values are given which show an extremely fast increase at the beginning, shortly after 1 ms with values of $df_v/dt \geq 4 \cdot 10^{-3}$ (cm^3/cm^3 s), the df_v/dt curve does change approximately proportional to f_v only when df_v/dt has fallen below 700 10^{-6} (s^{-1}). For this later part a k value of around 500 s^{-1} results.

The measurements of Graham et al. do show a special influence of temperature on the rate of soot formation.

We saw that the temperature dependence of the induction period for soot formation was similar for different hydrocarbons (for $C \approx 2 \cdot 10^{17}$ C atoms/cm^3). The τ values changed continuously with temperature. If the rate of soot formation would follow the same trend as the induction process or be independent of temperature, one would expect that the conversion of the fuel into soot would become faster with increasing temperature. There are, however, pronounced temperature effects which are not as simple. Graham et al. found that for measurements performed after 2.5 ms the total amount of soot measured first increased, passed through a maximum around 1800 K and then decreased with temperature for hydrocarbons like benzene, ethylbenzene, toluene, etc.

At lower temperatures these 2.5 ms (gas time) are close to the measured induction times. This side need not be considered further. The decrease towards higher temperatures could be due to a reduced rate of soot formation towards higher temperatures or, however, the rate remains the same or increases and there is less soot formed within the time available. Experimental results of Graham et al. point to the second possibility[9]. This is similar to the situation in premixed flames. A similar bahaviour of the soot yield at a given time as observed by Graham et al. has been found by Wang et al.[34]. They also showed that addition of hydrogen as well as oxygen reduced the soot yield. For a mixture of toluene with oxygen with a C/O ratio of 0.85 no more soot could be observed. Towards higher C/O ratios, a lower soot formation limit around T $\approx$ 1400 K could be seen, while

the upper limit moved to higher temperatures for increasing C/O
ratio. An apparent rate of soot formation which is not directly
df_v/dt seemed to have a maximum around 1800 K.

Graham et al. offer an explanation which in a simplified version
reads: "at low temperature molecules like benzene, etc. have their
route to soot. If the temperature becomes higher fragmentation of
these molecules into smaller particles becomes more important compared to condensation. The process comes closer to soot formation
from acetylene". Buckendahl[35] and Geck[36] performed measurements
about the growth of the soot mass by following the light emission
in shock waves in the temperature range from 1500 to 4000 K. As
mentioned above, the light emission is a good measure of f_v as long
as less than 10% of their hydrocarbon is transformed into soot[28,35,
36]. In most cases the turning point of the signals was close to the
beginning of the light emission. Therefore, the representation of
these signals (height h ~ f_v) by a first order rate law could be
used as a first approximation. A maximum value of the light emission
could usually be obtained from the plateau h_e of the light emission.
For the evaluation of k, $\ln(h_e - h)$ was plotted against time. Due to
the evaluation of Geck[36], these data for k can be represented in one
curve if they are divided by the total density. (It should be mentioned that the total density for these experiments were much higher,
around $2 \cdot 10^{-4}$ g/cm^3, than in the other experiments reported here.)
They increase with temperature up to a certain value where the temperature dependence seems to change sign. This range is, however,
not of interest here.

In the flame temperature regime, the temperature dependence of
their k values can be represented by an apparent energy of activation
of about 160 kJ/mol for C_2H_4 ($2 \cdot 10^{17}$ - $9 \cdot 10^{18}$ C atoms/cm^3) which is
similar to the activation energy for the induction period at high
pressures and to the activation energy for the pyrolysis of C_2H_4.
It is therefore not immediately obvious, whether this apparent energy
of activation is a property of the particle growth or is carried
along from the pyrolysis zone of the hydrocarbon. Towards higher
temperatures the formal first order rate constant seems to pass
through a maximum.

These measurements, including those of Tesner et al.,give many
details, nevertheless quantitative information about the rate of
growth of soot mass as a function of time in pyrolysis experiments
is still rare.

Soot Growth in Diffusion Flames

In flat flames and in shock waves the time axis can easily be
realized, the systems are pretty close to one-dimensional systems.
In diffusion flames the situation is not as simple because the stream
tubes do not follow straight lines. It is therefore necessary to

measure the soot properties as well as the properties of the gas
flow in two dimensions. This means more work to be done and more
possible sources of errors.

Normally the question of interest is whether a diffusion flame
emits soot or not. This, however, is a problem of a more or less
consecutive competition between soot formation and oxidation of soot
particles and their companions. The rate of soot formation is only
one influencing factor.

There have been several attempts to follow the rate of soot
formation in diffusion flames in the past. Probably the most complete
one was that of Kern and Spengler[37] which shows an increase of soot
above $4 \cdot 10^6$ g/cm^3, a mass that is about two orders of magnitude
larger than that of the identified polycyclic molecules larger than
napthalene, at their maximum. Rates of soot mass growth, however, can
only be given approximately. For the experiments of Gaydon et al.[38],
Hottel et al.[39], Kühn et al.[40], Kimugi and Jinno[41], Tesser et al.[42]
and Roper[43] the situation is similar.

In order to make the evaluation of the data simple, Haynes[44]
developed a two-dimensional burner, which could be used for laser
measurements. He could give profiles of N, d and f_v through the
flames. These experiments showed quantitatively how N, d and f_v vary
within the flame and how the particle number decreased away from
the flame zone towards the fuel side.

Kent[45] performed measurements of the soot and the flow field
which at least in principle can be used to evaluate the rate of
mass growth of soot along one stream tube. Some results are shown
in Fig. 7. From the flame surface the stream tube is slightly directed towards the fuel stream and bends at greater heights towards
the vertical direction, thus coming into cooler fuel rich domains.
The variation of the various quantities along this stream tube are
shown in Fig. 7. Particle size and soot mass fraction f_v increase
at first rapidly, then more slowly towards increasing flame height
and reaction time, while the temperature decreases continuously. The
rate of mass growth looks different from what we know from flames
and shock waves. Nevertheless, the general tendency: decrease of
df/dt with increasing f remains here too. The absolute values of
df/dt range up to about $2000 \cdot 10^{-7}$ s^{-1}. In different stream tubes
the situation will be different of course and the variation of f_v
respectively df_v/dt are not exlusively due to reactions of the substances entering the stream tube at the flame front. On the one
hand there is a fuel flow towards the flame and a burned gas flow
towards the center, on the other hand the soot particles move by
thermophoresis perpendicular to the stream tubes. Within the center,
stream tube pyrolysis, supported by oxygen starts while the temperature rises and another soot-forming process may start there if the
flames are sufficiently large.

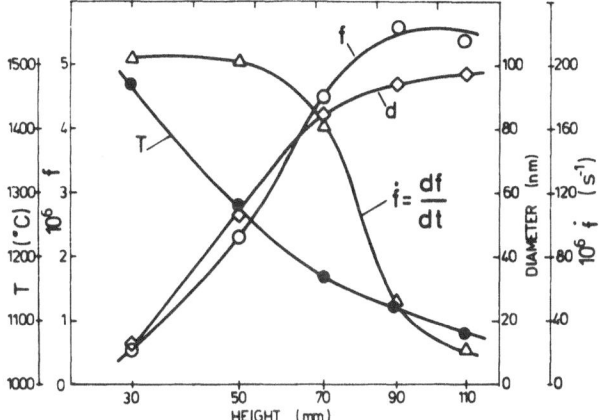

Fig. 7.   Soot volume fraction f_v, df_r/dt and temperature along a
streamline in an C_2H_4-air diffusion flame. Particle number
decreases left to right from 10^{12} cm^{-3} to about 10^{10} cm^{-3}.
(Kent)

The soot mass production rate, represented as lines of constant
df_v/dt, show a rather complicated picture for a two-dimensional diffusion flame. For turbulent diffusion flame the situation will become even more complex because there the data for f_v are already mean
values, the meaning of the various quantities measured from soot becomes even more diffuse as it is already.

DISCUSSION

The experimental results reported here show that most of the
mass growth, f_v, up to 95%, of soot takes place after particles with
around 20-30 Å mean diameter can be detected with number densities
around 10^{12} cm^{-3}. While the particle number density decreases to
around 10^{10} cm^{-3} or less their diameters grow to several hundred Å.

In that range the flame data for df_v/dt can be approximated by an interpolation formula

$$df_v/dt = k(f_M - f_v)^n$$

with the three parameters f_v, n and k. In flames the values of n = 1 fit best, k is around 100 s^{-1} (see Fig. 2) and f_v correlates directly with the maximum amount of soot measured within the experimental field of view. For the shock wave pyrolysis data, the approximation by this expression sometimes seems to be not as good. Here the values for the parameter k range up to above 1000 s^{-1} and n ≠ 1 gave sometimes better fits. These expressions give some information, how the mass growth process could happen, it seems to be more informative, however, to compare some models with this interpolation.

One simple way to find a theoretical estimate for possible growth rates of soot mass is to assume condensation of some species on the surface of the particles using eq. 5 or the corresponding equation for continuous flow conditions (see e.g. Hidy and Brock). The surface area F resp. the quantity R can be taken from the measurements and they do not change much along that growth time of the particle considered here. With the further assumption that the various condensing species can be represented by one, with the number concentration n_i and the volume v_i, where $S = n_i v_i$ is a convenient concentration variable, and that no reevaporation takes place, this expression reads $df/dt = k_M \cdot S$. For mean values of F and R from Fig.3 and 4, (one can take $F = 0.3$ cm^2/cm^3 resp. $4\pi R = 40 \cdot 10^4$ cm/cm^3) maximum values for k at 1800 K and 1 bar and species of molar mass m = 300 are under free molecular flow conditions $k_{MF} = 3.6 \cdot 10^3$ s^{-1} and under continuous flow condition $k_{MD} = 1.5 \cdot 10^5$ s^{-1}.

The shock tube results reported are in the range of small Knudsen numbers. In Graham's experiments (P ≈ 5 bar) the upper limit of k_{MD} for condensation is $k_{MD} = 3 \cdot 10^4$ s^{-1}. For the other shock tube experiments these values should be similar or even less[30-36]. The dependence of k_M on ρ[35,36] has been considered as an indication that diffusion may play an important role at high pressures. It is, however, not immediately obvious whether this is due to mass or heat transport to and from growing particles. From low pressure flames and from diffusion flames[40] there is evidence that the particle temperature in the soot mass growth zone is above gas temperature.

Accomodation coefficients often are between 1 and 0.1 so that for a normal condensation the upper bounds may even be somewhat lower.

For the atmospheric pressure flames the flow conditions vary. In the soot growth zone they are in or close to the transition

region to free molecular flow. The upper bound for the k_M values
should therefore be definitely below $k_{MD} = 1.5 \cdot 10^5 \text{ s}^{-1}$ (for m = 300,
$k_M \sim m^{-0.5}$).

The rate expression for this simple model has the same form as
the interpolation formula if the mass added to the particles $(f_M - f_v)$
is identified with S, the gas phase concentration of species that
can add to the particles. (In reality S should be higher.) A comparison of properly calculated k_M values (for reasonable molar masses
m of adding molecules) with those k values in Fig. 2 shows a rather
large difference, even when an accomodation coefficient of 0.1 is
used and isothermal reevaporation of the same species is considered.
This could be used as an argument against normal vapour condensation.
The fact, that the measured k values are always below the theoretical upper bounds given here, gives a certain confidence in the experimental results.

With this simple growth model one should expect a certain de-
dependence of k $(k_M \sim F$ or R) on experimental parameters like C/O. This
has not been found within the limits of experimental error, while
the fits for f_M and n in the interpolation formula (which considers
only the particles) work rather well.

The growth rate of soot mass decreases towards the hot gas,
even though there is a comparatively high concentration of hydrocarbon molecules left. This has been partly attributed to a reduced
surface activity of the particles, and various observations do indicate that tempering or ageing of soot particles at high temperatures
may strongly influence their properties. It has also been observed,
however, that at different C/O ratios the sizes of the elementary
soot spheres are different and the times, when the growth rate becomes very low, are also different. f_M and the mean particle volume
$\bar{v}$ increase strongly with increasing C/O ratio. In the same way the
amount of material available, which can readily add to the particles
may increase with increasing C/O ratio.

The total amount of hydrocarbons formed in the main reaction
zone of a hydrocarbon-air-flame increases strongly with increasing
C/O ratio, mainly because in a kinetic competition part of the
available oxygen is used to form CO_2 and H_2O which is later on,
especially at lower temperature, not accessible any more, in the times
available (see e.g. Warnatz). It is therefore not unreasonable to
assume that, when more hydrocarbons are formed the concentration of
those species, which can add readily, increases also with increasing
C/O ratio.

The soot particles loose hydrogen during their growth. This
can take place by the elimination of H_2 or of hydrocarbon molecules
which are relatively stable (acetylenes, smaller polycyclic aromates).

This simple model used here has therefore been modified by including the effects of reevaporation of species which are very probably different from those added. In addition, chemical reactions of molecules at the surface or within the particle have to be considered together with tempering processes and the diffusion of light species to the surface. Considering these effects changes and complicates the rate law to a certain extent. It is quite easy to derive expressions which are in agreement with experimental observations and the interpolation formula used. As long as the left side of the chemical equation, which describes the mass growth of soot (the species which add to the particles), is not better known, these expressions must remain rather speculative and shall not be considered here.

A rate law which is slightly different from the extrapolation formula and which seems to fit Graham's data better is the simple coagulation model. If one assumes that only particles above a certain diameter, say above 30 Å, can be seen by the experimental methods and that the whole matter appearing later-on as soot particles is already present, in the range below 30 Å, and ready to undergo sticky collisions, then a rate law for the soot mass growth in particles above 30 Å results in, for long times t

$$df_v/dt = k(n) \ f_M^{-1/2} \ (f_M - f_v)^{3/2}$$

For short times t, a factor with t^{n-1} appears on the right side. (The coagulation rate constant is assumed to be independent of the particle size, n is the number of elementary units in the smallest particle which can be seen). This rate law does not fit the flame data, it may well be, however, that a certain variation in the single rate constants could bring the exponent towards one.

These two simple models, which are connected with certain implications, show that very specific and careful measurements are necessary in order to sort out the various effects, influencing the process of soot mass growth.

I would like to thank Dr. Homer from Shell Research Thornton for making the induction times of their measurements available, Dr. Warnatz for calculations of hydrocarbon intermediates in C_2H_4- air flames.

REFERENCES

1.   B.S. Haynes, H.Gg. Wagner, Progress in Energy and Comb.Sciences,
        to be published;
     H. Gg. Wagner, Seventeenth Symposium (International) on
        Combustion, The Combustion Institute, Pittsburgh, p. 3-19
        (1979).

2. P. A. Tesner, Seventh Symposium (International) on Combustion,
    The Combustion Institute, p. 546 (1958).
3. J. Lahaye, and G. Prado., Chem. and Phys. of Carbon 14, p.168
    Marcel Dekker, New York (1978).
4. P.A. Tesner, Fiz Goreniya Vzryva 2:3 (1979).
    P.A. Tesner, A.E. Gorodetskii, T.D. Snegireva, E.F. Aref'eva,
    Dokl. Akad. Nauk SSSR 239:901 (1978).
5. B.L. Wersborg, J.B. Howard, G.C. Williams, Fourteenth Symposium
    (International) on Combustion, The Combustion Institute,
    Pittsburgh, p. 928 (1973); Faraday Symposium No.7:109 (1973).
6. K.H. Homann and H. Gg. Wagner, Ber. Bunsenges. Phys. Chem.
    69:20 (1965).
7. G.M. Hidy, J.R. Brock, The Dynamics of Aerocolloidal Systems,
    Pergamon Press (1970).
8. S.K. Friedländer, Smoke, Dust and Haze, Wiley Sons (1977).
9. S.C. Graham, J.B. Homer, J.L.J. Rosenfeld, Proc. Roy. Soc.
    A 344:259 (1975).
10. U. Bonne and H. Gg. Wagner, Ber. Bunsenges. Phys. Chem. 69:35
    (1965).
11. A. D'Alessio, A. Di Lorenzo, F. Beretta, C. Venitozzi, Four-
    teenth Symposium (International) on Combustion, The Combustion
    Institute, Pittsburgh, p. 941 (1973);
    F. Beretta, A. Borghese, A. D'Alessio, C. Venitozzi, Laser
    Measurement Methods in Combustion, Urbino, Italy 7-9 Sept.
    1977
12. B.S. Haynes, H. Jander, H. Gg. Wagner, Seventeenth Symposium
    (International) on Combustion, The Combustion Institute,
    Pittsburgh, p. 1365 (1979).
13. W. Morgenmeyer, Dissertation, Göttingen 1968.
14. B.L. Wersborg, L.K. Fox, J.B. Howard, Comb. Flame 24:1 (1975).
15. B.S. Haynes, H. Jander, H. Gg. Wagner, Ber. Bunsenges. Phys.
    Chem. 84:585 (1980).
16. K. Müller-Dethlefs, Doctoral Thesis, Imperial College 1979.
17. F. Zabel, Private communication 1968.
18. K.H. Meyer, Dissertation D17, Darmstadt 1979;
    H. Bockhorn, F. Fetting, U. Meyer, R. Reck, G. Wannemacher,
    Eighteenth Symposium (International) on Combustion, The
    Combustion Institute, Pittsburgh, p. 1137 (1981).
19. R. Reck, Dissertation D17, Darmstadt 1976.
20. G. Prado, J. Jagoda, K. Neoh, J. Lahaye, Eighteenth Symposium
    (International) on Combustion, The Combustion Institute,
    Pittsburgh, p. 1127 (1981).
21. M. Brei, private communication 1981.
22. S.C. Graham, Proc. Roy. Soc. A377:119 (1981).
23. K.H. Homann, W. Morgeneyer, H. Gg. Wagner, Comb. Institut
    Europ. Symp. 66, 394 (1973).
24. F.J. Wright, Twelfth Symposium (International) on Combustion,
    The Combustion Institute, Pittsburgh, p. 867 (1969);
    Comb. Flame 15:217 (1970).
25. B.S. Haynes, H. Mätzing, to be published.

26. A. D'Alessio, A. Di Lorenzo, A.F. Sarofim, F. Beretta, S. Masi, C. Venitozzi, Fifteenth Symposium (International) on Combustion, The Combustion Institute, Pittsburgh, p. 1427 (1975); A. D'Alessio, A. Di Lorenzo, A. Borghese, F. Beretta, S. Masi, Sixteenth Symposium (International) on Combustion, The Combustion Institute, Pittsburgh, p. 695 (1977).

27. E.E. Tompkins, R. Long, Twelfth Symposium (International) on Combustion, The Combustion Institute, Pittsburgh, p.1011 (1967).

28. J. Floßdorf, Diplomarbeit Göttingen 1965 J. Floßdorf, H. Gg. Wagner, Z. Phys. Chem., NF 54:8 (1967).

29. J.C. Street, A. Thomas, Fuel 34:4 (1955); Comb. Flame 6:46 (1962).

30. H. Jander, private communication (1980).

31. L. Baumgärtner, private communication (1981).

32. H. Bockhorn, DFG-Colloquium, Darmstadt (1980).

33. R.B. Cundall, D.E. Fussey, A.J. Harrison, D. Lampard, Eleventh Intern. Shock Tube Symposium, Seattle (1977); J.C.S. Faraday Trans 74: 1403 (1978); 75:1390 (1979).

34. T.S. Wang, R.A. Matula, R.C. Farmer, Eighteenth Symposium (International) on Combustion, The Combustion Institute, Pittsburgh, p. 1149 (1981).

35. W. Buckendahl, Diplomarbeit Göttingen 1970.

36. C.C. Geck, Diplomarbeit Göttingen 1975.

37. J. Kern, G. Spengler, Erdöl-Kohle-Erdgas-Petroleum, 23:813 (1970).

38. A.G. Gaydon, H.G. Wolfhard, Flames, Chapman and Hall, London (1970).

39. W.M. Dalzell, G.C. Williams, H.C. Hottel, Comb. Flames 14:161 (1976).

40. G. Kühn, R.S. Tankin, J. Quant. Spectroscop. Rad. Transf. 8:1281 (1968).

41. M. Kunugi, H. Jinno, Eleventh Symposium (International) on Combustion, The Combustion Institute, Pittsburgh, p. 257 (1967).

42. P.A. Tesner, E.I. Tsygankova, L.P. Guilazetdinov, V.P. Zuyev, G.V. Loskakova, Comb. Flame 17:279 (1971).

43. F.G. Roper, C. Smith, Comb. Flame 36:125 (1979).

44. B.S. Haynes, H.Gg. Wagner, Ber. Bunsenges. Phys. Chem. 84:499 (1980).

45. J.H. Kent, H. Jander, H. Gg. Wagner, Eighteenth Symposium (International) on Combustion, The Combustion Institute, Pittsburgh, p. 1117 (1981).

DISCUSSION

A.F. Sarofim (Massachusetts Institute of Technology)

One check on a diffusion limited growth mechanism is provided by the size dependence of growth rate. The approximate size

independence of the rate of soot mass addition at atmospheric
pressure rules out a diffusion-limited mechanism. At higher
pressures, where diffusion-limited growth rates are in approximate
agreement with measured values, does the dependence of rate of
particles size follow that expected from theory?

Wagner

   For the high pressure experiments, which are far away from
the transition to free molecular flow, particle size measurements
are not available. If one takes the data obtained by Geck, the
diffusion controlled coagulation rate, and assumes that more than
50% of the initial hydrocarbon is converted into soot, one can
calculate a growth rate of particle size. The dependence of the
rate on particle size does not seem to follow that expected from
theory. A definite statement, however, cannot be made.

A. Feugier (Institut Français du Pétrole)   .

   I would like to know how you explain the variation of chemical
composition of soot along the flame?

Wagner

   When soot particles grow older, their hydrogen content
decreases, they loose some of their radical character and the
structure of the cristallites becomes clearer (see Ref. 1).
During the establishment of the structure, hydrogen is released
from that domain, diffuses towards the surface. There, it is
released as hydrogen, or as hydrocarbon with relatively high
hydrogen content. At the same time an adsorption/desorption
process at the particle surface seems to take place which may
also shift the hydrogen content. Experiments on that problem
are under way.

IONIC MECHANISMS OF SOOT FORMATION

H. F. Calcote

AeroChem Research Laboratories, Inc.
P.O. Box 12
Princeton, NJ

INTRODUCTION

In a recent review[1] of soot nucleation mechanisms it was
demonstrated that mechanisms based upon neutral free radical
species are inadequate to explain soot formation in flames. Either
rates are too slow to account for the rapid rate of soot formation
or there are difficulties in accounting for the large numbers of
polycyclic rings observed in soot particles. These problems can
be overcome by assuming an ionic mechanism. Ion molecule reactions
are extremely fast compared to free radical reactions and ions
have a propensity to quickly rearrange to the most stable structure so that there is no difficulty in accounting for the observed
polycylcic structures. Much evidence has accumulated in the
literature indicating the importance of ions in sooting flames
and a number of workers have previously suggested ionic mechanisms.
For this, the reader is referred to a recent review.[1]

The literature on the soot formation process is, at times,
confusing because of the failure to recognize its complexity or
to overestimate it, and more specifically, a failure to recognize
that charged species or ions may enter into the soot formation
process in two completely separate steps. Thus, it seems useful
to define the specific steps involved in soot formation:

1. Formation of precursors – the generation of those free
    radicals or ions which are necessary for the initial
    stages of production of soot nuclei. Soot precursors may

grow by reaction with common, high concentration, flame
species, often acetylenes. These are called building
blocks.

2. Nucleation - the transformation from a molecular system to
    a particulate system, i.e., incipient soot particles in
    which the growing species take on the properties of parti-
    cles as opposed to large molecules. In soot formation this
    transformation occurs over a range of molecule/particle
    diameters unlike usual nucleation phenomena in which an
    abrupt transformation occurs because of the crossing of two
    rate-controlling steps.

3. Growth - the increase in size of the incipient soot particles
    by the further addition of molecular species.

4. Coagulation - the collision and coalescence of two particles
    of the same or different size into a single particle in
    which the identities of the two original particles are com-
    pletely lost.

5. Agglomeration - a process where a series of particles
    collide one at a time and adhere to each other to form a
    chain of individual particles which are still distinguish-
    able from one another.

6. Aggregation - a process where a series of particles collide
    one at a time to form a cluster of individual particles in
    which the particles are still distinguishable.

7. Oxidation - oxidative reactions of the growing nuclei or
    particles in any of the abovementioned forms to reduce the
    particle size and to reduce the H/C ratio.

We believe that charged species are important in steps 1, 2,
and 5. Much of the confusion in the literature arises from not
identifying observed effects of charged species with either step 2
or step 5. Since step 5 is much better understood than step 2 and
since there is considerable controversy as to the validity of an
ionic mechanism for step 2, we will concentrate this discussion on
soot nucleation, step 2.

After a brief review of the basic premise of ion molecule nucleation and some comments on the source of charged particles which
are important in step 5, the relationship between flame ion concentration and concentration of soot particles will be discussed in
some detail. This will be followed by a consideration of possible
alternative sources of the large molecular ions observed in flames,
i.e., alternative to the ion molecule nucleation hypothesis.

BASIC PREMISE

    Recent measurements using mass spectrometers to observe individual ion concentrations in flames at AeroChem[2] and by Michaud et al.[3] have demonstrated a dramatic change in the ions observed in nonsooting and sooting flames. This is demonstrated in Figs. 1 and 2 which are for a nonsooting and a sooting acetylene-oxygen flame at 2.0 kPa. The equivalence ratio at which acetylene soots in this flame is 2.6. It is interesting to note that even in the nonsooting flame, Fig. 1, very large ions, greater than 300 amu, appear very early in the flame and then disappear. In the sooting flame, Fig. 2, these large ions dominate and then disappear. Ions > 300 certainly do not include ions over 1000 amu because of the reduced sensitivity

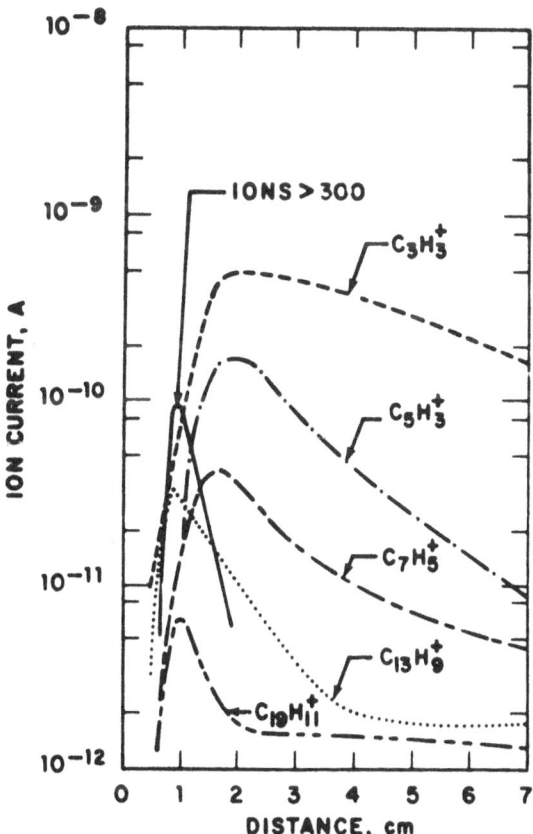

Fig. 1 Ion profiles in a nonsooting flame. Acetylene-oxygen, ϕ = 2.0, P = 2.0 kPa, T = 2300 K (Ref. 4).

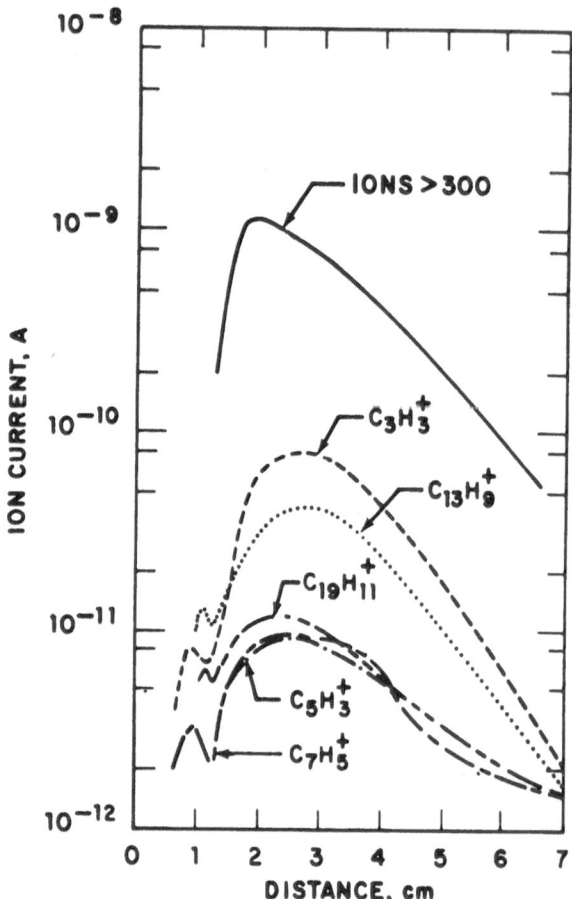

Fig. 2   Ion profiles in a sooting flame.   Acetylene-oxygen, ϕ = 2.7,
P = 2.0 kPa, T = 2170 K (Ref. 4).

of the mass spectrometer above 1000 amu.   Thus it is assumed that as
the > 300 amu ions disappear, larger ions are produced.   Figure 3
shows the effect of equivalence ratio on the peak mass spectrometer
ion current for selected ions.   The interesting feature here is that
as the equivalence ratio for the production of soot, indicated by
the shaded area, is approached the ions which are dominant just before
sooting decrease very rapidly and large ions, > 300, increase dramatically.   It is also interesting to note that the largest ion recorded
in this set of data $C_{19}H_{11}^{+}$ begins to increase in concentration as
the sooting equivalence ratio is approached and then disappears
rapidly presumably becoming a larger ion.

The basic premise by which we view the effect of ions and
charged particles on soot growth is summarized in Fig. 4.   Primary
flame ions produced by chemi-ionization react by rapid ion molecule

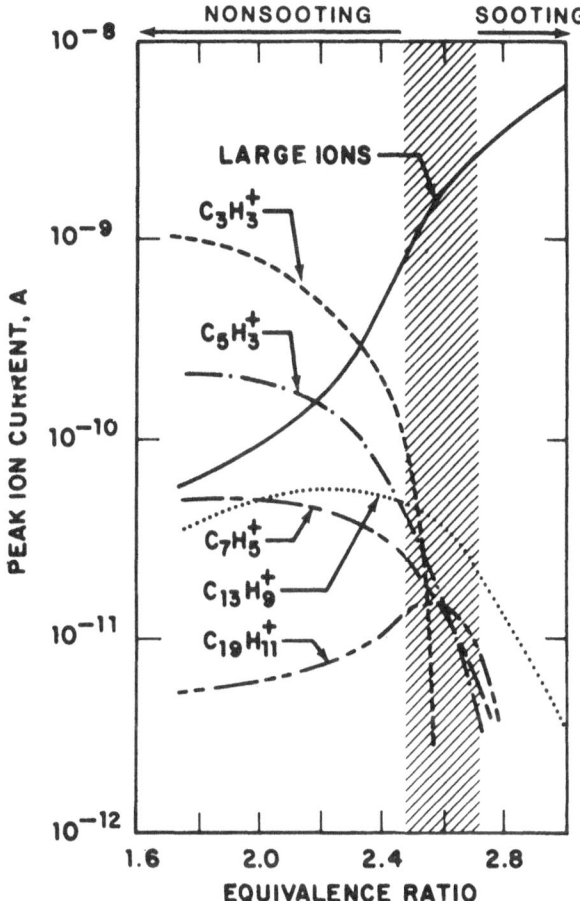

Fig. 3 Effect of equivalence ratio on ion profiles. Acetylene-
        oxygen, P = 2.0 kPa (Ref. 4).

reactions with neutral flame species, such as acetylene, polyacetylenes, and free radicals, to produce larger ions which rapidly
rearrange to produce even larger polycyclic aromatic ions. Some of
these ions are neutralized by recombination with electrons produced
in the primary flame ion reaction and become neutral incipient soot
particles; others grow to produce charged soot particles. These
particles then grow by surface addition as well as by coagulation to
form larger soot particles. As these particles reach a critical
size at a high enough temperature, their ionization potential
becomes sufficiently low that the particles are thermally ionized.
The charge on these particles determines their rate of agglomeration
and produces chains of individual particles distinct from aggregates.
It is also at this point that many of the effects of chemical additives or of electric fields are observed to affect the formation of

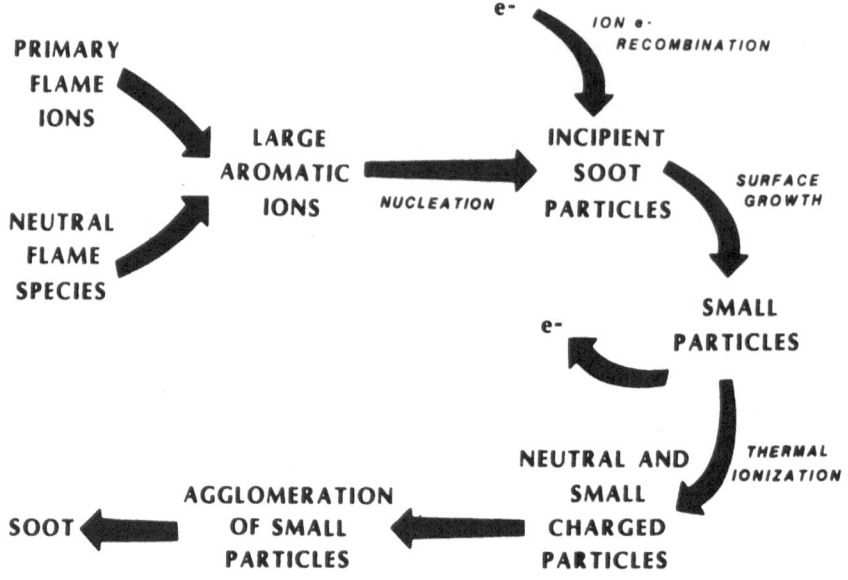

Fig. 4 Mechanism of soot formation.

soot in flames. Of course several of these steps are going on simultaneously; they are not necessarily sequential as indicated.

The formation of precursor ions in the proposed scheme is basically the same step observe in all hydrocarbon flames in which highly
nonequilibrium concentrations of ions are produced by chemi-
ionization. The basic reaction is

$$CH + O \rightarrow CHO^+ + e^- \tag{1}$$

The CHO^+ so produced rapidly transfers a proton to water

$$CHO^+ + H_2O \rightarrow H_3O^+ + CO \tag{2}$$

or by a series of ion molecule reactions produces the large numbers
of ions that are observed normally in flames. In particular $C_3H_3^+$
can be produced by a set of reactions such as

$$CHO^+ + CH_2O \rightarrow CO + CH_3O^+ \tag{3}$$

$$CH_3O^+ + C_2H_2 \rightarrow C_3H_3^+ + H_2O \tag{4}$$

There are, in fact, a large number of routes from CHO^+ through other
ion molecule reactions to $C_3H_3^+$. In rich hydrocarbon flames, $C_3H_3^+$

is the dominant ion while in leaner flames, i.e., equivalence ratios slightly greater than 1 to much less than 1, H_3O^+ is the dominant ion. Another proposed mechanism for the production of $C_3H_3^+$ is:

$$CH^* + C_2H_2 \rightarrow C_3H_3^+ \tag{5}$$

Evidence for Reaction (5) is fragile. It is also difficult to imagine that Reactions (1) through (4), leading to $C_3H_3^+$, might occur in very rich flames where the oxygen atom concentration can be safely assumed to be fairly low. Nevertheless, very large concentrations of $C_3H_3^+$ are observed in rich flames approaching sooting; therefore in our argument for an ionic mechanism of soot formation, at this stage, we accept an initial large concentration of $C_3H_3^+$ and assume that it is the initial precursor ion upon which ion molecule reactions build larger and larger ions. Further work is clearly needed to determine the mechanism by which $C_3H_3^+$ is produced in very rich flames.

Given $C_3H_3^+$ as the initial ion it is easy to write a series of ion molecule reactions in which the product ions continue to grow employing only those ions and neutral species which have been observed in sooting flames. Where thermodynamic data are available, all of the reactions employed are exothermic so that fast reaction rate coefficients are a good assumption. In fact, there is evidence that the rate coefficients of ion molecule reactions for a given homologous series are proportional to the exothermicity of the reaction.[5] Such a reasonable set of reactions, by no means inclusive, is summarized in Table 1. It is easy to envision the set of reactions in Table 1 continuing to produce larger and larger ions. Of course, while this growth process is proceeding, ion recombination with the electrons produced in the initial chemi-ionization reactions will neutralize some of the ions and thus reduce their concentrations. The rate coefficients for these recombination processes would be expected to increase with increasing molecular size. Thus the proposed ion molecule reactions naturally can be expected to rapidly produce large ions and incipient charged soot particles and, through ion/charged particle-electron recombination, incipient neutral soot particles. These incipient soot particles grow by surface addition of molecular species and by coagulation. At some point their work function or ionization potential decreases (Fig. 5) so that eventually they become thermally ionized. In fact it has been demonstrated that in the well-studied acetylene-oxygen 2.7 kPa flame the concentration of charged particles can be calculated by assuming thermal ionization and using the measured temperature and concentration of neutral soot particles.[1]

ION CONCENTRATION FOR SOOT NUCLEATION

It is frequently assumed that the ion concentration in sooting flames is too small to account for the observed concentration of soot.

Table 1. Proposed Ion-Molecule Soot Nucleation Reactions
(from Ref. 4)

$$C_3H_3^+ + \begin{cases} C_2H_2 &= C_5H_5^+ \\ C_4H_2 &= C_5H_3^+ + C_2H_2 \\ C_4H_2 &= C_7H_5^+ \end{cases}$$

$$C_5H_3^+ + \begin{cases} C_2H_2 &= C_7H_5^+ \\ C_4H_2 &= C_7H_3^+ + C_2H_2 \\ C_4H_2 &= C_9H_5^+ \end{cases}$$

$$C_5H_5^+ + \begin{cases} C_2H_2 &= C_7H_7^+ \\ C_4H_2 &= C_7H_5^+ + C_2H_2 \\ C_4H_2 &= C_9H_7^+ \end{cases}$$

$$C_7H_5^+ + \begin{cases} C_2H_2 &= C_9H_7^+ \\ C_4H_2 &= C_9H_5^+ + C_2H_2 \\ C_4H_2 &= C_{11}H_7^+ \end{cases}$$

$$C_7H_7^+ + \begin{cases} C_2H_2 &= C_9H_9^+ \\ C_4H_2 &= C_9H_7^+ + C_2H_2 \\ C_4H_2 &= C_{11}H_9^+ \end{cases}$$

$$C_9H_7^+ + \begin{cases} C_2H_2 &= C_{11}H_9^+ \\ C_4H_2 &= C_{11}H_7^+ + C_2H_2 \\ C_4H_2 &= C_{13}H_9^+ \end{cases}$$

etc.

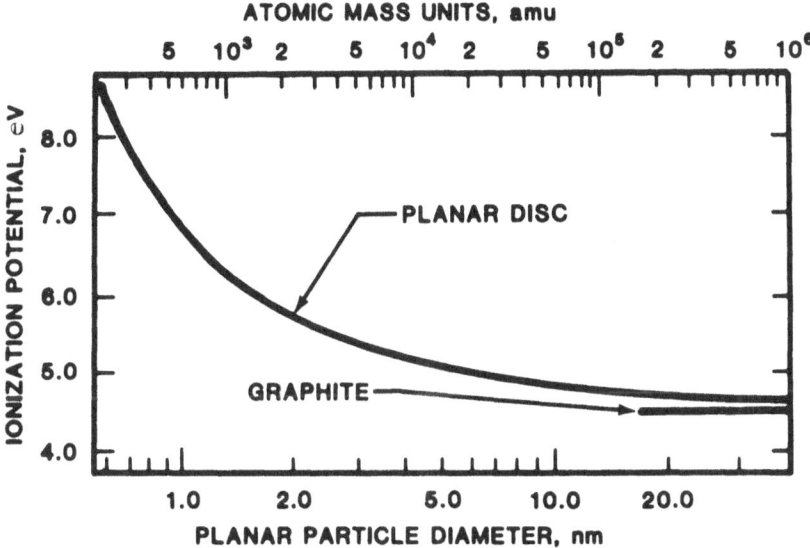

Fig. 5   Ionization potentials of soot particles.

This question arises for two reasons: (1) the available ion concentration data are not definitive and (2) the argument has not been
thoroughly analyzed.   Several measurements of ionization in sooting
flames have been reported but, unfortunately, these have not always
been performed under comparable experimental conditions with those
for soot concentrations.   Some data for a well-studied premixed flame
are shown in Fig. 6.   The flame is an acetylene-oxygen flame at an
equivalence ratio ϕ = 3.0, a pressure of 2.7 kPa, and an unburned
linear gas feed velocity of 50 cm s^{-1}, except as noted.   The curve
entitled "soot" is a combination of data obtained by Bonne et al.[7] and
Wersborg et al.[6]   The data marked "charged soot" are from Prado and
Howard.[8]   The curve marked "Yeung '73"[6,8] was obtained using a molecular beam sampling system similar to those used in molecular beam
mass spectrometer sampling of flames with the ions collected by a
Faraday cage.   The curve marked "Homann '79" was obtained by a similar technique.[9]   The probe curve is a recently obtained set of data
from AeroChem[10] obtained by a Langmuir probe technique using the
continuum electrostatic probe theory of Clements and Smy[11] which in
the past has given very reliable absolute ion concentrations in non-
sooting flames.   In obtaining this curve we have used the molecular
weight increase through the flame measured by Homann.[9]   The close

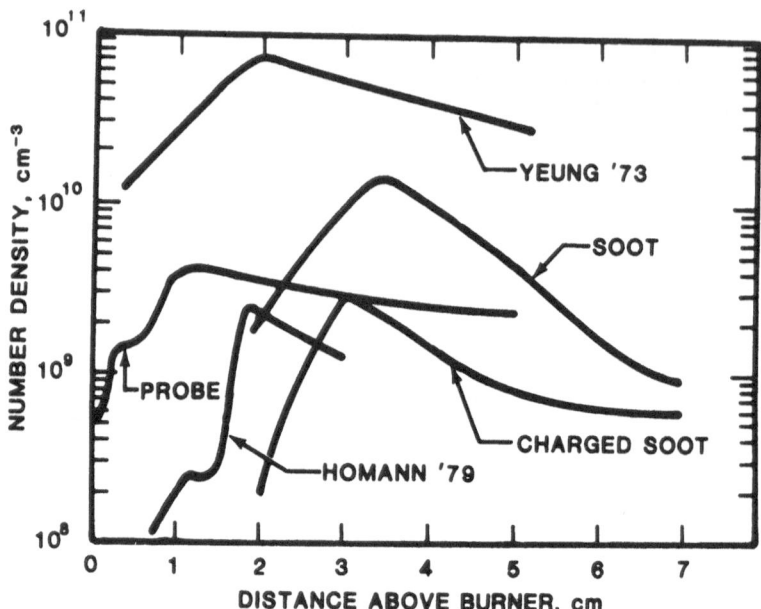

Fig. 6   Number density of ions, soot, and charged soot particle pro-
         files. Acetylene-oxygen, ϕ = 3.0, P = 2.7 kPa, u = 50 cm s^{-1},
         (except where noted). Soot from Wersborg et al.[6] and Bonne
         et al.[7]; charged soot particles from Prado and Howard[8]; Yeung
         '73 represents a molecular beam total ion sampling[6,8] for u =
         38 cm s^{-1}; Homann '79 is a similar type measurement[9] for ϕ =
         2.9, 44 cm s^{-1}; the probe curve represents total ion concen-
         tration measured by Gill et al.[10] with a Langmuir probe.

agreement between the AeroChem probe data and the molecular beam data
of Homann are encouraging. Certainly it remains an important task to
identify the absolute concentration of individual ions and small
charged particles as a function of position in the flames and equivalence ratio.

     Diffusion flames also contain large concentrations of ions, Fig.
7. This set of data was taken on a 3.3 kPa ethylene-oxygen flame.
The soot is as visually observed; the temperature was measured with
a radiation corrected coated thermocouple and the curve indicated as
"probe current" is the actual probe current obtained with a Langmuir
probe 0.305 cm long and 0.0127 cm radius. The indicated ion concentrations were deduced from the probe current assuming an average mass
of 50 amu at 1 cm and 1000 amu at 3 cm. The ions > or < 300 amu were
obtained from mass spectrometric flame sampling. The peaks in the
C_2 and CH emission have been indicated as obtained photographically
using a combination of light filters and film to isolate their

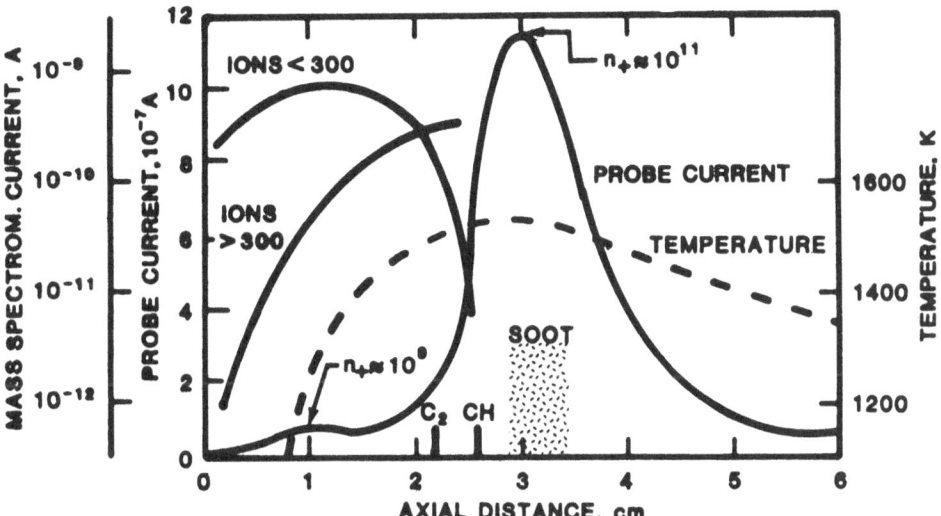

Fig. 7   Ionization in sooting diffusion flames. Ethylene-oxygen,
         P = 3.3 kPa; fuel burner diam = 2.4 cm; fuel flow =
         39 cm s^{-1}; oxygen flow = 50 cm s^{-1} (Ref. 10).

emission spectra. Some striking features of this set of data are
their close similarity to those from premixed flames, the large
amount of reaction occurring in the center of a diffusion flame (note,
however, the low pressure), and the relatively large concentration of
ions at low temperature.

We now address the problem of the relationship between ion
concentration and soot number density observed in flames with respect
to the premise that ions are the key element in the nucleation step.
Consider one of the first reactions in the proposed scheme

$$C_3H_3^+ + C_2H_2 \rightarrow C_5H_3^+ + H_2 \tag{6}$$

If this is the first reaction in an ion-molecule growth sequence, we
can deduce the rate of formation of $C_3H_3^+$ from the rate of its
disappearance; thus

$$\frac{d(C_5H_3^+)}{dt} = -\frac{d(C_3H_3^+)}{dt} = k(C_3H_3^+)(C_2H_2) \tag{7}$$

For this reaction we will assume, for the sake of argument, an ion
concentration of $C_3H_3^+$ of 10^8 ions cm^{-3} which is conservatively low
compared with the values reported, for example, in Fig. 6. The rate
coefficient k has recently been measured by Ausloos and Lias[12] to be
10^{-9} cm^3 s^{-1} and the temperature coefficient was estimated to be

negligible. The acetylene concentration is taken from the work of
Homann and Wagner[13] in a flame similar to that described in Fig. 6.
This equation would lead to a rate of $C_3H_3^+$ formation (and disappearance) of about 10^{15} ions cm^{-3} s^{-1}. In the most simplistic interpretation of our theory the $C_5H_3^+$ ion would then sequentially add acetylene molecules in a large number of steps to produce charged soot
particles of (assumed for this discussion) 5 nm diameter. This would
require approximately 10^3 steps:

$$C_5H_3^+ \xrightarrow[]{\quad C_2H_2 \quad} \rightarrow \rightarrow \rightarrow \rightarrow \rightarrow P^+(5 \text{ nm})\tag{8}$$

This overestimates the number of steps but ignores some loss of ions
through ion-recombination reactions. If we assume the rate coefficient for each step is the same as that for Reaction (6), then the
total time required for the 10^3 steps will be approximately 6 x
10^{-5} s. This is a short time on the time scale of the flame and is
consistent with the observation that very large ions are observed
early in the flame front.

The charged soot particles will be neutralized by recombination
with free electrons

$$P^+(5 \text{ nm}) + e^- \rightarrow P(5 \text{ nm})\tag{9}$$

The rate coefficient for this reaction is[14]:

$$k = \frac{\pi d^2}{4} \bar{c}_e S\left(1 - \frac{e^2}{CkT_e}\right)\tag{10}$$

where d is the particle diameter; $\bar{c}_e$ is the mean electron velocity,
S is a sticking coefficient assumed to be 1; e is the charge on an
electron, C the electrical capacitance of a planar disk (the assumed
shape for small particles); $C = 4\varepsilon_0 d$, where ε_0 is the dielectric
constant of free space and T_e is the electron temperature assumed to
be equal to the gas temperature. For particles of 5 nm diameter, k
is approximately 10^{-5} cm^3 s^{-1}. Thus, assuming that the concentration
of charged particles is equal to the initial concentration of ions
because of the rapidity of Reaction (8) and neglecting any recombination of ions in this chain, the characteristic time for Reaction (9)
is about 10^{-3} s, again a fairly short time but not nearly so short
as the time at which the ions grow to produce charged particles.
The rate of charged particle recombination increases rapidly with
charged particle diameter. The rate of particle production through
this sequence of reactions would be equal to the initial ion production rate which gives 10^{15} particles cm^{-3} s^{-1}. Wersborg et al.[15]
observed a particle production rate in the same flame of 10^{13} particles cm^{-3} s^{-1} and a particle concentration of 10^{10} cm^{-3}. Thus it is
possible, using reasonable values to have a considerably lower ion
than particle concentration and still have the ions be the primary

source of particles. For example, in the above calculation
$(C_3H_3^+) = 10^8$ cm^{-3} which is much less than $(P) = 10^{10}$ cm^{-3}. The
above example clearly demonstrates that particle concentrations and
ion concentrations per se are not important for the argument but that
the rate of ion formation and the rate of particle formation are
really the important parameters which should be addressed. The ion
concentration is, of course, important in that it is rate controlling
for some of the steps in the process.

ALTERNATIVE SOURCES OF LARGE MOLECULAR IONS

    In our interpretation of the observation of large molecular ions
in sooting and nearly sooting flames, we have assumed that these ions
were produced by a continuing sequence of ion molecule reactions producing larger ions. To substantiate that argument it is necessary to
eliminate other possible mechanisms by which these large molecular
ions could be produced. We attempt to do that in this section for
six different mechanisms:

1. Thermal ionization of large molecules.

2. Charge transfer from particles to molecules.

3. Proton transfer from particles to molecules.

4. Hydride transfer from molecules to particles.

5. Chemi-ionization of large molecules.

6. Ion-molecule equilibria.

    1. Thermal ionization can be calculated by Saha's equation which
for both large molecules and particles takes the form[14]

$$K = \frac{N_p^+ \, N_e}{N_p} = G \left(\frac{2\Pi \, m_e \, kT}{h^2} \right)^{3/2} \exp\left(-\frac{\phi_p}{kT} \right) \qquad (11)$$

where G is taken as 2 because the statistical weights of the initial
and final states of the particles or large molecules and ions will be
equal, N_p^+, N_p, and N_e are the concentrations of charged particles or
molecules, neutral particles or molecules and electrons, respectively;
m_e is the mass of an electron, k is Botzmann's constant; h is Planck's
constant, and ϕ_p is the ionization potential of a large molecule or the
equivalent work function for a particle where ϕ_p for a particle is
given by $\phi + e^2/2c$, with ϕ the work function of graphite. ϕ_p is
displayed in Fig. 5 as a function of particle diameter. As already
pointed out, it is possible to account for the concentration of
charged particles as being in thermal equilibrium with neutral

particles by the use of this equation. However, it is clear that
this cannot be true for most of the large molecules especially in
lower temperature flames such as the diffusion flame reported in
Fig. 7. Some typical ionization potentials of large polycyclic molecules are: 7.97 eV for phenanthrene $C_{14}H_{10}$; 7.52 eV for pyrene
$C_{16}H_{10}$; and 7.43 eV for coronene $C_{24}H_{12}$.[16] For the same flame as in
Fig. 6 the mass concentration of large molecules peaks at about
2×10^{-8} g cm^{-3} at about 2.5 cm above the burner where the flame temperature is 2120 K.[1] Assuming the molecular weight of coronene this
is equivalent to 4×10^{13} molecules cm^{-3}. If ion and electron concentrations are equal Eq. (11) gives 3×10^7 ions cm^{-3}. This is
considerably less than the observed ion concentration, Fig. 6.

Another compelling reason for rejecting thermal ionization of
large molecules is that the ions observed in flames have odd numbers
of hydrogen atoms and the neutral species observed have even numbers.[1]
Of course this does not eliminate the possibility that there may be
large concentrations of free radicals with odd numbers of hydrogen
atoms. Assuming these are in about the same concentration as the
large molecules and that the ionization potential is roughly the same
as for large molecules, a reasonable assumption, then the above equilibrium calculation shows that only minor amounts of ions containing
an odd number of hydrogen atoms could be produced by thermal
ionization.

2. Charge transfer from a particle to a molecule

$$P^+ + M \rightarrow P + M^+ \tag{12}$$

can also be easily eliminated by reference to Fig. 5 in which it can
be seen that as particle size increases the ionization potential
decreases so one would expect the equilibrium to be heavily weighted
to the left. Again, another compelling argument against this mechanism is that quoted above: the ions have odd numbers of hydrogen
atoms and the observed large molecules have even numbers.

3. It is also conceivable that a proton could be transferred
from a particle to a molecule

$$PH^+ + M \rightarrow P + MH^+ \tag{13}$$

This possibility can be eliminated by reference to Fig. 8 which shows
the calculated proton affinity as a function of a particle diameter.[14]
The validity of this calculation is supported by observing that for
small particle diameters the curve extrapolates very satisfactorily
through the proton affinities[16] for a number of large polycyclic
aromatic compounds indicated on the figure. For this extrapolation
the diameter of the large molecule was estimated as the diameter of
a circle with the equivalent cross sectional area as the large molecule. In addition, for the large diameter particles the curve can

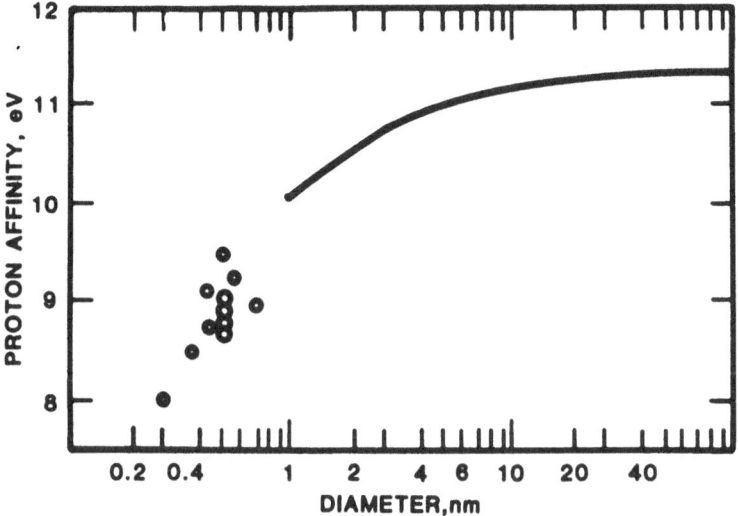

Fig. 8 Effect of particle diameter on proton affinity. (Ref. 14).

be smoothly extrapolated to the bulk graphite value derived by Meot-Ner.[16] From Fig. 8 it is easy to see that Reaction (13) will be highly biased towards having the proton remain on the particle rather than on the molecule because of the very rapidly decreasing proton affinity as a molecule or particle increases in size.

    4. Hydride transfer from a molecule to a particle represented by

$$P^+ + MH \rightarrow PH + M^+ \tag{14}$$

can also be considered as a possible mechanism for production of large molecules by summarizing the individual steps which thermodynamically make up this process:

						ΔH, kJ
MH		$\rightarrow$	M	+	H	420
M		$\rightarrow$	M^+	+	e^-	680
P^+	+ e^-	$\rightarrow$	P			-440
P	+ H	$\rightarrow$	PH			-420
MH	+ P^+	$\rightarrow$	M^+	+	PH	240

The large endothermicity of this process argues against its being an important mechanism for large molecular ion formation.

   5. Another possible mechanism for production of large ions is chemi-ionization of large molecules. This mechanism cannot be eliminated out-of-hand because the following reactions involving large molecules can readily be written down:

$$\Delta H, \text{ kJ mol}^{-1}$$

$$C_3H_7 + C_{10}H_2 \rightarrow C_{13}H_9^+ + e^- \qquad -370 \qquad (15)$$

$$C_5H_7 + C_8H_2 \rightarrow C_{13}H_9^+ + e^- \qquad -290 \qquad (16)$$

$$C_4H_6 + C_9H_3 \rightarrow C_{13}H_9^+ + e^- \qquad -210 \qquad (17)$$

$$C_6H + C_{13}H_{10} \rightarrow C_{19}H_{11}^+ + e^- \qquad -160 \qquad (18)$$

$$C_6H + C_7H_8 \rightarrow C_{13}H_9^+ + e^- \qquad -30 \qquad (19)$$

It is somewhat difficult, however, to envision the energy from such large molecular reactions involving complex rearrangements ultimately residing in the electronic state needed to ionize the molecule. Nevertheless, because of the large exothermicities involved in some of these reactions they will require further consideration. For example, how does the rate production of ions by such processes compare with the measured rate of ion production?

   6. Ion molecule equilibria might also allow for production of larger ions. In fact, it would be just such reactions that we consider in our scheme to produce the larger ions. There are several types of ion molecule equilibrium which one might consider. The most obvious is, of course, proton transfer

$$mH^+ + M \rightleftharpoons m + MH^+ \qquad\qquad (20)$$

where m represents a small molecule and M a large molecule.

   Two examples of such equilibria are

$$C_3H_3^+ + C_{12}H_8 \rightleftharpoons C_3H_2 + C_{12}H_9^+ \qquad K \approx 10^{-7} \qquad (21)$$

$$C_4H_3^+ + C_{12}H_8 \rightleftharpoons C_4H_2 + C_{12}H_9^+ \qquad K \approx 10^{-2} \qquad (22)$$

In both of these examples the equilibrium constant is estimated to be far to the reactant side.

   Another class of ion molecule equilibria would be hydride transfer from the large molecule to a small ion

$$m^+ + MH \rightleftharpoons mH + M^+ \qquad\qquad (23)$$

Two examples of this type of reaction are:

$$C_3H_3^+ + C_8H_8 \rightleftarrows C_3H_4 + C_8H_7^+ \qquad K \sim 10^{-5} \qquad (24)$$

$$C_3H_3^+ + C_{12}H_{10} \rightleftarrows C_3H_4 + C_{12}H_9^+ \qquad K \sim 10^{-3} \qquad (25)$$

Again the equilibrium constants indicate that the reaction would be favored toward the left.

Another type of ion molecule equilibria would involve the transfer of C_3H^+. For example in the equilibrium

$$C_3H_3^+ + C_2H_2 \rightleftarrows C_5H_3^+ + H_2 \qquad (26)$$

This reaction, in fact, is one of the assumed steps in our proposed mechanism. The heat of formation of $C_5H_3^+$ is not well known; assuming the heat of formation derived from Ref. 17, the equilibrium constant would be 10^{-1} while using the heat of formation recently derived from flame measurements by Michaud et al.[3] the equilibrium constant would be 10^3. This reaction has been observed to be very rapid in the forward direction when the linear isomer of $C_3H_3^+$ is assumed.[12] Another type of ion molecule reaction which again is a part of our assumed mechanism is the CH^+ transfer which is typified by reactions:

$$C_3H_3^+ + C_3H_4 \rightleftarrows C_4H_5^+ + C_2H_2 \qquad K \sim 10^3 \qquad (27)$$

$$C_3H_3^+ + C_{12}H_8 \rightleftarrows C_{13}H_9^+ + C_2H_2 \qquad K \sim 10^8 \qquad (28)$$

In both of these reactions the equilibrium is favored toward the larger product ion.

In summarizing this discussion of alternate sources of large molecular ions, none of them seem nearly as attractive and reasonable as the assumption that the large molecular ions arise by ion molecule reactions in which the product mass exceeds the reactive mass of the ion, i.e., the scheme we have championed for the nucleation step in soot formation.

SUMMARY AND CONCLUSIONS

In summary: ionic mechanisms appear to be important in the nucleation step in which chemi-ions react by ion molecule reactions to produce increasingly larger ions and in the coagulation step where charged particles are produced by thermal ionization. It has also been demonstrated that comparison of the absolute concentration of ions and particles is not a significant argument for or against an ionic mechanism except insofar as these concentrations indicate rates of reaction. It is important to demonstrate that the rate of ion production is equal to or greater than the rate of particle production.

It is difficult to explain the appearance of large ions except by ion molecule growth reactions in which the molecular weight of the product ion exceeds the molecular weight of the reactant ion.

ACKNOWLEDGEMENTS

The work reported herein was partially supported by the Air Force Office of Scientific Research (AFSC) under Contract F49620-81-C-0030 and the National Bureau of Standards Center for Fire Research under Contract NB80NADA1038. The United States Government is authorized to reproduce and distribute reprints for governmental purposes notwithstanding any copyright notation hereon. The author is also indebted to his colleagues, especially Drs. D.B. Olson, R.J. Gill, and W. Felder for many helpful discussions and criticisms on the manuscript.

REFERENCES

1. H.F. Calcote, Mechanisms of Soot Nucleation in Flames - A Critical Review, Combust. Flame, in press.
2. D.B. Olson and H.F. Calcote, Ions in Fuel Rich and Sooting Acetylene and Benzene Flames, in: "Eighteenth Symposium (International) on Combustion," The Combustion Institute, Pittsburgh (1981).
3. P. Michaud, J.L. Delfau, and A. Barassin, The Positive Ion Chemistry in the Post-Combustion Zone of Sooting Premixed Acetylene Low Pressure Flat Flames, in: "Eighteenth Symposium (International) on Combustion," The Combustion Institute, Pittsburgh (1981).
4. D.B. Olson and H.F. Calcote, Ionic Mechanisms of Soot Nucleation in Premixed Flames, in: "Particulate Carbon: Formation During Combustion," Plenum Press, New York (in press).
5. P. Ausloos, J.A.A. Jackson, and S.G. Lias, Reactions of Benzyl Ions with Alkanes, Alkenes, and Aromatic Compounds, Int. J. Mass Spectr. Ion Phys. 33:269 (1980).
6. B.L. Wersborg, A.C. Yeung, and J.B. Howard, Concentration and Mass Distribution of Charged Species in Sooting Flames, in: "Fifteenth Symposium (International) on Combustion," The Combustion Institute, Pittsburgh (1975).
7. U. Bonne, K.H. Homann, and H.Gg. Wagner, Carbon Formation in Premixed Flames, in: "Tenth Symposium (International) on Combustion," The Combustion Institute, Pittsburgh (1965).
8. G.P. Prado and J.B. Howard, Formation of Large Hydrocarbon Ions in Sooting Flames, in: "Evaporation-Combustion of Fuels," J.T. Zung, ed., American Chemical Society, Washington, DC (1978).
9. K.H. Homann, Charged Particles in Sooting Flames. I. Determination of Mass Distributions and Number Densities in $C_2H_2-O_2$ Flames, Ber. Bunsen. Gesell. Phys. Chem. 83:738 (1979).

10. R.J. Gill, D.B. Olson, and H.F. Calcote, work in progress.
11. R.M. Clements and P.R. Smy, Ion Current from a Collision-Dominated Flowing Plasma to a Cylindrical Electrode Surrounded by a Thin Sheath, $\underline{J}$. $\underline{Appl}$. $\underline{Phys}$. 41:3745 (1970).
12. K.C. Smyth, S.G. Lias, and P. Ausloos, Can $C_3H_3{}^+$ be an Important Soot Precursor?, to be presented at Eastern States Combustion Symposium, Pittsburgh (1981).
13. K.H. Homann and H.Gg. Wagner, Some New Aspects of the Mechanism of Carbon Formation in Premixed Flames, $\underline{in}$: "Eleventh Symposium (International) on Combustion," The Combustion Institute, Pittsburgh (1967).
14. H.F. Calcote, Ionic Mechanisms of Soot Formation in Flames, $\underline{Progr}$. $\underline{Energy}$ $\underline{Combust}$. $\underline{Sci}$., in preparation.
15. B.L. Wersborg, J.B. Howard, and G.C. Williams, Physical Mechanisms in Carbon Formation in Flames, $\underline{in}$: "Fourteenth Symposium (International) on Combustion," The Combustion Institute, Pittsburgh (1973).
16. M. Meot-Ner (Mautner), Ion Thermochemistry of Low-Volatility Compounds in the Gas Phase. 3. Polycyclic Aromatics: Ionization Energies, Proton and Hydrogen Affinities. Extrapolation to Graphite, $\underline{J}$. $\underline{Phys}$. $\underline{Chem}$. 84:2716 (1980).
17. H.M. Rosenstock, K. Draxl, B.W. Steiner, and J.T. Herron, Energetics of Gaseous Ions, $\underline{J}$. $\underline{Phys}$. $\underline{Chem}$. $\underline{Ref}$. $\underline{Data}$ 6:Suppl. 1 (1977).

DISCUSSION

D. Rivin (Cabot Corporation)

The number of $C_{19}H_{11}{}^+$ ions (239 amμ) in the acetylene oxygen flame rapidly decreases near the sooting limit, while ions of 300-1000 amμ continue to increase in concentration at much higher equivalence ratios. Why do ions with a small difference in mass exhibit such divergent behaviour?

Calcote

The smaller ions are replaced by larger ions as they grow by the addition of electrically neutral species, e.g., C_2H_2 and C_4H_2. It is this rapid growth of ions which we think is responsible for soot nucleation.

CHARGED SOOT PARTICLES IN UNSEEDED AND SEEDED FLAMES

Klaus H. Homann and Eckhard Ströfer

Institut für Physikalische Chemie
Technische Hochschule Darmstadt
6100 Darmstadt

INTRODUCTION

The fact that a fraction of the soot particles in flames carry a natural charge and that the overall concentration of charge carriers can be influenced by additives has stimulated numerous investigations. Of the many questions that arise from those works two major ones are: How can soot formation be controlled either by electrical means or additives or by both? What is the role of the different charge carriers in carbon formation, and how is the mutual interaction between them and the soot particles? This paper attempts to make a contribution to the second problem and tries to give some answers.

After introducing a simple method which allows to measure concentrations and mass distributions of larger positively and negatively charged particles in low pressure flames the questions to be dealt with are: What are the number densities and the masses of larger ionic species in sooting hydrocarbon flames and how do they depend on fuel and burning conditions? What is the influence of inorganic salt additives on these larger charged species and how does the presence of large particles in the flame in turn interact with the ionization of the metal component of the salt? The mechanism of ionization will be discussed.

This paper deals with the analysis of charged particles in a mass range from about 100 to about $6 \cdot 10^4$ mass units. Since sooting flames stabilized on cooled burners are very sensitive to the burner and the burning conditions we used the same (or nearly so) low pressure flat flames that had been analyzed before by different workers using mass spectrometry, optical spectroscopy and different

types of particle analysis[1,2,3,4]. Thus, the temperature, the con-
concentration profiles of higher hydrocarbons and the number density
and size of soot particles are known better than in any other sooting flame.

EXPERIMENTAL METHOD

The Burner and the Seeding System (Fig. 1.)

The flat premixed C_2H_2-O_2 and C_6H_6-O_2 flames burned on a water-
cooled sintered plate burner (7.0 cm diameter, cooling coil within
the plate) which is vertically movable in a burner housing and has
been described previously[3,5]. The burning pressure was 20 or 30 Torr
(26.6 and 40 mbar). For the flames seeded with a spray of salt solution the sintered plate burner was exchanged for one with a silver
plate (5 mm thick, 7.0 cm diameter) with uniformly distributed
holes (ca. 400) of 1 mm diameter. It was water-cooled from the peri-
phery. This burner could be isolated electrically from the burner
housing by means of a teflon insertion between shaft and burner head.

The input of acetylene and oxygen was controlled in the usual
way. For benzene flames pure oxygen was saturated with benzene
vapor and then mixed with further oxygen to get the desired ratio.
For the seeded flames an aerosol of salt solution and unburned gas
was prepared by means of an atomizer (Nebulizer Tantalum Variable,
Varian Techtron) which was driven directly by the premixed unburned
gas mixture. The pressure difference across the atomizer was 80mbar.
In a settling chamber (2 l volume, 2 s residence time) between atomizer and burner ca. 93% of the sprayed solution separated from the
aerosol which was then led into the burner. The unburned gas mixture and the atomizer had to be heated to 40°C in order to avoid
freezing of the nozzle. Flash back of the flame into the settling
chamber was never observed with burning pressures below 50 mbar.

The Sampling Nozzle and the Mass Analyzer (Fig. 1)

The sampling orifice on top of a conical quartz probe (height
60 mm, base diameter 50 mm) served as a source for a supersonic
nozzle beam from the flame gases. This beam was expanded into a
vacuum chamber evacuated by a 1000 $l \cdot s^{-1}$ diffusion pump at a background pressure of about 0.1 Pa ($7.5 \cdot 10^{-4}$ Torr). Since charged particles were to be sampled platinum paint was burned onto the inner
and outer surface of the probe so that it could be kept at ground
potential as was the burner housing. The diverging beam from the
sooting flame source represents a "seeded nozzle beam" which accel-
erates the heavy particles in it until the flow has become free
molecular. The beam passes through three grids. The first is a
shielding grid kept at ground potential which confines a field free

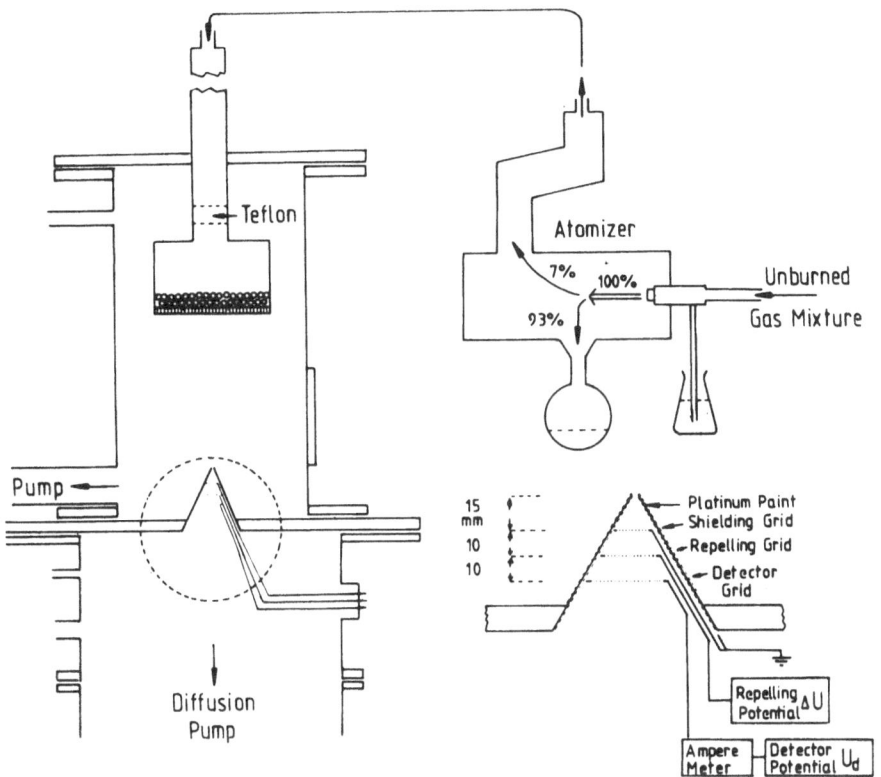

Fig. 1.   Burner, atomizer and mass analyzer for the study of heavy
          ions in sooting low-pressure flames

space in the tip of the cone. It is placed where the beam has already
reached its free molecular flow region. An energy separation of
the charged particles is obtained by applying a repelling voltage
ΔU to the second grid. This leaves through only those particles for
which $1/2 \ mv^2 \geq q \ \Delta U$, where m, v, and q are mass, velocity and charge
of the particle, respectively.

The third grid serves as the detector. For positive ions it is
biased negatively and vice versa for negative charge carriers. An
integral energy spectrum is obtained when the detector current is
monitored as a function of the repelling potential ΔU. This spectrum
is then differentiated. To convert it into a mass spectrum the velocity and the charge of the particles must be known. The method of mass
separation has been described previously[5].

This mass separator which naturally does not have the resolving
power of a conventional mass spectrometer is especially suited for
particles of masses between 500 and $5 \cdot 10^4$ mass units. The "spectrum"

of a single ion of comparatively low mass Cs$^+$, M = 133 u, which was
obtained from the burned gas of a CsCl seeded flame is shown in
Fig. 2.

To reduce the scattering of the ions by the background gas and
to improve the resolution, the distance between the orifice and the
grids had been diminished. A third grid was used as a detector instead of a Faraday cup as in a previous work[5].

Reproducible results were obtained when the sampling orifice,
the grids and the insulation were kept clean. This demanded frequent
cleaning and changing of grids when sampling from strongly sooting
flames.

RESULTS

Determination of the Mass of Heavy Species in the Beam;
Scattering Effects

The differential energy spectrum is converted into a mass spectrum by assuming that the particles carry one elementary charge and
applying the results of the "seeded beam"-theory for determination
of their velocity[6,7]. The terminal gas velocity after the flow
acceleration is obtained from the final Mach number of the flow
which may be calculated as a function of the source Knudsen number.
There is a considerable velocity slip for heavy particles[8]. As a

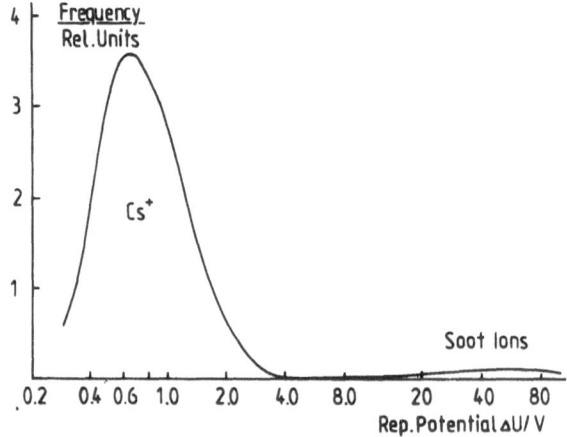

Fig. 2.   Cs$^+$ mass peak from a CsCl seeded C_2H_2-O_2 flame.
The repelling voltage at the maximum corresponds to
a velocity of 970 m s^{-1}.

consequence, the heavier the particles are, the smaller is the ratio
of their terminal particle velocity to that of the gas. This ratio
is taken from the work of Schwartz and Andres[8] where the results
for different heavy gaseous components and for suspended latex particles are given. Details of these results applied to nozzle beam
sampling of flames have been published elsewhere[5].

Scattering of the charged beam particles by the background gas,
however, diminishes their velocity component in forward direction
before they reach the repelling field. Experiments with different
background pressures have shown that this has to be taken into consideration when estimating the mass of the charged soot particles.
Since the effect of scattering on the particle velocity cannot be
rigorously calculated for beams from a flame source it is estimated
by experiments with different background pressure in the vacuum
chamber. The procedure is illustrated on Fig. 3. It was found that
the logarithm of the repelling voltage at the second maximum of the
differential energy distribution, ΔU_{max} (see Fig. 4), is a linear
function of the background pressure or the area of the sampling
orifice, respectively. Fig. 3 shows how this repelling voltage decreases when the background pressure increases for samples from
flames at two different burning pressures and three different
sampling heights. The four samples comprise four different mass
distributions of larger charged particles. Yet, the slope of the
straight (solid) lines seems to be little dependent on the particle
mass, but the negative slope is proportional to the burning pressure of the flame.

The straight lines can be described by the relationship

$$\lg \frac{\Delta U_{max}(A \to 0)}{\Delta U_{max}} = b \cdot p_v \propto b \cdot A \cdot p$$

where p_v and p are the pressures in the vacuum chamber and in the
flame, respectively. A is the orifice area and b is a proportionality factor characterizing the loss of translational energy in the
direction of the beam axis. That this factor is little dependent on
the particle mass m suggests that it is proportional to the geometric cross section of the large particle, σ_p, and inversely proportional to its kinetic energy when the particle has reached its
terminal velocity $v_{p,\infty}$:

$$b \propto \frac{\sigma_p}{\frac{1}{2} m v_{p,\infty}^2}$$

Taking the soot particles as spheres of density ρ_p and diameter d_p
we have:

$$\sigma_p = \pi\left(\frac{d}{2}\right)^2 = \frac{3}{2}\frac{m}{d\varsigma_p} \quad \text{and}$$

$$b \propto \frac{m}{d\,m\,v_{p,\infty}^2} \propto m^{-\frac{1}{3}}\,v_{p,\infty}^{-2}$$

For typical relative masses M and terminal particle velocities in the beam the product gives

M/u	$v_{p,\infty}/m\,s^{-1}$	$M^{-\frac{1}{3}}\,v_{p,\infty}^{-2}$
10 000	1 200	$3.2 \cdot 10^{-8}$
2 000	1 700	$2.7 \cdot 10^{-8}$
1 000	1 800	$3.1 \cdot 10^{-8}$
100	1 900	$6.0 \cdot 10^{-8}$
20	2 100	$8.3 \cdot 10^{-8}$

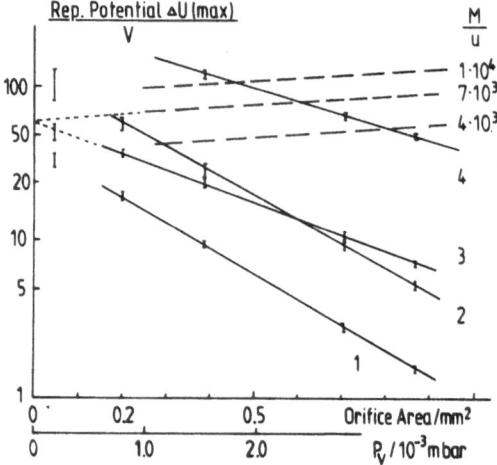

Fig. 3. Influence of scattering on the kinetic energy of ions: The solid lines connect experimental data of repelling potentials at the maxima of different kinetic energy distributions of C_2H_2 flames for various background pressures in the vacuum chamber.
1: $C/O = 1.06$, 40 mbar, $v_0 = 42$ cm s^{-1}, $h = 14$ mm;
2: $h = 20$ mm
3: $C/O = 1.12$, 27 mbar, $v_0 = 42$ cm s^{-1}, $h = 20$ mm;
4: $h = 35$ mm. Broken lines are calculated (see text).

b increases for masses below about 1000 u. That means that the part
of the differential energy spectrum corresponding to smaller mass
particles is shifted more strongly to smaller repeller voltages than
that of larger mass particles ($u \gtrsim 1000$).

A correlation between the mass of the particles and the repelling voltage is found by an extrapolation using Fig. 3. The broken
lines have been obtained from seeded beam theory by plotting the
repelling voltage corresponding to the terminal particle velocity
for zero background pressure versus the nozzle area with the particle mass as parameter. An extrapolation, for example, of line 3
to $p_v \rightarrow 0$ gives 63 V. Since the broken line for M = 7000 u also
extrapolates to 63 V this is approximately the mass corresponding
to the maximum in the energy spectrum of the respective sample.

A similar extrapolation from another repelling potential of
the same spectrum using the same slope then leads to the respective
mass, etc. A calibration curve so obtained can be extended to lower
masses by using an ion of known mass such as Cs^+ (M = 133). Scattering also causes a loss of intensity in the current due to smaller
mass ions (see below).

Charged Particles in Acetylene Flames

Fig. 4 shows mass distributions of positive ions from different
heights in a C_2H_2-O_2 flame.
A short qualitative description of this flame is as follows[1,3,9]:
Maxima of chemiluminescence of OH^*, CH^*, C_2^*: 4-7 mm above the burner
Temperature maximum (2180 K): 11 mm
Maximum concentration of polyacetylenes(mole fraction = 0.018):10 mm
Complete consumption of O_2: 13 mm
Mass fraction of soot in burned gas: $0.7 \cdot 10^{-4}$

Until the end of the oxidation zone mainly ions with masses
$\lesssim 300$ relative mass units (u) are detected. These ions are formed
in the oxidation zone by chemi-ionization. They have also been analyzed by conventional mass spectrometry[10]. Their number density decreases rapidly at larger heights. At 15 mm a broad shoulder appears
in the mass spectrum which is due to larger particles. The mass
distribution of these newly formed soot ions first is rather asym-
metrical with a steep increase at masses around 500 u and a long
smoothly sloping extension to about 5000. This asymmetry is still
obvious at 17 mm. The heavier charged particles increase strongly
in number density up to 20 mm. The strongest increase is found for
particles of masses between about 10^3 and $3 \cdot 10^3 u$. The shoulder now
has developed into a second strong maximum in the spectrum. With
increasing height the distribution broadens while the maximum is
shifted to higher masses. A mass of $10^4 u$ corresponds to the limit
size of particles that have been found by electron microscopy[3,11].

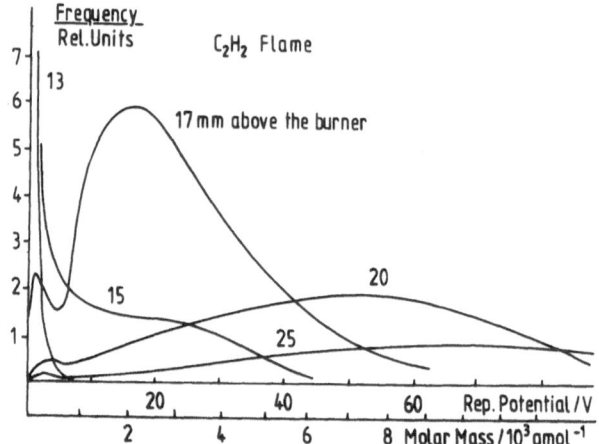

Fig. 4.   Mass distribution of positive ions in a sooting C_2H_2-O_2
flame. C/O = 1.12, p = 27 mbar, v_o = 42 cm s^{-1}

Fig. 5 shows the differential kinetic energy (or mass) spectra for
negatively charged particles from different heights in the same flame.

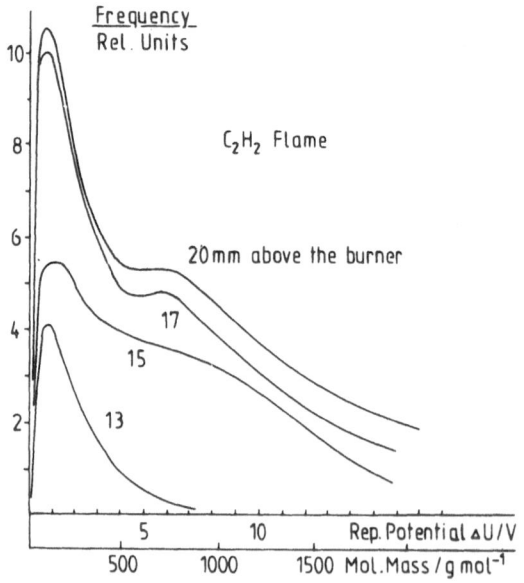

Fig. 5.   Mass distribution of negative ions in a sooting C_2H_2-O_2
flame.
(Same as in Fig. 4.)

At 13 mm the distribution is similar in shape and extends to about
the same masses as for the positive ions. The development at larger
heights, however, is different. At 15 mm the increase in negative
ionization is mainly due to ions of masses between about 300 and
1500 u giving rise to a broad shoulder in the distribution. This
develops again into a second maximum, which, however, remains nearly
in the same mass region around 1000 u. In contrast to small positive
ions the number density of small negative ions increases with the
general increase of soot ionization between 13 and 15 mm. The growth
of larger negative soot ions after the first burst is much less pronounced than for positive soot ions. Their masses remain between
about 400 and 2-3000 u. This is also true for larger heights.

   The formation of positive and negative soot ions is accompanied by a strong increase in total ionization, shown for this flame
in Fig. 6 where the detector current is plotted versus the height
in the flame. The estimation of number densities is described below.
Above 20 mm recombination of charges dominates while the positive
soot ions are still growing. Since the flame gases are almost elec-
troneutral the profile for negative ionization should coincide with
that of positive ions. The reason why this is not the case is the
stronger scattering of negatively charged particles before they
reach the detector. Negative charge carriers, on the average, have
a smaller mass than positive ones in this flame.

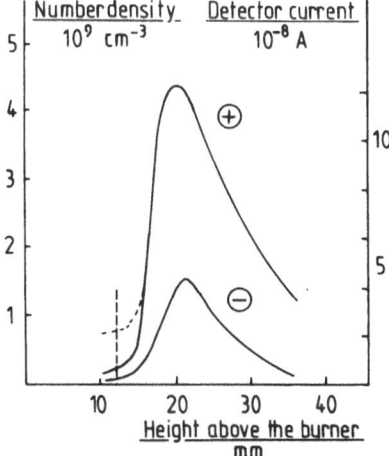

Fig. 6. Profiles of positive and negative ionization.
         (Same flame as in Fig. 4.)

Most of the electrons will escape detection . If the repelling grid
which is nearer to the orifice is used as the detector by biasing
it positively and thereby drawing almost all negative charges onto
it and vice versa for the positive ions the ionization profiles of
positive and negative particles coincide better as demonstrated in
Fig. 9. The broken line in Fig. 6 allows for the scattering of low
mass ions (see below). The ionization in a sooting C_2H_2-O_2 flame
increases with increasing C/O ratio between 1.0 and 1.3. It also
increases with decreasing initial flow velocity of the unburned gas
(60 – 25 cm s^{-1})[5],[12].

Charged Particles in Benzene-Oxygen Flames

Fig. 7 shows mass distributions of charged particles from dif-
ferent heights in a sooting benzene-oxygen flame, C/O = 0.82, p =
27 mbar, v_0 = 44 cm s^{-1}. Its structure can be described qualitative-
ly by the additional data[4],[13]:
Maxima of chemiluminescence (CH*, OH*, C_2*): 7.5 – 8.5 mm
Maximum temperature (2200 K): 15 mm, O_2 consumed at 15 mm
Maximum concentration of polyacetylenes (x ≈ 0.004): 11 mm
Maximum of polycyclic aromatic hydrocarbons (2-4 rings): 9 – 11 mm
Main region of soot formation and particle growth: 11 – 30 mm
Mass fraction of soot in burned gas: 1.9 · 10^{-3}.

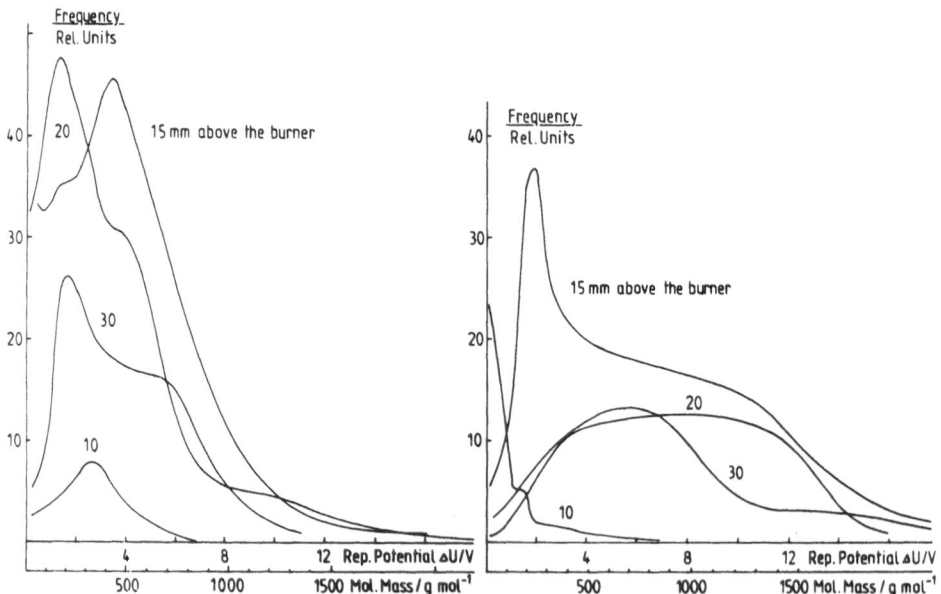

Fig. 7. Mass distribution of positive (right) and negative ions
          from different heights in a sooting benzene-oxygen flame.

The mass distributions of the positively charged particles show
similarities to those of acetylene flames but also some striking
differences. There is again a strong increase in soot ionization
giving rise to a very asymmetric mass distribution of soot ions at
15 mm. This distribution increases steeply at masses below about
250 u and declines more smoothly to about 2000 u. At heights larger
than 15 mm the peak at about 250 u has disappeared and the shoulder
has developed into a more symmetrical distribution with a mean mass
of about 10^3u. Positive ions apparently do not continue to grow but
decrease in mass and number density.

The mass distributions of negative ions in the benzene flame
also show a similar development as in the C_2H_2 flame. After the beginning of the strong ionization two groups of ions can be distinguished: One with an average mass around 180 u and the other with
its mass distribution peaked at 5-600 u. The number density of the
higher mass group appears to increase earlier than that of the lower
masses. Between 15 and 20 mm the fraction of higher mass ions de-

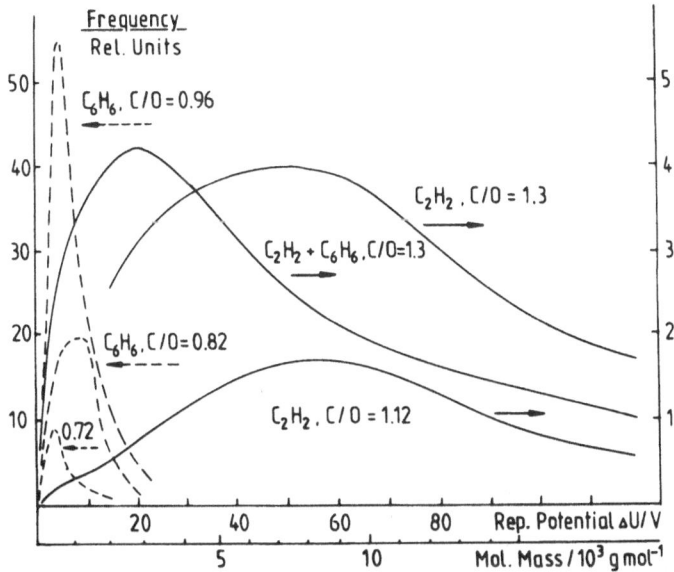

Fig. 8.   Mass distributions in acetylene and benzene flames;
          $p = 27$ mbar, $v_o = 43$ cm s^{-1}, $h = 20$ mm.
          Flames of C_2H_2 (C/O = 1.30) and C_6H_6 (C/O = 0.82) have
          about equal soot concentration in the burned gas.

creases while that of lower mass increases. There is hardly any
growth of negative soot ions above 15 mm of height from the burner.

Comparison between C_2H_2 and C_6H_6 Flame of Equal Soot Concentration

According to Morgeneyer et al.[4,14] a C_2H_2 flame with a C/O=1.30
has about the same soot concentration in the burned gas as the benzene flame described above. For these two flames the mass distribution of positive ions at a height of 20 mm are shown together on
Fig. 8. The number densities at this height of the two flames are
about equal (see Fig. 9.). The mass distribution, however, in the
benzene flame is narrow, peaked at about $1 \cdot 10^3$ u, while that in the
C_2H_2 flame is broad with the maximum at about $7 \cdot 10^3$ u.

This discrepancy is even stronger at larger heights. The number densitiy of positive soot ions with a mass larger than about
$2 \cdot 10^3$ u cannot be appreciable in the benzene flame. At any rate, it
is very much smaller than that in a C_2H_2 flame of the same amount
of soot in the burned gas. It should be mentioned that at a height
of 20 mm the temperature of the benzene flame is about 70 K higher
than that of the acetylene flame.

Another mass distribution of Fig. 8 demonstrates the influence
of benzene addition to the C_2H_2 flame with an original C/O = 1.12
raising it to 1.31. This leads to a relatively stronger increase in
the ionization of lower mass particles giving rise to an asymmetric
distribution. The other curves show the influence of the C/O ratio
in sooting flames of pure C_6H_6 and C_2H_2, respectively. It changes
the number density rather than the mass distribution.

Fig. 9 presents the profiles of total current due to positive
and negative charge carriers in the C_2H_2 and the C_6H_6 flame. The
total current has been measured with the repeller grid as detector
applying a biasing voltage of $\pm$ 300 V for negative or positive
charge carriers, respectively. Under these conditions there is a
field penetration through the first shielding grid which causes
more scattered particles to be drawn to the detector. There is much
less difference now between the positive and negative currents.

In the C_6H_6 flame the maximum ionization is larger by a factor
of about two, the peak is sharper and is shifted towards the oxidation zone bei 5 mm. The profile for negative charges lies somewhat
above that for positive ions. This is the opposite way for C_2H_2
flames. The ionization wiht C_6H_6 starts 6-7 mm before the temperature maximum is reached and the increase up to the maximum lies
within the oxidation zone of the flame. In contrast to this, the
strong ionization in the C_2H_2 flame starts when O_2 is consumed. This
different behavior is in line with an overlap of the oxidation zone
and the soot forming region in C_6H_6 flames whereas in C_2H_2 flames
these zones are much more separated[4].

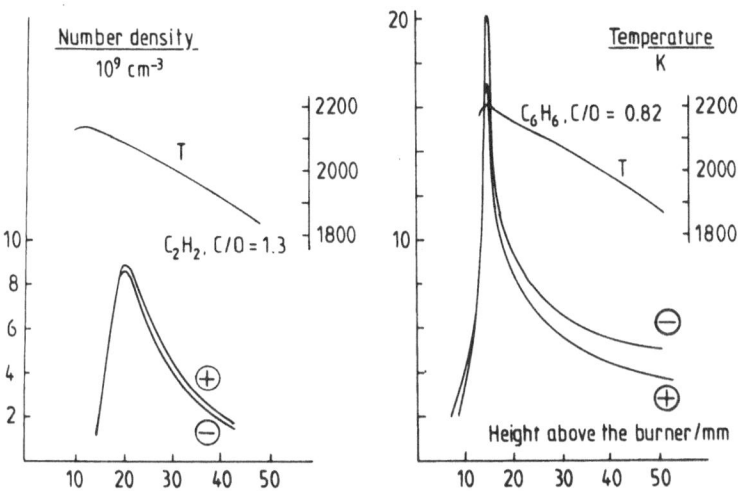

Fig. 9.   Ionization and temperature profiles of a C_2H_2-O_2 (C/O =
1.3) and a C_6H_6-O_2 flame (C/O = 0.82) having equal concentration of soot in the burned gas.

Determination of Ion Number Densities

The knowledge of the absolute number densities of positive and
negative ions in the zone of beginning soot formation is important
for the discussion of the mechanism of ionization and of their role
in carbon formation. The ion number density $C_{\pm,o}$ is estimated from
the respective total detector current $I_\pm$ measured at the repeller
grid, which for this purpose is biased with opposite polarity so
that the current is measured under saturation conditions. If it is
assumed that there is sonic flow at the nozzle throat and an isen-
tropic change of state, then

$$C_{\pm,o} = \frac{I_\pm \left(\frac{\varkappa+1}{2}\right)^{\left(\frac{1}{\varkappa-1}\right)} \left[\frac{2\,\varkappa\,RT_0}{(\varkappa+1)\,\overline{M}_g}\right]^{-\frac{1}{2}}}{e \cdot A \cdot f}$$

$\varkappa$ is the ratio C_p/C_v of the gas mixture, T_0 the stagnation temperature, $\overline{M}_g$ the average molar mass of the gas, e the elementary charge
and A the nozzle throat area. The coefficient f contains a factor
0.9 (estimated) by which the effective throat area is smaller through
a boundary layer and allows also for the transparency of the grids
used (0.8).

The formula neglects any loss of ions due to scattering by background gas. This is difficult to calculate quantitatively. From the
work of Fenn and Anderson[15] on argon beams it can be estimated
that the loss is probably no more than 50% under our conditions for
ions of a mass comparable to that of the gas molecules. For heavier
ions it will be less. Although the ion-molecule scattering cross sections are larger than those of neutral particles, the electric field
probably overcompensates this effect.

Scattering is strongest for ions of low mass. This can be estimated from experiments with flames seeded with alkali ions of different mass: Cs, K, Na. When the flame is seeded so that the number
densities of the different ions can be assumed to be about equal
(see below) the ratios of the ion currents are $I(K,M=39)/I(Cs,M =
133) \approx 1/3$ and $I(Na,M= 23)/I(Cs,M= 133) \approx 1/4$. Thus, if a loss of Cs^+
of 50% is admitted the currents of ions of masses in the range 20-
40 u are measured too low by a factor of 8-6 at the most. Even with
this presumably overestimated factor there is an appreciable extra
ionization in the soot forming zone which cannot be due to a growth
of chemi-ions from the oxidation zone. The general appearance of the
ionization profiles remained the same no matter how far the detector grid was moved into the tip of the probe.

Seeding of Sooting Flames with Metal Salts

It has been reported that metal additives to sooting flames
inhibit or promote soot formation depending on where they are admitted to the flame[16]. Another effect is that they inhibit particle
growth[17,18]. The mass separator used for unseeded flames is well
suited to study the influence of different metal atoms on number
density and mass distribution of small, charged soot particles.
Since the salts were admitted through an atomized aqueous solution
to the unburned gas the effect of water vapor was first determined.
In the following all results obtained with solutions were compared
to those with pure water.

Fig. 10 shows ionization profiles of a strongly sooting C_2H_2-
O_2 flame compared to a flame of the same C_2H_2/O_2 ratio with 16% by
volume (relative to the total volume of unburned gas) of water
vapor. The addition of H_2O reduces the flame temperature by about
100 K and also reduces the amount of soot formed. The maximum number
density of positive soot ions decreases by a factor of 18 and is
reached at a larger height from the burner. If a $2 \cdot 10^{-4}$ m CsCl solution is used instead the ionization increases again reaching its
maximum at a slightly larger height than with pure water. If a mass
distribution from the flame with H_2O is compared to that of the
pure C_2H_2 flame for the same height it is shifted towards lower
masses and is reduced in frequency.
Fig. 11 shows the influence of CsCl on the mass distribution of positive soot ions at three different heights in the flame. Their con-

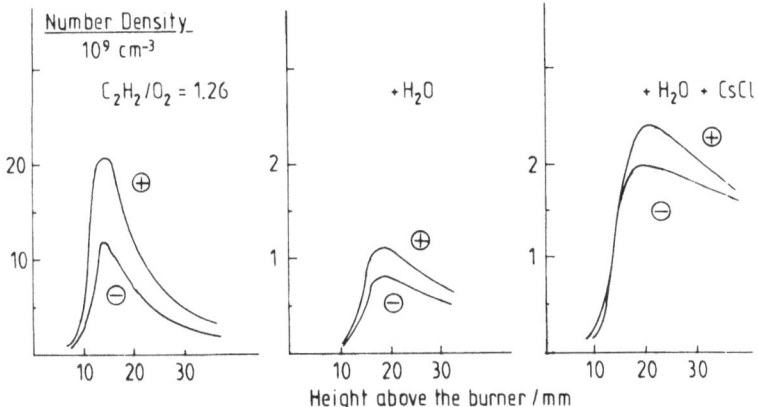

Fig.   10.   Influence of water and $2 \cdot 10^{-4}$mCsCl solution on the ioni-
             zation profiles of a sooting acetylene flame.
             C_2H_2/O_2 = 1.26, p = 40 mbar, v_0 = 25 cm s^{-1} (seeded
             flame 30 cm s^{-1}).

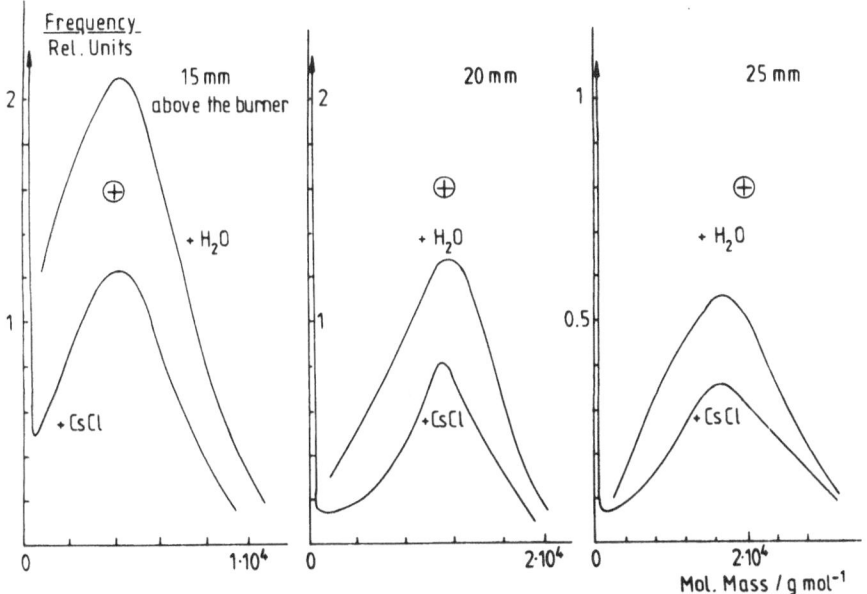

Fig.   11.   Effect of $2 \cdot 10^{-4}$m CsCl solution on the mass distribution
             of positive soot ions.

centration decreases while the distribution curves remain relatively
unaffected. Their maxima are found in the same mass region as with
pure water. The positive ions continue to grow with increasing height.
The steep increase of the mass distributions at very low masses is
due to Cs^+ ions.

   If the concentration of salt is increased the suppression of
positive soot ions is reinforced but the mass distribution is still
unaffected. Seeding with a 0.1 m CsCl solution causes the positive
soot ions to disappear completely. This behavior is shown on Fig.12
for a height of 20 mm. The amount of soot formed as judged from the
luminosity is hardly influenced by the salt addition.

   The behavior of negatively charged particles under the same
conditions as for Fig. 11 is given in Fig. 13. Their concentration
increases through addition of CsCl in about the same ratio throughout the mass range. No appreciable change in the relative mass
distribution is noticeable compared to H_2O addition. As with unseeded flames there appears to be only a very slight growth of negative soot ions with height from the burner.

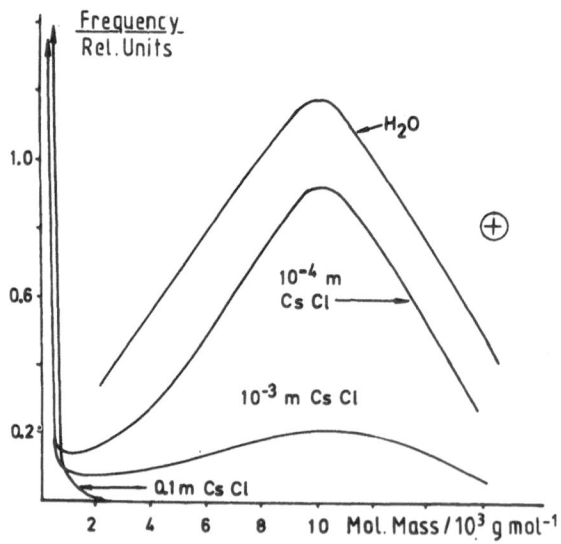

Fig. 12. Suppression of positive soot ions by CsCl solution of
           various concentration.

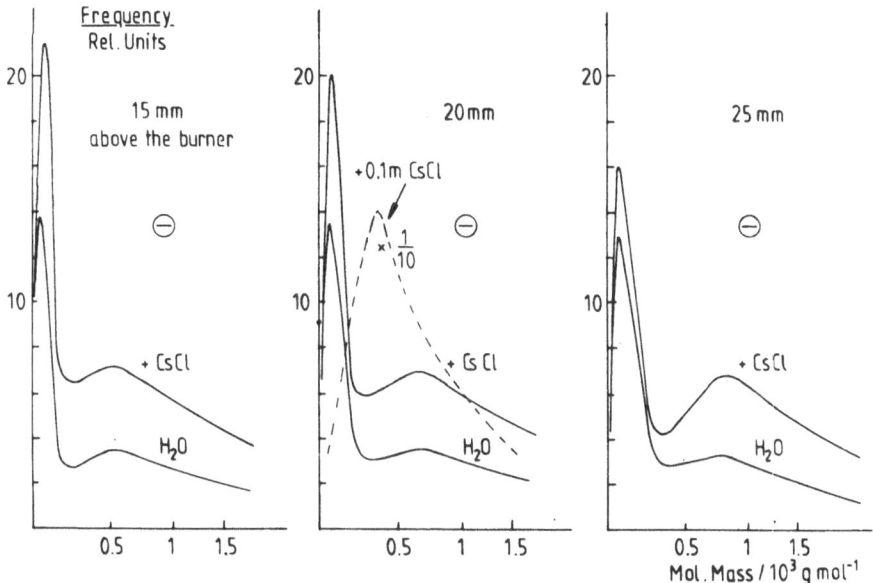

Fig. 13. Effect of $2 \cdot 10^{-4}$m CsCl solution on the mass distribution of negative soot ions.

If the flame is seeded with 0.1m CsCl the concentration of negatively charged particles increases by a factor of about 30. In this case, the (second) maximum in the distribution is shifted from about 700 to 400 u.

Effect of Other Metal Salts

Of the alkali salts other than CsCl the compounds KCl, KNO_3, K_2CO_3, NaCl, NaOH, Na_2CO_3, Na_2WO_4 and LiCl were tested and also the earth alkaline salts: $BaCl_2$, $SrCl_2$, and $CaCl_2$. All these salts have in principle the same effect on the soot ions as has CsCl. The anion is immaterial. To obtain the same quantitative reduction of positive and gain of negative ions, however, the concentrations of the salt solution had to be increased in correlation with the ionization potential of the metal component if this was higher than that of potassium. Fig. 14 shows a plot of the logarithm of the concentrations of alkali and alkaline earth halide solutions that have the same effect as a $2 \cdot 10^{-4}$m CsCl solution versus the ionization potential. The alkali metals can be connected by a curve showing a sudden change in slope between 4.5 and 5 eV. Two straight lines drawn through the point of Cs and K and through Na and Li intersect at about 4.8 eV. The alkaline earth metals do not fall onto the same curve.

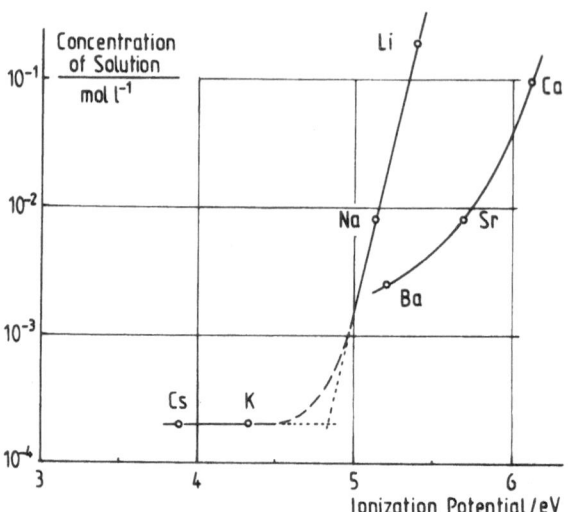

Fig. 14. Correlation between concentrations of salt solutions of equal efficiency and the ionization potential of the metal.

Other metal salts such as $CrCl_3$, $(NH_4)_6Mo_7O_{24}$, $Pb(NO_3)_2$, $MnCl_2$, $CuCl_2$, $FeCl_3$, $CoCl_2$, $(UO_2)(CH_3COO)_2$ have no effect, not even in saturated solution. The ionization potential of the metal component in each case is higher than 6.7 eV. An exception is uranium (≈ 4 eV). Therefore, the conclusion may be allowed that the flame did not contain free uranium atoms.

DISCUSSION

Estimation of Ionization Equilibrium

It has been argued that the charge concentration in the burned gas of a sooting flame corresponds to an ionization equilibrium[5], [19,20]. It is indeed possible to get a satisfactory agreement between calculated and experimental ion number densities. The calculation, however, is not free from some uncertainties. The work function of soot was set equal to 5.0 eV which seems not unreasonable in view of the effect of different alkali metals (Fig. 14, see also below). The electron concentration was assumed to be 1/10 of that of soot ions. If a soot ion is taken as a little sphere the extra energy necessary to remove an electron from it causes the maximum in the calculated profile of Fig. 15. Admitting these estimates and regarding the number densities only it seems reasonable to explain the charge concentration by an ionization equilibrium.

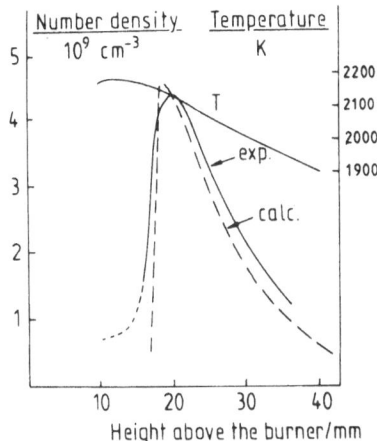

Fig. 15. Measured and calculated ionization profiles in a
C_2H_2-O_2 flame (same flame as in Fig. 4.).

     There is evidence, however, that the increase of ionization is
not controlled by a rapidly established equilibrium. Moreover, it
must be assumed that not even at larger heights there is a true
equilibrium under the conditions in low pressure flames. This is
based on the observation that the size distribution of charged particles is quite different from that of all particles. The charged
particles are smaller on the average. This discrepancy is extreme
in benzene flames. It shows that processes such as thermionic
emission from large particles and charge exchange which would lead
to complete equilibrium are too slow under these conditions.

Non-Thermal Effects in Ionization of Soot Particles

     The lower average mass of charged soot particles in the burned
gas is probably correlated to the fact that they acquire their
charge when they are still very small. In acetylene flames the
strongest increase in ionization takes place when soot ions with
masses between $1 \cdot 10^3$ and $3 \cdot 10^3$ u are formed. In benzene flames this
range is even lower, 500-1500 u. In this flame zone both the concentrations of polycyclic hydrocarbons and polyacetylenes are

relatively high[3,13,21]. In C_2H_2 flames the maxima of polyacetylene
concentrations and temperature lie at 10-11 mm from the burner. The
increase in ionization, however, is strongest at 15-17 mm where
polycyclic hydrocarbons have formed and the concentration of polyacetylenes is still high. In the benzene flame the polycyclic hydrocarbons appear earlier and simultaneously with polyacetylenes (10-
11 mm) before the temperature maximum is reached. This is in line
with an earlier increase in ionization. It falls off rapidly when
both polyacetylenes and polycyclic hydrocarbons decrease in concentration at the end of the oxidation zone although the maximum
temperature has just been reached. It should be mentioned that in
both flames the region of maximum ionization coincides with the
maximum of a transient absorption profile which was attributed to
larger polycyclic hydrocarbons[1,4].

A qualitative description of the ionization process could be
as follows: Neutral polycyclic radicals grow rapidly by exothermic
addition reactions with polyacetylenes and small radicals. The product species keep a fraction of this exothermicity and become "hot"
particles. The ionization potential ϕ falls as the particles are
growing to 500-3000 mass units. Ionization sets in when ϕ has become
low enough and the particles do not have acquired too much mass for
the extra heating to become inefficient. The rate of ionization decreases when the concentration of partners for exothermic addition
reactions falls and the particles become too large and probably
less reactive.

In benzene flames there is an overlapping of oxidation and
soot forming zone. As a consequence, a large fraction of the carbon
that is found in the soot at larger heights in the flame is concentrated in a comparatively large number of small particles at the
end of the oxidation zone[4]. Consequently, the soot ions are also
small. In the following the carbon particles grow mainly through
collisions between these smaller particles without taking much more
carbon from the gas phase. Since these sticky collisions are hardly
exothermic and are slow compared to recombination of charged particles with electrons different, mass distributions of neutral and
charged particles result. It is amazing, however, and not completely
understood why there are only very small soot ions in benzene
flames. One should expect a certain growth by sticky collisions
between soot ions and neutral soot particles.

In acetylene flames there is little overlap of oxidation and
soot forming zone. The concentration of polyacetylenes do not decrease so drastically behind the oxidation zone as in benzene flames.
A smaller fraction of the final soot is found in the small particles
at the end of the oxidation zone. These small particles grow larger
not only through sticky collisions between each other but also
through addition of e.g. polyacetylene and acetylene from the gas
phase. Thereby, larger charged particles are formed. Recombination

with negative charges which is kinetically more favorable with larger soot ions contributes to the difference in mass distribution between charged and neutral particles. Comparing the mass distributions of charged particles in flames of different C/O ratio there is evidence that also in C_2H_2 flames soot ions grow relatively more through the addition of gas phase species than neutral particles. The growth of the latter to sizes corresponding to 10^6-10^7 mass units comes mainly from sticky collisions between the particles.

Negative Soot Ions

     The formation of negative soot ions is difficult to discuss without knowing the profile of free electron concentration in the flame. The development of the mass distributions of negative soot ions show that the strong ionization takes place simultaneously with the formation of negative ions in the mass range 300-1500 u in acetylene flames and around 600 in benzene flames. At larger heights relatively more smaller negative ions are formed with both fuels, and there is no appreciable growth of negative ions. This behavior could be explained if parallel to a soot ionization generating free electrons there is a mechanism by which in an elementary step negative ions of some hundred mass units are formed together with the somewhat larger positive ions. Thermodynamically this could be more favorable than emission of an electron. The mechanism which causes the shift of the average mass of negative ions to smaller values is not known. This can happen via free electrons but also by splitting of smaller negative ions from larger ones.The fact that there is no appreciable growth of negative ions both in flames of acetylene and benzene shows that a sticky collision of a larger negative ion and a neutral particle to form a larger negatively charged soot particle is an inefficient process.

The Effect of Salt Addition

     The addition of metal salts leads to a decrease of positive and an increase of negative soot ion concentration. Thus, if the growth of soot particles is inhibited by these additives it must be through the negative soot ions rather than the positive ones. It is indeed possible to suppress the positive soot and hydrocarbon ions completely. Since this has no dramatic effect on the total amount of soot formed positive hydrocarbon and soot ions cannot play an important role as nuclei for soot particles[20].

     The number density of negative ions can be very much increased by using more concentrated solutions. As in unseeded flames these ions do not grow larger. If it could be shown that this increase goes parallel with the decrease of the neutral particle number density this would be a direct proof that the formation of a large number density of small negative soot ions inhibits the growth of soot particles. Experiments along these lines are being prepared.

The rate of metal ion formation in seeded flames increases enormously when the flame becomes sooty. Kinetic considerations led to the conclusion that this can only be explained through a mechanism

$$Me + soot\ particle \longrightarrow Me^+ + (soot\ ion)^-.$$

This mechanism is similar to a Langmuir ionization of the metal atom on a hot metal surface having a larger work function. The charge transfer, $Me + (soot\ ion)^+ \longrightarrow Me^+ + soot\ particle$, can be efficient for reducing the positive soot ion concentration but not for the observed production of Me^+. This Langmuir mechanism also explains the fact that K is equally effective as Cs (shown in Fig. 14.) and that the strong curvature of the plot takes place at an ionization potential equal to the work function of the small soot particles. A quantitative description of these processes will be subject of a further publication.

We are indebted to the Deutsche Forschungsgemeinschaft and the Fonds der Chemischen Industrie for financial support of this work which is gratefully appreciated.

REFERENCES

1. U. Bonne and H.Gg. Wagner, Ber.Bunsenges.Phys.Chem. 69: 35 (1965)
2. B.L. Wersborg, L.K. Fox and J.B. Howard, Comb.and Flame, 24:1 (1975)
3. K.H. Homann and H.Gg. Wagner, Ber.Bunsenges.Phys.Chem., 69: 20 (1965)
4. K.H. Homann and W. Morgeneyer and H.Gg. Wagner, Combustion Institute Europ.Symp., p. 394, Academic Press, London(1973)
5. K.H. Homann, Ber.Bunsenges.Phys.Chem. 83: 738(1979)
6. J.B. Anderson, R.P. Andres and J.B. Fenn, Adv.Chem.Physics (Ed.J. Ross) 10, 275, Interscience, New York(1966)
7. N. Abuaf, J.B. Anderson, R.P. Andres, J.B. Fenn and D.R.Miller, Rarefied Gas Dynamics, Adv.Appl. Mechanics,Suppl.4, Vol. II (ed.C.L. Brundin), p.1317, Adacemic Press, New York(1967)
8. M.H. Schwartz and R.P. Andres, Rarefied Gas Dynamics, 10th Symp., p.135 (J.L. Potter ed.), Progr. in Astronaut. and Aeronaut., Vol. 51, Part I, Aspen/Col.(1976)
9. U.v. Pidoll, Diplomarbeit, Technische Hochschule Darmstadt(1981)
10. D.B. Olson and H.F. Calcote, 18th Symp.(Internat.) on Combustion, p.453, The Combustion Institute, Pittsburgh(1981)

11. B.L. Wersborg, J.B. Howard and G.C. Williams, 14th Symp.(Internat.) on Combustion, p.929, The Combustion Institute,Pittsburgh (1973)
12. B.L. Wersborg, A.C. Yeung and J.B. Howard, 15th Symp.(Internat.) on Combustion, p. 1439, The Combustion Institute, Pittsburgh (1975)
13. G. Sticha, Dissertation, Technische Hochschule Darmstadt (1977)
14. W. Morgeneyer, Dissertation, Univ.Göttingen (1968)
15. J.B. Anderson and J.B. Fenn, 4th Symp.on Rarefied Gas Dynamics (J.H. Leeuw ed.) Vol. II, p.311, Academic Press New York(1966)
16. K.C. Salooja, Combustion Institute Europ.Symp. p.400, Academic Press, London (1.973)
17. E. Bartholomé and H. Sachse, Z.Elektrochem.angew.phys.Chem. 53: 326(1949)
18. B.S. Haynes, H.Jander and H.Gg. Wagner, 17th Symp.(Internat.) on Combustion, p. 1365, The Combustion Institute, Pittsburgh (1979)
19. D.R.Hardesty and F.J. Weinberg, 14th Symp.(Internat.) on Combustion, p.907, The Combustion Institute, Pittsburgh(1973)
20. J.B. Howard, 12th Symp.(Internat.) on Combustion, p.877, The Combustion Institute, Pittsburgh (1969)
21. K.H. Homann and H.Gg. Wagner, 11th Symp.(Internat.) on Combustion, p.371, The Combustion Institute, Pittsburgh (1967)

DISCUSSION

A. Feugier (Institut Français du Pétrole)

Can one expect that the ionization potential of soot has a constant value, independently of its chemical composition or of its condition of formation?

Homann

The ionization potential of large polycyclic hydrocarbons found in sooting flames is about 7 to 8 eV while that of graphite is 4.4 eV. An interpretation of the increase in ionization on the basis of an ionization equilibrium suggests that there is a decrease in the ionization potential to a value of about 5 eV while the particles grow to a mass of 2000 to 3000 u. It can be expected that the ionization potential depends on the hydrogen content and on the internal structure of the particles but it is unknown whether a value near that for graphite is reached in the burned gas of low pressure flames. It is even conceivable that soot particles have a work function lower than that of graphite because of their structure defects and dislocations and their radical character as compared to a graphite crystal.

R. Delbourgo (Centre National de la Recherche Scientifique)

Adding alkaline or alkaline earth metals has an inhibition effect on flame properties such as flame velocity, and flame temperature.

Isn't the effect observed on soot merely reflecting that influence?

Homann

An inhibitory effect of the salt additives on these strongly stabilized low pressure flames could not be observed when compared to those to which pure water was added. The temperature profile did not change when salt solutions were used. From experiments on the influence of inhibitors at various burning pressures [1] it can be concluded that an influence at such a low pressure is negligible. Moreover, Haynes et al. [2] who studied the influence of alkali and alkaline earth metal salts on soot formation in premixed ethylene-air flames at 1 atm do not report an inhibitory effect of the salts added.

H.F. Calcote (AeroChem Research Laboratories, Inc.)

1. In your explanation for the source of ions in sooting flames, you assume the energy of exothermic addition reactions accumulates in the growing species until it becomes "hot" enough to thermally emit electrons. This would require that the incipient soot particles have a temperature several hundred degrees above the gas temperature. Is there any evidence for this? Would you not expect that radiation, and collisions of the particle with the flame gases would prevent such a large temperature difference from existing? If you could account for a non-equilibrium concentration of incipient soot particles by this mechanism, what mechanism would you invoke to explain the non-equilibrium concentrations of ions smaller than the ions produced by this mechanism? Incidently, between 500 and 3000 amu the ionization potential of a soot particle is still much higher than that for graphite, compare 7.1 eV to 5.6 eV for 500 to 3000 amu particles with the work function of graphite, 4.6 eV (see my paper this workshop).

2. You show that the addition of CsCl to a sooting flame greatly suppresses the large positive ion concentration and comment that "the rate of metal ion formation in seeded flames increases enormously when the flame becomes sooty." No rate data are reported so I assume you deduce the relative rates from the observed concentrations. Charge transfer could account for the larger concentration of metal ions because the recombination coefficient of metal ions is much less than for soot ions. Thus the set of reactions:

$$Soot^+ + Me \rightarrow Soot + Me^+ \qquad very\ fast$$

$$Me^+ + e \xrightarrow{M} Me \qquad k \sim 10^{-9}\ cm^3 s^{-1}$$

$$Soot^+ + e \rightarrow Soot \qquad k \sim 10^{-5}\ cm^3 s^{-1}$$

with a fixed rate of $soot^+$ production would lead to a higher observed concentration of Me^+ in the presence of soot than when no soot is present. You also note that "the amount of soot formed as judged from the luminosity is hardly influenced by the salt addition."

Did you determine the change in particle size? Often the effect of chemical additives on soot formation is to change the particle size distribution but not the quantity of soot produced.

Homann

1. Several experiments indicate that the relatively strong ionization of small soot particles with masses of a few thousand atomic mass units or less is a non-equilibrium process. If one tries to describe this process by an equilibrium formula such as the Saha equation corrected for the finite size of the particles it appears reasonable to assume a higher temperature or a lower work function for the particles (or both) compared to the gas temperature and to graphite, respectively. I agree with Dr. Calcote that in doing so there are many uncertainties. If one assumes that the amount of carbon grows through the addition of polyacetylenes to small particles a temperature difference between the particles and the gas up to about 150 K can be accounted for. This depends on the estimation of the unknown accomodation coefficient between the gas and the particles. If the addition of energy rich unsaturated radicals contributes to the growth of particles the exothermicity and the temperature difference is even larger. On the other hand, it is known from ESR measurements that very young carbon particles have radical character and are certainly not made up of graphite crystals. Thus, their work function may be less than that calculated on the basis of that for graphite.

The non-equilibrium concentration of smaller ions can be explained by a growth of chemi-ions from the oxidation zone. But this mechanism cannot account for the formation of small charged soot particles since the number of charges increases considerably whereas a decrease would be expected if only ion-molecule and recombination reactions were considered to be responsible for the formation of larger ions.

2. Rate data for the formation of metal ions when CsCl is added to a sooting flame will be published in a forthcoming paper. The

lower recombination coefficient of metal ions with free electrons will cause a relatively higher level of M^+ concentration. But even if recombination is totally neglected an unreasonably high reaction cross section for the charge transfer from a soot particle to a metal atom would have to be assumed since the concentration of charged soot particles relative to that of uncharged ones is still very small in the flame zone of strong metal ionization.

We did not yet determine the particle size of neutral soot particles. This will be one of the subjects of further studies on this seeded flame.

ELECTRICAL INTERVENTION IN THE SOOTING OF FLAMES

Felix J. Weinberg

Imperial College
London SW 7
England

INTRODUCTION

There is much evidence that both electrical and chemical
effects are important in the formation of soot in flames. It
is difficult to distinguish between the two by non-invasive
methods since the chemical and electrical properties of species
are closely related. This paper reviews and discusses the con-
sequences of gross electrical interactions with sooting flames.
The literature of this subject is very extensive; see for example
references 1 to 13.

It must be borne in mind that in much of this work the primary
aim was, for example, control of sooting in practical systems
rather than the elucidation of mechanism. Indeed, such gross
electrical intervention leaves much to be desired as a diagnostic
tool because, although the application of electric fields inter-
acts selectively with charged species, it also modifies the system
under study by transposing these charged particles with respect to
the flames structure. Thus it will be shown that large DC electric fields effectively remove any small ions which might act as
nuclei for the pyrolysis so that this particular process could
not be studied easily by such means.

Since the object of this paper is to initiate discussion at
a workshop, I will attempt to review the subject in such a manner
as to provide the basis for re-appraising the crucial issues. In
view of the extensive literature, such a focused review necessarily
has to be very selective and somewhat over-simplified. In particular previous experimental results will be considered from the
point of view of the light they shed on renewed attempts to link

243

nucleation of soot directly with ionisation - see Dr H.F. Calcote's contribution to this meeting.

EFFECTS OF LARGE DC FIELDS

It has been known for many years that control of charged particle movement by electric fields can act in at least three distinct ways, depending on the stage in the particles growth. The first is manipulation of fully formed particles - especially their removal to electrodes. The second is the control of trajectories, and hence residence times, during the process of growth or burn up, which allows interaction with the reaction directly. The third is the transposition or removal of the highly mobile charged nuclei on which particulates are thought to grow.

The first of these is useful though not perhaps very interesting from the point of view of the mechanism of soot formation. It is akin to electrostatic precipitation except in that the charges are not provided by a corona discharge or other external source. They can derive from chemi-ionisation in the flame reaction zone, from thermionic emission by the soot and from any growth on ionic nuclei.   By subjecting the flame to a field such that the collecting electrode is negative, it is easy to arrange[3] for any or all these processes to reinforce so that all the soot particles are positively charged. (This will tend to happen automatically, unless the field and burner geometry are so arranged that soot particles are subjected to a flux of negative charge from the flame, in which case diffusion and/or bombardment charging will cause some, and eventually all of the soot to acquire negative charge).

Apart from allowing soot to be collected on a cold, or burnt off on a hot electrode, the structure of the carbon black so collected may also be modified. Since electric field lines converge on points of high curvature due to particles deposited initially these tend to collect particles arriving subsequently which might otherwise have travelled to adjacent sites. The longer such a filament grows, the larger the surrounding area which it protects from further deposition. Soot deposited on an electrode under these conditions therefore occurs in branching structures with bald patches between them resulting in much reduced bulk densities. This is helpful when the soot is to be burnt off by causing it to deposit on an electrode heated by hot combustion products. It is of course also possible to use the field to prevent deposition on a specific surface - i.e. prevent rather than promote electrostatic precipitation - by making the surface the positive electrode.

Such work[3] has been carried out with fairly conventional diffusion flames on tubular burners. For the detailed study of the effect of direct and alternating electric fields on processes beyond the deposition of fully formed particles, the flat, counter flow, diffusion flame [14] has been extensively used. This is because the matrices which streamline the approach flow in the two burner mouths can be used as electrodes parallel to the flat flame which are cooled by transpiration of the reactants. Depending on the direction of the applied field, positive or negative charge carriers from the flame zone can be made to pass through the pyrolysis zone and bring down any charged carbon particles on the matrices which can be covered by light gauzes for subsequent removal and weighing of the deposit or its examination under an electron microscope. In addition to providing a nearly uni-dimensional configuration the system has been favoured because of its advantage of minimizing ionic wind effects as these occur in opposition to the gas flows.

Because of the large volume, age and previous exposure of this work, a terse summary of how it has been interpreted will have to suffice: All soot particles were found to be charged, at least in the presence of a field. Except in the case when the pyrolysis zone was subjected to a flux of negative charge, all particles were charged positively. Even in the latter case some positively charged particles were collected, though they had to cross the flame to deposit on the electrode, on the opposite side. Although a sufficiently large flux of negative charge would eventually result in charging the soot negatively, contrary to its normal tendency, I thought it sensible to concentrate here on conditions producing only positively charged particles.

At any one field strength, the particles were of fairly equal size. This size decreased rapidly with increasing field strength, by a factor of about 5. At the same time the mass of soot formed (= deposited) fell to about 2% of the amount in the absence of the field. Note that this implies an increase, with increasing field strength, of the number of particles col-lected if their density remains constant. The minimum particle radius is 5 nm. This is independent of whether of not the flame is seeded with a caesium salt and/or whether the flux of charge is positive or negative. It is reached at a relatively low applied potential (approximately one-third of the maximum of 6 kV) and remains sensibly constant thereafter as the voltage is increased. Mobility measurements, to which this burner system also readily lends itself simply by incorporating a second gauze further up the burner tube and maintaining a control field between these two gauzes, range from 10^{-3} to 3×10^{-2} cm^2 s^{-1} V^{-1}, depending on the applied potential. Taken together with size measurements from the electron micrographs, these values show that each particle carries unit charge over practically the entire experimental range, the majority of the current being carried by smaller charge carriers.

The above conclusions, which apply with only minor variations

for a variety of conditions (including flames with oxygen, and with
caesium chloride seeding considerably increasing the current drawn)
seem quite incontrovertible.   They do not involve any hypothesis
or speculation, are subject only to relatively minor experimental
errors and, time permitting, will be documented by experimental
records at the meeting.   What follows is their interpretation;
this led to further experiments which provided confirmation of the
basic concepts.

     Since all the particles are charged, application of the field
causes them to be removed from the pyrolysis zone at a rate which
depends on their size (mobility) and the applied potential.   This
shortens their residence, and hence growth time in the pyrolysis
zone, resulting in smaller particles and reduced mass.   The consequence manifests itself even to the casual observer by a very
great reduction in flame luminosity.   By the time the applied
potential has reached 1 kV, the yellow luminosity due to carbon
particles in the flame has virtually disappeared, leaving only the
blue flame zone.   Another experimental test which immediately
suggests itself is to attempt to increase the residence time by
holding charged particles stationary against the gas flow by means
of an applied field.   When this critical value is reached, macro-
scopic filaments and particles suddenly appear in the pyrolysis
zone and eventually deposit on the burner flanges to produce a
network of carbon filaments extending right into the flame.   The
field strength at which this occurs is relatively small;   at any
appreciable field strength the drift velocities of the particles
greatly exceed the flow velocity of the gas.   In this electrically distorted system the particles are immersed in unipolar clouds
of charge.   Thus there are virtually no electrons on the positive
side and the positively charged soot particles head rapidly for the
nearest electrode where they arrive having all experienced similar
histories and having attained similar sizes (depending on the
particular potential) without much chance for agglomeration into
chains.

     The deviation from electrically unperturbed flames is even
more extreme when we consider nucleation.   The mobility of molecular ions is of the order of a hundred times greater than that of
the soot particles.   It follows that at any appreciable field
strength the concentration of ionic nuclei is reduced to negligible
proportions.   It emerges from the results that the rate of arrival
of ions at the electrode is very much larger (from 14 to 100 times)
than that of soot particles.   Since under all conditions of
appreciable potential the minimum particle size is constant at
9.2 nm it is concluded that these particles acquired their charge
when they reached this size.   In other words, if large electric
fields are used to remove ions, then soot particles will grow on
uncharged nuclei.   This agrees well with the theory of diffusion
charging. [1]   Indeed a model based on a constant rate of
nucleation in the absence of ions,   followed by growth with

reasonable rate constants for agglomeration and surface accretion gives remarkably good agreement[7] with experiments. It predicts correctly all the major features of the mass deposition, particle sizes and numbers as a function of applied potential (the latter increase with voltage simply because increasing fields allow less time for aggregation).

Clearly this is not an argument against nucleation by charged precursors - it simply accounts for the behaviour of a system in which these have been removed by a large field. To look for charged nucleation, it is necessary to apply much smaller potentials. If it is assumed that the majority of chemi-ions are formed in the main reaction zone, the calculated ion density in the pyrolysis zone peaks at a few hundred volts, rather than the several thousand considered hitherto. Electron micrographs taken at these very low potentials indeed show a very different picture. Instead of all particles being the same size there is a range of small sizes corresponding to particles growing on charged nuclei or acquiring charge by attaching to ions early in their lives, thereupon being immediately removed to the electrode. The max-imum size is shared by a family of large particles, evidently produced by the process of uncharged nucleation, followed by sub-sequent charge acquisition as described above. It is remarkable that, because of the very rapid withdrawal and deposition by the field of small charged particles, it is actually possible to identify the history of each individual particle which appears in such a photograph. By plotting the percentage of particles smaller than half the average diameter against the calculated ion concen-centration in the pyrolysis zone, divided by the number of particles collected, a correlation is obtained; this percentage varies from as little as 2% to 60% at the low fields which allow appreciable ion concentrations.

Accordingly it was concluded that charged nucleation probably is very important in systems not involving large uni-directional fields. When such fields are applied, however, the experiments prove conclusively that nucleation by uncharged species takes over and, when growth occurs on uncharged nuclei, the resulting particles acquire charge by other means when their radius reaches approxi-mately 5 nm.

Although this review deals primarily with electrical inter-vention in flames, it is appropriate to mention that DC fields have been applied also behind reflected shock waves (e.g. [15-17]). Sooting has generally been inferred in such work from optical or spectroscopic measurements. Several of the findings are of interest here. In the absence of any oxygen[15] chemi-ionisation was never-theless shown to persist to a very low temperature by saturation current measurements and an alternative mechanism not involving species containing oxygen was accordingly proposed (not as a rival

to that generally accepted but as an alternative in the absence of
oxygen). For fuel-rich methane/oxygen mixtures electric fields
were found[16] to suppress soot formation if methane was only in
limited excess but to enhance sooting for very rich mixtures. It
must be borne in mind that the geometry here is totally different
from that relevant to flames: the pyrolysis zone here does not occur
in a thin region adjacent to a flame but rather occupies the
entire volume of the tube behind the reflected shock. Nevertheless
the conclusions drawn from this research are almost identical to
those arrived at above. Lester and Wittig inferred [16] from their
results a dual nucleation mechanism involving both charged and
uncharged nuclei. Theoretical modelling based on this concept
predicted trends found to be in good agreement with their experimental results.

EFFECT OF ELECTRON EMITTERS AND SMALL DC FIELDS

        There is a very extensive literature (see for example[18] to
[28]) dealing with the effect of additives on soot formation which is
not the subject of this review. However, some of it is relevant
background because the effectiveness of many of the additives
seems to be associated with their readiness to ionize, or to
emit electrons in virtue of a low work function. Beyond that the
picture is complicated in that such additives can act either to
suppress or to enhance soot formation depending on where they are
introduced in the flame, on their amount and on the composition of
the reactants. Various mechanisms have been proposed to account for
these phenomena, based on steps involving charge carriers - for
example reversible charge exchange reactions between the metallic
ion and an organic nucleus for soot formation.[23] Electric
fields enter this picture because of attempts [29] to suppress or
enhance electron emission from such additives when coated onto an
electrode surface. The potentials involved here are one to two
orders of magnitude smaller than those discussed hitherto and they
are not capable of producing a major distortion in the electrical
and aerodynamic structure of the flames. The original object of
the research was to test the hypothesis that the effect of ionising
additives is largely due to the electrons they emit. To this end
small electric fields were applied to promote or retard the emission
of electrons from the surfaces of small wires coated with barium
salts,or even just carbon, and maintained hot by contact with flame
gases. Effects on soot formation at various positive and negative
potentials were examined in relation to the current - potential
characteristics obtained with coated and uncoated wires: at negative potential coated wires act effectively like the cathode of a
diode, freely emitting electrons into the flame gases. The sequence
of experiments followed some work of Salooja[28] who showed that
when an additive on a wire is introduced at the base of the flame
it leads to suppression,whereas introduction higher up leads to
promotion of soot formation. Experiments at each of these stations

were repeated with positive, negative and zero potentials applied
to barium-, soot-coated,and uncoated surfaces. Results in the absence of an applied potential allowed non-electrical effects - such
as vaporisation of the additive - to be assessed,while experimenting
with uncoated wires tests the importance of just placing a material obstacle in that part of the flame. In each case a negative
potential accentuated the effect characteristic of an additive and
a positive potential suppressed it. In other words,the behaviour
associated with the additive is manifested by the electrons emitted
from it , irrespective of whether sooting is being promoted or
suppressed.

The discussion of these observations[29] whilst recognising
the possible role played by chemical effects including those involving nucleation by charged precursors, deliberately avoided
invoking such effects in order to demonstrate that the mechanism
could be accounted for by physical processes such as aggregation.
Where all soot particles are charged positively, electrostatic
repulsion would be a barrier to aggregation. Electrons would then
act as a "glue" by neutralising some of the positive charges. It
has been shown[30,31]* that such processes involving charge are
very important in coagulation and chain formation. In regions
where more than half of the charge has been neutralised and further
accretion occurs by dipole action, electron neutralisation of positive charges at the end of chains could have the opposite effect.
The point about the size of the agglomerate is that smaller particles are much more likely to burn up in the flame than larger
aggregates and hence the total yield would be affected.

If we extend speculation to take into accound ionic nucleation,
alternative interpretations suggest themselves. Neutralisation of
nuclei would be expected to have a much more direct effect on soot
reduction if the charge itself plays an important part in the nucleation process. Even the above mentioned mechanism proposed by
Bulewicz et al.[23] is not ruled out by supplying electrons alone
since positive ions formed in reversible equilibrium reactions along
with an electron would tend to be suppressed when the equilibrium
is driven into reverse by increasing electron concentration.

This method is unique in that it generates only one kind of
charge carrier. Part of the interest it has aroused is undoubtedly
due to its practical implications rather than any light that it

---

*Much useful work on the electrical characteristics of soot parti-
cles and their precursors has been carried out by Howard et al.
(e.g.[30,31,32]) by studying such particles using mobility or
Faraday cage methods following sampling by microphobes. Although
the methods involve the application of electric fields, these fields
are not applied to flames and therefore they do not come within the
scope of this review, though the results which are relevant are quoted.

sheds on the mechanism of soot formation. This is because of the
implication that electrons so emitted from hot surfaces coated with
additives could be used in place of injecting those additives into
combustors via the fuel and thereafter into the atmosphere. Additionally it has been suggested that this would make the effect
subject to control by varying electron emission, using small
retarding or promoting potentials applied to the surfaces.

EFFECT OF AC FIELDS

The first application of alternating potentials dates back to
1969.[10] Radial fields were applied to sooting turbulent diffusion flames causing them to become shorter, broader and less
luminous. These results are much less informative than might at
first sight appear, for two reasons. Firstly, the maximum frequency
of 400 Hz makes such fields effectively unidirectional, insofar as
ions and electrons are concerned, since these charge carriers will
reach their respective electrodes during a period of oscillation.
Even large soot particles would be expected to be drawn out of the
pyrolysis zone and the authors' surprise at the insensitivity of
the results to the AC frequency appears unjustified. The other
major complication is due to ionic (or Chattock) wind effects.
These have not been discussed before because the use of the counter
flow diffusion flame makes them relatively unimportant. In open
flames with appreciable electrode separation the aerodynamic disturbances due to the induced gas flow [1, 33]   become dominant.
Because of its effect on soot formation through altered mixing of
reactants and flame distortion, such geometries do not give results
conducive to analysing the underlying processes of growth and
have therefore not been given prominence in this review.

A very important recent piece of work which could greatly
advance the subject[13] has applied AC fields over a wide range of
frequencies, as well as DC fields in both directions, to flat
counterflow diffusion flames and to premixed flames using acetylene
as fuel. The AC frequencies ranged from zero to 10^7 Hz, field
strength up to 3 kV cm^{-1} and the effects of these variables on flame
luminosity and on the amount of soot carried in the products was
studied. In addition to the very considerable reductions in mass
and luminosity previously observed[1,4,7] when DC fields are
applied to counterflow diffusion flames, luminosity _increased_ at
higher frequencies. The largest increases of approx. 230% over
the unperturbed value occurred for premixed flames on cylindrical
burners subjected to radial fields. Although these are less tract-
able from a fundamental point of view than the counterflow diffusion
flames discussed below, the practical potential of this observation
should not be overlooked.

As regards the mass of soot collected from the exhaust, large
decreases due to applying DC fields in either direction begin to

vanish at increasing frequencies, their effects having diminished
appreciably already even below 50 Hz.   (These threshold frequencies are very important to the discussion below; the increases in
luminosity further considered later do not set in till frequencies
increase by more than two orders of magnitude.) By the time the
frequency has reached 10^5 Hz any change on the unperturbed rate of
soot production becomes immeasurably small.

   Luminosity variation shows a more intricate pattern due to an
increase at higher frequencies following the decrease for low and
zero frequency fields.   The exact values depend on the potential
applied, flow velocities and fuel/air ratio but, in round numbers,
the increase sets in at about 10^4 Hz , peaks at around 10^6 Hz and
has not disappeared even at the very highest frequency, for any
appreciable field. A small extrapolation suggests that at high
fields it would still be appreciable at 10^8 Hz !

   The authors account[13] for these results in terms of the
theory of Mayo and Weinberg[7] on the basis of surface growth
mechanism of small carbon particles. They suggest that as the particles are made to oscillate in an AC electric field they sweep
a larger gas volume in the pyrolysis zone and, due to their increased relative velocity, increase their rate of surface growth
The consequent increase in the volume fraction is deemed to lead
to the increase in luminosity, which reaches about 20% in the
counterflow flame. The authors calculate the amplitude of these
oscillations using the mobility of a singly charged particle of
0.5 nm radius and compare this with the dimensions of the pyrolysis zone obtained, along with the flame location, from photo-
graphic records of the counterflow flame. They also calculate the
mean drift velocity of such "particles" in the field and demonstrate that above around 10^4 V. cm^{-1} this velocity induced by
the forced oscillation exceeds that due to thermal motion according to the kinetic theory. The enhancement of luminosity is thought
to be limited, at the lower frequencies, by excursions beyond the
pyrolysis zone and, at the higher frequencies, by the fall in the
mean drift velocity.

DISCUSSION

   Before we turn to speculation about the mechanism of sooting,
it seems right to acknowledge that much of this work was carried
out with the object of offering methods for the manipulation
of soot formation in practical processes.   To the ability of
DC fields to modify deposition, aggregation, growth and nucleation
we must now add the effects of electron injection from captive
emitting coatings controlled by small potentials and of high
frequency fields which can increase radiation from sooting flames
without other major disturbances. It is worth recalling that,
given a particular flame with fixed flows of fuel and oxidant, we

have no other method of comparable power and versatility for controlling processes involving soot formation - only additives approach
anywhere near this aim and their use seems very haphazard by comparison.

Turning now to mechanism, I am anxious not to pre-empt the
discussion which is the object of the workshop and I would therefore
like to confine myself to leading the way in suggesting how we might
re-examine some of the results presented above, particularly with a
view to the role of nucleation by ions. Judging from the abstracts
I received, Dr. Calcote's contribution at this meeting will emphasize ionisation as an essential process in the growth of carbon, at
the cost of nucleation by neutral radicals. Dr. Homann, on the
other hand, whilst recognising the importance of charge effects on
soot growth, suggests that this process has its own charging mechanism, independent of normal chemi-ionisation in flames.

Work with DC fields proves conclusively that when large uni-
directional fields sweep the pyrolysis zone clean of ions - i.e.
reduce ion concentrations to negligible proportions - soot grows on
uncharged nuclei. The particles generated under these conditions
then acquire charge by other means when they grow to 5 nm. All the
subsequent phenomena which can be produced by large field strengths
can then be fully accounted for in terms of the movement and transposition of charged particles in and out of pyrolysis zones, manipulation of the residence times, and so forth. This proof that
uncharged nucleation is possible has perhaps led to an undue tendency to explain other electrical effects in purely physical terms
when it would be more logical simply to recognise the obvious fact
that systems in which small charged particles are immediately
removed to electrodes are not suitable for the study of nucleation
by ions. The appearance of tiny charged particles when very small
fields are applied, as well as soot reduction by electron injection,
under appropriate conditions, are most easily and logically explained
by growth on charged nuclei, as has been indicated above. I would
suggest, however, that the evidence of experiments with alternating
fields, when correctly interpreted, is even more conclusive.

This is because the mobilities of the charge carriers of
interest: particles, ions and electrons, differ so greatly from
one another that, even with the crudest assumptions, it is possible
to define which band of frequencies will cause each to oscillate
within certain limits - e.g. so as to drift to the electrodes and
be removed, go beyond the boundaries of the pyrolysis zone, or
barely at all. Figure 1 shows such a rough monogram based on
the data of Kono et al. [13]   It is drawn for a mean field (peak
to peak) of 1 kV cm^{-1} which is quite a high value bearing in mind
the distribution of field intensity[9] and the elevated temperature. It is quite clear from this that AC fields will behave like
DC fields up to frequencies of over 10^3 Hz if the effect concerns

an ion but under 10^2 Hz if we are looking at particle effects. At the other end of the scale, frequencies of the order of 10^7 Hz will simply not shift a particle, whereas they will greatly perturb the behaviour of ions and electrons. Once this is accepted, we are led to the conclusion that the observation that fields of frequencies in excess of 10^7 Hz still greatly enhance luminosity cannot be due to a mechanism involving particles – as distinct from ions.

To some extent the issue of when an "ion" becomes a "particle" is a question of semantics. In this discussion the term "particle" is applied to the range of sizes of what electric fields remove from flames when ions are not involved as nuclei – the smallest radius being 5 nm at which the first charge is acquired and for which the mobilities have been measured. The radius of 0.5 nm used by Kono et al.[13], however, is that of a molecule (e.g. molecular diameters from viscosity measurements are .46 nm and .75 nm for carbon dioxide and benzene, respectively). Presumably those values were chosen because otherwise the effect of higher frequencies would not be explicable (although the factor is only 10 in terms of diameters, it is, of course, 10^3 in terms of mass).

However, not all the effects observed are due to ions. The mass of soot collected, for example, is quite different for a DC field and one oscillating at 50 Hz. Yet it is clear from Fig.1 that, to an ion, the two are similar since its amplitude of oscillation at 50 Hz is at least one order of magnitude greater than the distance to the nearest electrode. Accordingly the mass of soot carried by the products is determined by electrical inter-actions with the particles. As would be expected this effect disappears at the high frequencies.

Let us now consider why luminosity should increase at higher frequencies if the amount of carbon produced does not. I do not want to discount higher growth rates due to increased relative velocities, even if the charge carrier responsible looks more like an ionic nucleus than a particle. However, during the process of growth, the number of particles will not increase; they can only coalesce and will be aided in this by any oscillation. Moreover, the oscillation will only affect the smallest particles which must form a minute fraction of the total mass and any increase in their growth rate due to their movement will only decrease their propor-tion. In case this does not seem a very plausible mechanism to you, I would tentatively propose that we are seeing an increased number of smaller particles due to increased nucleation. This would allow the number density to increase without having to result in an increased mass eventually. Several mechanisms by which the application of a high frequency field to the initial charge carriers could increase their number suggest themselves. Firstly, alternating fields decrease recombination because the clouds of opposite charge carriers which oscillate with respect to one

another overlap over smaller volumes.   In equilibrium, decreased
recombination results in further ionisation.   Secondly, if the
main zone of chemi-ionisation does not coincide with the pyrolysis
zone, alternating fields would provide an effective means of transposing ionic nuclei at regular intervals.   Thirdly, secondary ionisation might create additional nuclei, especially at the higher
fields.   In this context it is worth noting that the excess over
thermal velocities has been calculated only for ions, not for
electrons.

     It may be that increased nucleation could be detected by elec-
tron microscopy of the soot particles finally collected and such
an experiment has been proposed.   I would like to leave the final
conclusions to the discussion at the workshop but the whole prospect
of using oscillating fields of varying frequency as a diagnostic
tool seems very exciting – especially at frequencies so high that
neither particle motion nor ionic winds can affect the results.
Any effects then observed must be due to field interaction with
ions, and/or electrons, and I would have thought it most probable
that we are looking at artificially increased ionic nucleation in
this instance.

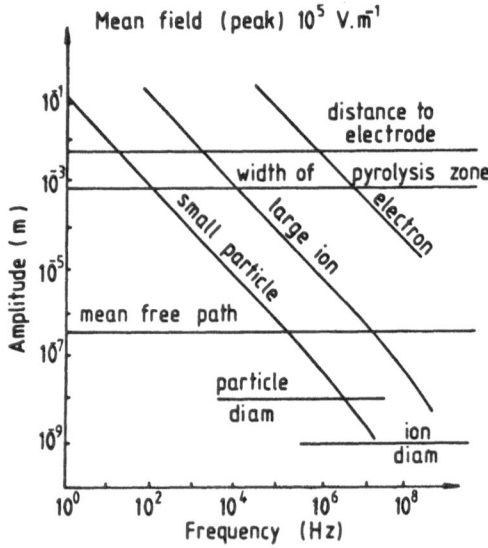

Fig. 1.   Approximate excursions of main charge carriers
          as function of frequency (based on data from
          refs 13 and 7).

ACKNOWLEDGMENT

    I am indebted to Professor T. Takeno and, through him, to
Professor N. Kono for helpful discussions.

REFERENCES

1. J. Lawton and F. J. Weinberg, "Electrical Aspects of Combustion",
    Clarendon Press, Oxford (1969).
2. W. T. Brande, The Bakerian lecture: on some new electrochemical
    phenomena, Phil. Trans. Roy. Soc. 104: 51 (1814)$\frac{1}{2}$
3. K. G. Payne and F. J. Weinberg, A preliminary investigation of
    field-induced ion movement in flame gases and its applica-
    tions, Proc. Roy. Soc. A 250: 316 (1959).
4. E. R. Place and F. J. Weinberg, Electrical control of flame
    carbon, Proc. Roy. Soc. A 289: 192 (1965).
5. E. R. Place and F. J. Weinberg, The nucleation of flame carbon
    by ions and the effect of electric fields, in 11th Symposium
    (International) on Combustion, The Combustion Institute,
    Pittsburgh, p.245 (1967).
6. F. J. Weinberg, Electrical aspects of aerosol formation and
    control, Proc. Roy. Soc. A 307: 195 (1968).
7. P. J. Mayo and F. J. Weinberg, On the size, charge and number-
    rate of formation of carbon particles in flames subjected
    to electric fields, Proc. Roy. Soc. A 319: 351 (1970).
8. D. R. Hardesty and F. J. Weinberg, Electrical control of
    particulate pollutants from flames, in 14th Symposium
    (International) on Combustion, The Combustion Institute,
    Pittsburgh, p.907 (Invited paper). (1973).
9. F. J. Weinberg, Smokes, droplets, flames and electric fields,
    Faraday Symposia of The Chemical Society, No.7: 120 (1973).
10. J. E. Mitchell and F. J. Wright, Effects in diffusion flames
    by radial electric fields, Combustion & Flame, 13: 413 (1969).
11. R. J. Heinsohn and P. M. Becker, in "Combustion Technology:
    Some Modern Developments", H. B. Palmer and J. M. Beer, eds.,
    Academic Press, New York, p.239 (1974).
12. B. S. Chittawadgi and R. S. Mate, Effect of electric field on
    diffusion flame, 6th National Conference on Internal Combus-
    tion Engines and Combustion, Indian Institute of Technology,
    Bombay, Paper FG-9/79 (1979).
13. M. Kono, K. Iinuma and S. Kumagai, The effect of DC to 10 MHz
    electric field on flame luminosity and carbon formation,
    in 18th Symposium (International) on Combustion, The Com-
    bustion Institute, Pittsburgh, p.1167 (1981).
14. T. P. Pandya and F. J. Weinberg, The structure of flat counter-
    flow diffusion flames, Proc. Roy. Soc. A 279: 544 (1964).
15. R. J. Bowser and F. J. Weinberg, Chemi-ionisation during pyro-
    lysis, Combustion & Flame, 27: 21 (1976).
16. T. W. Lester and S. L. K. Wittig, Soot nucleation kinetics in

premixed methane combustion, in 16th Symposium (International) on Combustion, The Combustion Institute, Pittsburgh, p.671 (1977).

17. S. L. K. Wittig and T. W. Lester, in "Evaporation-Combustion of Fuels", J. T. Zung, ed., American Chemical Society, Advances in Chemistry series 166, Washington, p.167 (1978).

18. E. Bartholome and H. Sachsse, Katalytische Erscheinungen an Aerosolen, Z. Elektrochem. 53: 326 (1949).

19. K. S. B. Addecott and C. W. Nutt, Mechanism of smoke reduction by metal compounds, presented at American Chemical Society meeting, New York City, September 7-12 (1969).

20. D. H. Cotton, N. J. Friswell and D. R. Jenkins, The suppression of soot emission from flames by metal additives, Combustion & Flame, 17: 87 (1971).

21. A. Feugier, The effect of alkali metals on the amount of soot emitted by premixed hydrocarbon flames, in Combustion Institute European Symposium 1973, F. J. Weinberg, ed., Academic Press, London, p.406 (1973).

22. K. C. Salooja, Carbon formation in flames; control by novel catalytic means, in Combustion Institute European Symposium 1973, F. J. Weinberg, ed., Academic Press, London, p.400 (1973).

23. E. M. Bulewicz, D. G. Evans and P. J. Padley, Effect of metallic additives on soot formation processes in flames, in 15th Symposium (International) on Combustion, The .Combustion Institute, Pittsburgh, p.1461 (1975).

24. A. Feugier, Effect of metal additives on the amount of soot emitted by premixed hydrocarbon flames, in 2nd European Symposium on Combustion, vol.I, The Combustion Institute, Orleans, France, p.362 (1975).

25. A. Feugier, in "Evaporation-Combustion of Fuels", J. T. Zung, ed., American Chemical Society, Advances in Chemistry series 166, Washington, DC, p.178 (1978).

26. J. M. Goodings, C-W. Ng and D. K. Bohme, Int. J. Mass Spectr. Ion Phys. 29: 57 (1979).

27. B. S. Haynes, H. Jander and H. Gg. Wagner, The effect of metal additives on the formation of soot in premixed flames, in 17th Symposium (International) on Combustion, The Combustion Institute, Pittsburgh, p.1365 (1979).

28. K. C. Salooja, Combustion control by novel catalytic means, Nature, 240: 350 (1972).

29. R. J. Bowser and F. J. Weinberg, Electrons and the emission of soot from flames, Nature, 249: 339 (1974).

30. R. T. Ball and J. B. Howard, Electric charge of carbon particles in flames, in 13th Symposium (International) on Combustion, The Combustion Institute, Pittsburgh, p.353 (1971).

31. B. L. Wersborg, J. B. Howard and G. C. Williams, Physical mechanisms in carbon formation in flames, in 14th Symposium (International) on Combustion, The Combustion Institute, Pittsburgh, p.929 (1973).

32. B. L. Wersborg, A. C. Yeung and J. B. Howard, Concentration
     and mass distribution of charged species in sooting flames,
     15th Symposium (International) on Combustion, The Combustion
     Institute, Pittsburgh, p.1439 (1975).
33. J. Lawton, P. J. Mayo and F. J. Weinberg, Electrical control
     of gas flows in combustion processes, Proc. Roy. Soc. A
     303: 275 (1968).

BURNOUT OF SOOT PARTICLES

Jean-Baptiste Donnet and Jacques Lahaye

Université de Haute-Alsace, 61 rue A. Camus
68093 Mulhouse cedex, France   and
Centre de Recherches sur la Physico-Chimie des
Surfaces Solides, 24 av. du Prés. Kennedy
68200 Mulhouse, France

Soot in flames is often a necessary component as it makes possible an efficient heat transfer by radiation. Naturally, it must be subsequently destroyed to avoid pollution. The present paper is focussed on soot oxidation by gaseous reagents.

Oxidation of different varieties of carbon has been examined for centuries. Around 1875 Marcellin Berthelot carried out research work on combustion of carbon. "More recently" (1907), Henry Chatelier wrote a wonderful book "Leçons sur le carbone et la combustion" (Lectures on Carbon and Combustion). In 1965, Letort[1] et al. edited a comprehensive review on oxidation of carbon. These authors emphasized the difficulty to compare results obtained on materials all named carbons but as different as coal, coke, pyrocarbon, graphite etc.

These materials frequently exhibit unexpected behavior. A striking illustration is given in Figure 1 where the rate of oxidation by oxygen at low pressure of non graphitized carbon filament is plotted versus temperature[2]. The presence of a maximum when temperature increases is unusual in heterogeneous kinetics.

An abnormal or unexpected result in scientific investigation often means that the variables used to describe the system are not properly selected. In the cases of combustion of soot in flames, it is particularly important to well define the constituents of the system: carbon and gaseous environment.

259

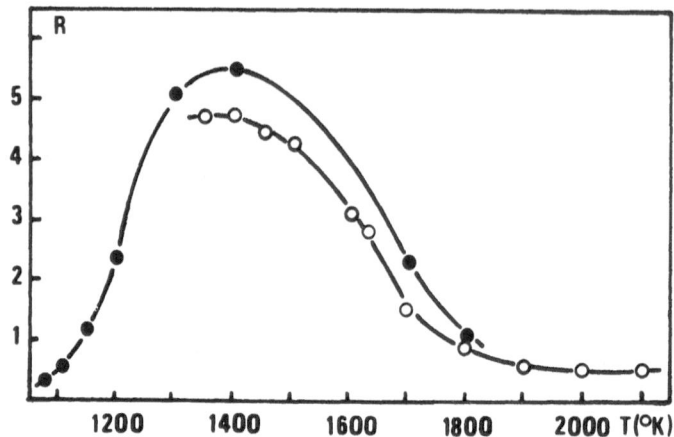

Fig. 1. Reaction rate R (arbitrary unit) of C + O_2 as function
         of temperature (two curves corresponding to different
         pressures and filaments).[1]

OXIDATION OF CARBON IN THERMAL SYSTEM (at low temperatures)

    In a first approximation, and considering reactivity of soot,
we shall consider soot and carbon blacks as equivalent materials
from the standpoints of morphology and internal structure. It has
been shown recently that soot can be significantly different from
carbon blacks.[3]

    Carbon blacks or soot are carbonaceous materials obtained from
the gas phase generally by thermal decomposition or incomplete
combustion of carbonaceous gases. They are made of aggregates of
up to several tens of pseudospherical individual particles. For
particles formed in flames, the diameter of the primary spheres
does not depend much on the conditions of formation, and is usually
in the range 10 to 40 nm. An aggregate is a single entity, made up
of carbon layers which are continuous from one particle to the next.
From the point of view of the final materials the concept of individual spherical particles for most of the blacks appears to be
irrelevant.

    Crystalline structure and morphology of carbonaceous materials
are important parameters to understand their reactivity.

    In graphite, the carbon atoms are located in planes made of
adjacent hexagons (Figure 2). The different planes are parallel
and graphite exhibits a three dimensional structure. Degenerated
structures correspond to lower degree of symmetry. In a turbostratic structure, the stacking of the layers remains, but the
layers are disoriented by arbitrary rotation in their planes.

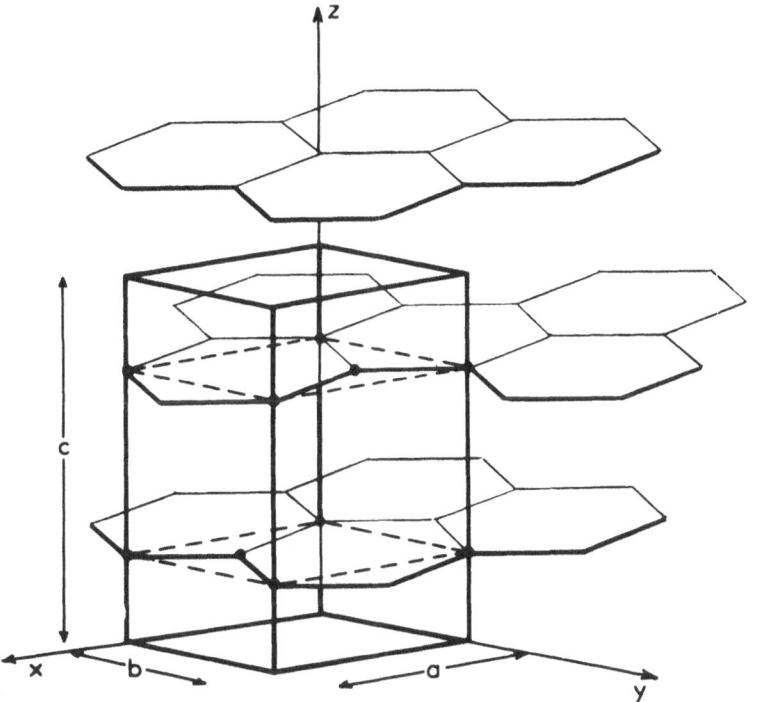

Fig. 2.   Schematic of graphite structure.

Transmission electron microscopy of oxidized carbon blacks[4],
dark field and phase contrast electron microscopy[5,6] led to describe carbon blacks internal structure as made of turbostratic
graphitic carbon layers preferentially oriented tangentially
to particle surface and around centers (of nucleation or crystal-
lisation).

The reactivity of carbon is strongly dependent on structure
as one of us has pointed out earlier in a comprehensive review. For
oxidative agent which do not proceed by intercalation, and not
taking into account the planes or structural defects which are
preferential sites, two types of carbon atoms can be considered:
those located in basal planes and the border atoms located in the
prismatic planes of the graphitic structure. The second ones are
expected to be more reactive than the first ones. A systematic
study of Laine et al.[8], though carried out at low temperature
illustrates the difference of reactivity between both types of
carbon atoms. Graphon, a graphitized carbon black, was oxidized
up to 35 % burn-off. After outgassing of surface oxides, the
sample is oxidized at 300°C during 24 h using an initial oxygen
pressure of 0.5 torr. It appears that k_e/k_b = 2000 where k_e and k_b
represent the reaction rate constants for attack at the edge and

basal plane carbon atoms, respectively. No systematic study has
been performed for other molecular or atomic species. But we
can expect the reactivity of edge carbon atoms to be higher than
those of basal carbon atoms with the consequence that the highest
reactive parts of the material will be preferentially oxidized.
If the reaction temperature is sufficiently high, so that the life
time of surface oxides is very limited, pores can be formed.

A clear example of the difference of reactivity of carbon
atoms is given by J.B. Donnet et al.[4] Furnace and thermal carbon
blacks during oxidation either by liquid oxidant (nitric acid)
or by air can be converted into concentric shells or empty
spheres (Figure 3) due to the difference of reactivity between
carbon atoms in the outermost and innermost layers of particles.

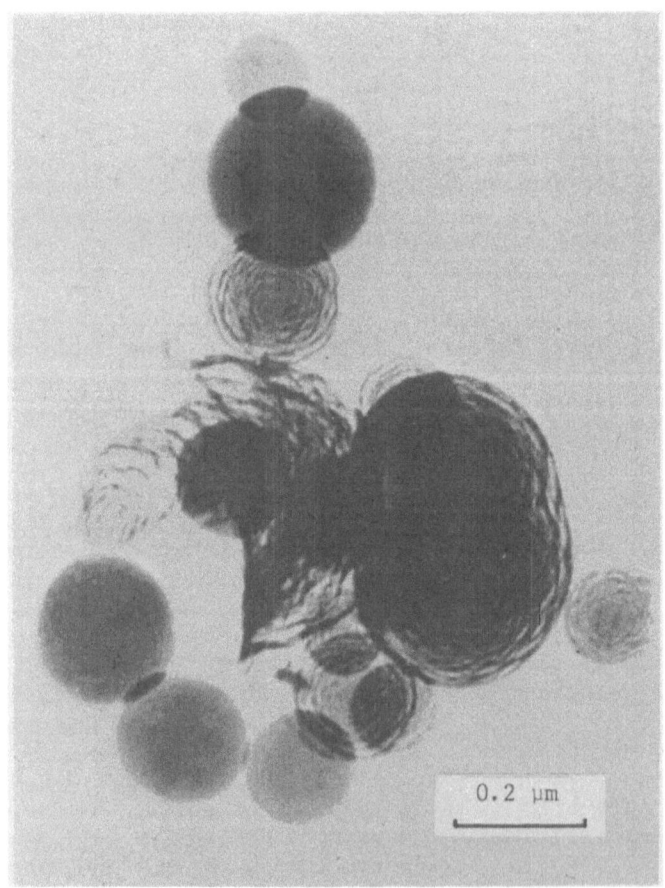

Fig. 3. FT carbon black oxidized with nitric acid.[4]

For very reactive reagents (e.g. excited species) k_e and k_p are both very high. Particles will be progressively used from the outside; no porosity is developed during burn out.

Oxidation of soot or carbon blacks by oxidizing gases, in the absence of flame environment, has been examined. The published results are mainly concerned with molecular[9-12] and atomic[13] oxygen.

In 1962, an important paper on "oxidation of carbon between 1000-2000°C" has been published by J. Nagle and R. F. Strickland-Constable[14]. These authors studied the oxidation by oxygen of artificial carbon, reactor quality graphite and pyrographite at high flow rate and at pressures of about 20 kPa. Above 2000 K the oxidation rate becomes fairly constant. Referring to the theory developped by Blyholder, Binford and Eyring[15], it was assumed that there are two types of sites on the carbon surface, namely A, a more reactive type and B, a less reactive type. Oxidation has been summarized by:

$$A + O_2 \rightarrow A + 2CO$$
$$B + O_2 \rightarrow A + 2CO$$
$$A \rightarrow B \qquad \text{(thermal rearrangement)}$$

Rate equations were developped. Values of the different constants were computed in order to fit experimental results and theoretical expressions. The most significant part of the agreement lies in the correct prediction of the maxima in the rate curves.

The work of Nagle et al. is slightly anterior to the paper of Laine et al.[8]; no hypothesis is made on the nature of A and B sites. It is reasonable to identify A and B sites as edge and basal carbon atoms, the thermal transformation of A into B corresponding to a graphitisation process or annealing.

Surface oxidation rates of two types of carbon black, which are considered to be representative of soot formed during the combustion of hydrocarbon fuels, have been measured by Park and Appleton[16] in a shock tube over the range of temperature of 1700-4000 K and of pressure 5-1300 kPa of oxygen. The results illustrate that the specific surface reaction rate is nearly the same as the one measured for the oxidation of pyrolytic graphite samples and can be approximately correlated by the formula proposed by Nagle and Strickland-Constable.

OXIDATION OF SOOT IN FLAME ENVIRONMENT

The work of Nagle et al. had a major impact on the scientific community so that during several years all papers concerned with soot oxidation, even in flame environment, were referring to Nagle and Strickland-Constable analysis.

Almost simultaneously with the Nagle et al. publication, Lee, Thring and Beér[17] studied the rate of combustion of soot in a laminar soot flame. The change in size of the soot particles (the initial particle size was about 400 Å) was followed by electron micrographs. A semi-empirical rate equation was worked out from the experimental data and gave the specific rate of reaction as a function of the temperature and the partial pressure of the oxygen.

In fact, in a flame, due to high temperature and the presence of oxygen and hydrogen elements, excited species are present at fairly high concentrations. As an example, Figure 4 [18] represents the mole fractions of O, OH and HO_2 for a lean CH_4/O_2 flame (9.5 % CH_4, 90.5 % O_2), burning at 5.3 kPa. The cold gas stream velocity is 67 cm/s. The concentrations of CO_2, CO, O_2 and H_2O in such a flame are one or two orders of magnitude higher than O, OH and HO_2 species[16] but the reactivity of radical species is much higher than those of molecular species.

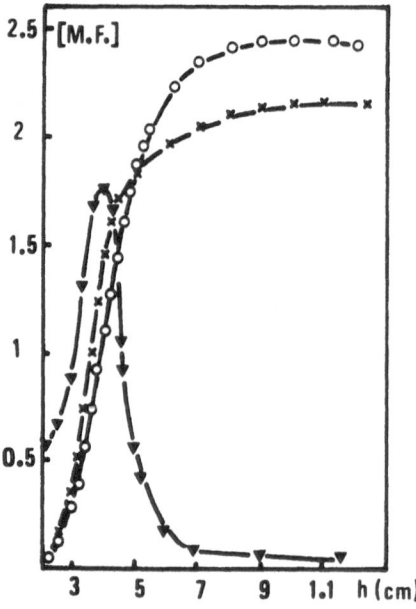

Fig. 4.   Mole fraction [M.F.] O (O), OH (×) and HO_2 (▲) versus distance h from burner in lean methane-oxygen flame (9.5 % CH_4 - 90.5 % O_2). The HO_2 profile should be shifted 0.05 cm to the left.[18]

To our knowledge, it is only in 1967 that OH was proposed as
a main oxidant in a premixed flame.[19]   In the Fenimore and Jones
experiments[19] soot produced by burning a premixed rich mixture of
C_2H_4, O_2 and Ar is burnt in a second burner.   Results showed that
the oxidation rate of soot depends only slightly on molecular oxygen
concentration.

In order to give an acceptable interpretation of their results,
they assume that OH radicals were the main oxidizing agent of soot
in flames.   An experimental evidence of the role of OH radicals
was given in 1980 in the remarkable work of K. Neoh, J. B. Howard
and A. F. Sarofim.[20,21]

In the work of K. Neoh, soot is produced by burning premixed
CH_4 and O_2 with a fuel equivalence ratio of 2.1.   After partial
cooling, soot is passed in a secondary mixing chamber into which
O_2 and N_2 are injected;   the mixture, including combustion primary
gases, is ignited.   The concentrations of N_2, CO, CO_2, H_2 and hydrocarbons in the system are determined by probe and gas chromatography.
H concentrations are measured by atomic adsorption with Li/LiOH
method.[22]   OH and O concentrations are estimated, using partial
equilibrium consideration from H, H_2O and H_2 concentrations in rich
flames and from H, H_2O and O_2 concentrations in lean flames.

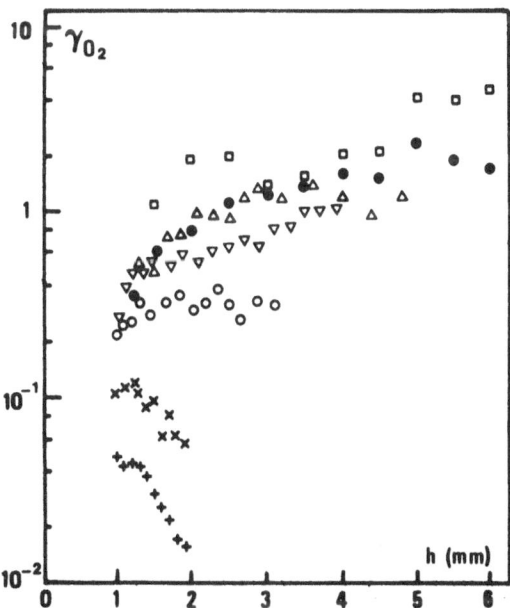

Fig. 5.   Collision efficiency γ_{O2} versus height above burner h
for different fuel equivalence ratios: + 0.85, ×0.95, ○ 1.05
● 1.10, □ 1.15, △ 1.15 (+ CH_4), ▽ 1.15 (+ CH_4 + CO_2).[21]

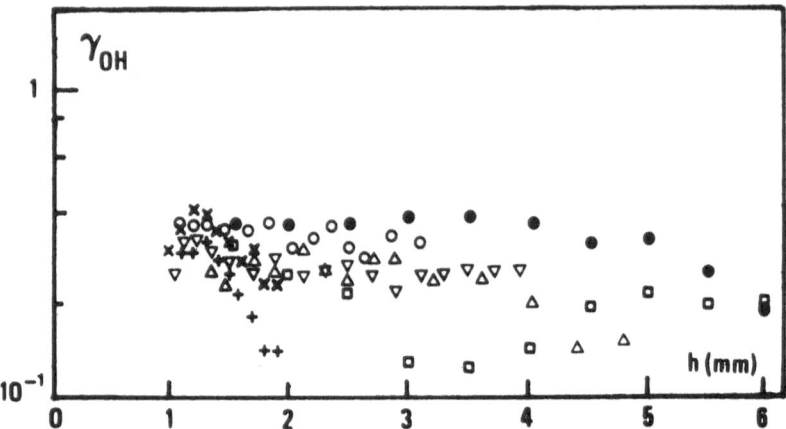

Fig. 6. Collision efficiency γ_{OH} versus height above burner h
for different fueld equivalence ratios: + 0.85, × 0.95,
○ 1.05, ● 1.10, □ 1.15, △ 1.15 (+ CH_4), ▽ 1.15 (+ CH_4
+ CO_2).[21]

The soot burn-out rate is mainly followed by measuring the
soot size and concentration using optical methods similar to those
of D'Alessio et al.[23] and J. Jagoda et al.[24]

In order to determine which species (H_2O, CO_2, O, O_2 or OH)
is the main oxidant, the collision efficiency γ_i, required for
each species, acting alone to account for the observed burnout
rate, was computed (γ_i is defined as the probability of removal
of a carbon atom for one collision by an oxidant species).

It is shown that the role of H_2O, CO_2 and O can be disregarded
because of the absolute value of γ_i ($\gamma_i > 1$) and/or their variation
with height above burner is incompatible with the variation of temperature in the flame.

The values of γ_{O_2} (Figure 5)[21] are increasing over two orders
of magnitude from the leanest to the richest flame: the values
obtained for the two leanest flames are not unrealistic but the
high values of γ_{O_2} in rich flames indicated that O_2 acting alone
could not account for the observed burnout rates. On the contrary,
γ_{OH} (Figure 6)[21] variation with equivalence ratio and position in
flame is fairly small; the average value is close to 0.28 when
calculated on the basis of an optical diameter. The authors are
brought to the reasonable conclusion that OH radicals are the main
oxidizing agent while molecular oxygen can be neglected except for
fuel lean conditions.

Let us finally mention an important observation of K. G. Neoh[21]
(Figure 7). In lean flames, the number density of the soot was

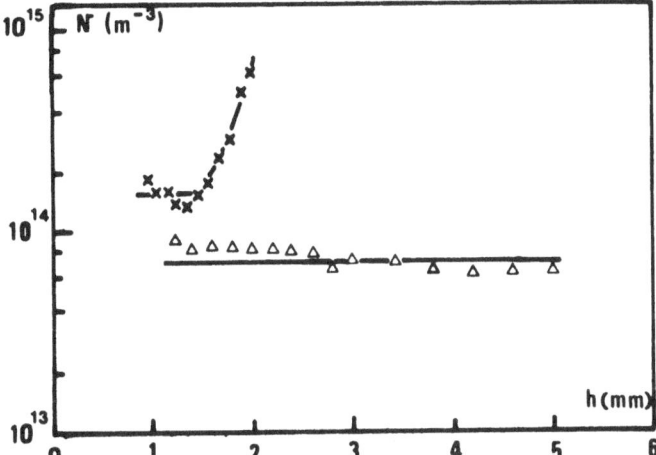

Fig. 7. Profiles of number concentration N for fuel equivalence
         ratio of 0.95 (**x**) and 1.15 (+ CH4) (**Δ**) versus height
         above burner h.[21]

observed to increase sharply with height above burner while no
variation was observed for rich flames.

DISCUSSION AND CONCLUSION

We have shown, 20 years ago,[4] that oxygen is able to oxidize
the core of carbon black particles and, therefore, to break
aggregates. In lean flame where molecular oxygen is a significant
agent of oxidation, increase of particle number density (Figure 7)
corresponds to breaking of aggregates. The collision efficiency
of O_2 is sufficiently low so that oxygen can diffuse into particles
and undergoes internal oxidation.

In rich fuel flames, the OH radical is the main oxidizing
species and its reactivity is high. OH species react with carbon
of the soot surface; they do not diffuse into particles: aggregates are not broken during oxidation.

According to Neoh's computation, the collision efficiency of
OH species γ_{OH} lies between c.a. 0.1 to 0.3. It can be assumed
that only the border carbon atoms are reactive even if that concept
is not totally clear in highly turbostratic graphitic materials
like soot. A crude approximation of the proportion N_b of border
atoms with respect to the total number of carbon atoms has been
given by Studebacker[25]

$$N_b = \frac{4}{L_a}$$

where L_a (in $\overset{\circ}{A}$) is the length of coherence of the graphitic structure in the direction parallel to graphitic planes.

For commercial furnace black L_a lies between 12 and 20 $\overset{\circ}{A}$ [25] and therefore N_b goes from 0.33 to 0.20.

The comparable order of magnitude of γ_{OH} and N_b may be fortuitous, and a systematic investigation might clarify the relation between γ_{OH} and crystalline structure of soot.

The collision efficiency of reactive species γ_i can change during the oxidation process of soot. A reagent capable of oxidizing the core of the particle and to develop surface area of the material may present a large increase of γ_i during oxidation process though it is unlikely that γ_i may change over one or two orders of magnitude.

It appears that a better understanding of soot oxidation and, more generally, carbonaceous particulates (e.g. coal) will be achieved by taking into consideration not only flame data and size or surface area of the solid but also the variation of internal structure and reactivity of carbon during burn-out.

REFERENCES

1. M. Letort, L. Bonnetain, G. Hoynant and H. Guerin Les carbones, A. Pacault ed., Mason et Cie, 234 (1965)
2. F. Boulangier, X. Duval et M. Letort, Proceed. 3rd Conf. on Carbon (Pergamon Press), 257 (1959)
3. A. I. Medalia and D. Rivin, Ext. Abst. 15th Biennial Conf. on Carbon, 480 (1981)
4. J. B. Donnet and J. C. Bouland, Rev. Gen. Caout. 41(3):407 (1964)
5. A. Oberlin, Carbon 17:7 (1964)
6. R. D. Heidenreich, W. M. Hess and L. L. Ban, J. Appl. Cryst. 1:1 (1968)
7. J. B. Donnet, George Skakel Memorial Award Conference, Carbon Conference, Philadelphia, June 1981, to be published.
8. N. R. Laine, F. J. Vastola and P.L. Walker, Jr., J. Phys. Chem. 67:2030 (1963)
9. C. W. Snow, D. R. Wallace, L. L. Lyon and G. R. Crocker, Ext. Abst. 3rd Biennial Conf. on Carbon, 279 (1959)
10. C. W. Snow, D. R. Wallace, L. L. Lyon and G. R. Crocker, Ext. Abst. 4th Biennial Conf. on Carbon, 79 (1961)
11. H. L. Riley, Fuel 24:8 (1945)
12. J. D. Watt and R. E. Franklin, Nature 180:1190 (1957)
13. J. D. Blackwood and F. K. Mac Taggart, Austral. J. Chem. 12:114 (1959)

14. J. Nagle and R.F. Strickland-Constable, Ext. Abst. 5th
     Conf. on Carbon, 1:154 (1963)

15. G. Blyholder, J. S. Binford and H. Eyring, J. Phys. Chem.
     62:263 (1958)

16. C. Park and J. P. Appleton, Comb. and Flame 20:369
     (1973)

17. K. B. Lee, M. W. Thring and J. M. Beér, Comb. and
     Flame 6:137 (1962)

18. J. Peeters and G. Mahnem, 14th Symp. (Int'l) on Combustion,
     The Combustion Institute, p.133 (1973)

19. C. P. Fenimore and G. W. Jones, J. Phys. Chem. 71:593
     (1967)

20. K. G. Neoh, "Soot Burnout in Flames", Sc.D. Thesis,
     Department of Chemical Engineering, Massachusetts
     Institute of Technology (October 1980)

21. K. G. Neoh, J. B. Howard and A. F. Sarofin, Soot Oxidation
     in Flames, in: "Particulate Carbon: Formation During
     Combustion", D. C. Siegla and G. W. Smith, Eds,
     Plenum Press, New York (1981)

22. M. J. Mc Evan and L. F. Phillips, Comb. and Flame
     9:420 (1965)

23. A. D'Alessio, A. Dilorenzo, A. F. Sarofim, F. Beretta,
     S. Masi and C. Venitozzi, 15th Symp. (Int'l) on
     Combustion, The Combustion Institute, p.1427 (1975)

24. J. Jagoda, G. Prado and J. Lahaye, Comb. and Flame
     37(3):251 (1980)

25. M. L. Studebaker, Rubber Chem. and Technol. 30:1400 (1957)

DISCUSSION

A. Feugier (Institut Français du Pétrole)

In diesel engines, combustion of soot is not rapid enough, and
it would be useful to add some suitable catalyst, which can be
incorporated in the diesel fuel or which can be impregnated into
the trap structure located in the exhaust pipe. For example, it
is known that lead increases drastically the oxidation of graphite.
Were they any similar experiments with soot particles?

Donnet

The effect of metal additives on soot emission from a flame is
not only dependent on the nature of the metal but also on the
location of introduction into the flame.

As far as I know there is no clear information on the role
of metals or metallic compounds on soot emission in diesel engines.

P. Cadman (The University College of Wales)

    Could Prof. Donnet comment on the reason for the maximum in the slide of the oxidation rate versus temperature shown at about 1600-1700 K. Certainly, we have evidence from an X-photoelectron spectroscopy study of O_2 and O on the basal plane of single crystal graphite that surface oxides have been desorbed (> 99%) by 1200 K so that this process cannot be the main reason for the 1700 K maximum.

Donnet

    During carbon oxidation at high temperature there is a competition between the oxidation itself and graphitization of the carbonaceous material. At high temperature, oxidation being diffusion limited, its rate increases moderately with temperature, whereas the rate of graphitization of carbon becomes quite significant. As graphitized carbons are much less reactive than ungraphitized materials (except when oxidation proceeds through intercalation) one can expect a maximum of the global rate.

P. Cadman

    Studies at Aberystwyth using X-ray photoelectron spectroscopy on the rate of attack of O_2 and O on the basal planes of single graphite has suggested that O atoms can pull out carbon atoms from the basal plane at relatively low temperatures. This then creates defects which can then expand and act like the prismatic edge, with its enhance reactivity. As O_2 at high temperatures contains increasing equilibrium amounts of O atoms it is difficult to separate out the reactivity of these species in temperature studies. In addition, the large difference in reactivity quoted for reaction of O_2 with the prismatic edge compared with the basal plane is probably not true at the higher temperatures, when O_2 contains an increasing proportion of the very reactive species, oxygen atoms?

Donnet

    You are right, oxygen (atomic or molecular) may act as an "activator" of carbon material. Do not forget, however, the opposite effect of annealing by graphitization.

K. H. Homann (Technische Hochschule, Darmstadt)

    Is there experimental evidence that the density in the middle of a soot spherule is less than in the surface-near layers? Can one assume that small spherules have a smaller density than large ones?

Donnet

The core of the particle is generally more sensitive to oxidation than the surface-near layers. As mentionned earlier in the
discussion, oxidation of carbon is strongly dependent on the degree
of graphitization of the material and, therefore, on its density;
reactivity is an indirect proof of the low density of the core of
soot particles.

The organized carbon layers are more easily built when the
curvature radius of the particle is high. For structural reasons
large particles are expected to have higher density than small
ones. This is experimentally verified.

D. Rivin (Cabot Corporation)

I suggest that the microstructure of carbon produced by high
temperature pyrolysis or flames differ in some important respects
from the simple microcrystal model. The particles are composed of
large, overlapping turbostratic layers oriented parallel to the
surface. These layers have many basal plane defects such that the
X-ray diameter (L_a) and surface group concentration indicate much
smaller planes layers. Aromatic layer size and order decrease in
going from the surface to the interior of the particle.

Donnet

It is, indeed, another way of answering Prof. Homann's question.
Anyway, I do not like much the concept of microcrystals whatever
the temperature of carbon formation. It was convenient some 15
years ago but, as suggested by X-ray analysis, and demonstrated by
phase contrast electron microscopy, it is not relevant to carbon
black crystalline structure.

D.W. Smith (General Motors Research Laboratories)

From a practical point of view studies of catalytic burnout,
either via inclusion of catalytic species in fuel or via substrates,
would be useful.

Donnet

Yes.

AEROTHERMOCHEMISTRY OF DIFFUSION FLAMES

Marcel Barrère

Office National d'Etudes et de Recherches
Aérospatiales, 92320 Châtillon, France

INTRODUCTION

Soot formation in combustion zones concerns many scientific
and technical fields, e.g.:
- flame pollution
- increase, in some applications, of the flame radiation due
  to soot formation,
- utilization of soot in some fabrications ("carbon black").

Among the important parameters governing this formation, we
should mention the temperature level, the adequate composition and
the residence time necessary for not limiting the production in
the reactor. For instance, in a premixed flame the temperature is
a function of equivalence ratio, being maximum around the stoichiometric value, and soot formation increases as the mixture is more
fuel-rich, so there exists an equivalent ratio for an optimum
production, compromise between temperature and composition[1]
(Fig. 1).

Soot production may be noticeably increased by using heterogeneous flames where stoichiometric, hence high temperature masses
coexist with masses with a large excess of fuel, i.e. carbon-rich.
Diffusion flames constitute a first configuration of these heterogeneous flames, hence the importance of the flow organization in
this type of combustion, object of this paper.

The structure of diffusion flames and organization of the flow
depend on the physical state of the fuel and oxidiser; that is why
we propose the following classification of diffusion flames:

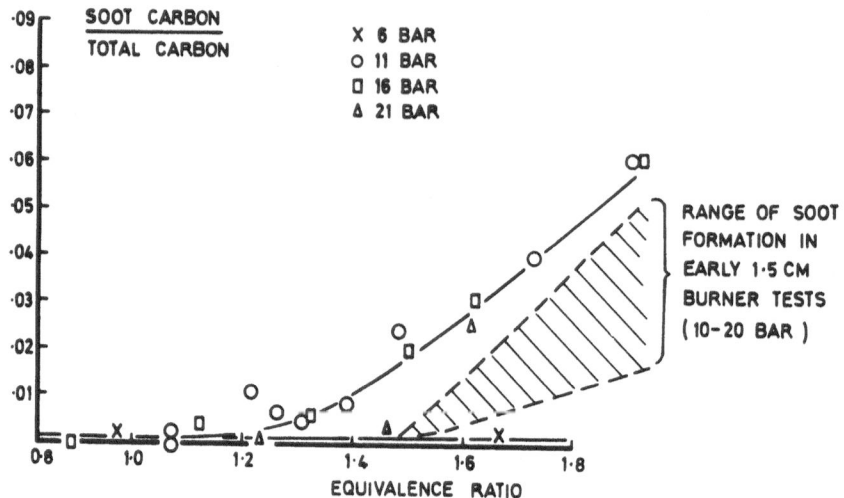

Fig. 1. Soot formation premixed kerosine vapour/air.[1]

- gas fuel and gas oxidiser: parallel flow with plane or cylindrical
  symmetry, laminar or turbulent; applications: torchs, combustors,
  etc...
- liquid fuel and gas oxidiser: plane and spherical symmetry,
  combustion of drops and spray; applications: combustors, reactors,
  fires, etc...
- solid fuel and gas oxidiser: plane, cylindrical, spherical
  symmetry; applications: combustion of plastics, coal, hybrid
  propulsion, etc...
- solid fuel and liquid oxidiser: cylindrical symmetry; applications:
  hybrid propulsion, etc...
- solid fuel and solid oxidiser: complex geometry; applications:
  solid propellant propulsion, etc...

     Thus, there exist a broad variety of possibilities and we
shall limit this paper to gaseous systems: diffusion flames with
plane or cylindrical symmetry, the flow being laminar or turbulent,
combustion of drops, and investigation of the multiphase flows
often encountered in this type of problem.

DIFFUSION FLAME WITH PLANE OR CYLINDRICAL SYMMETRY IN LAMINAR FLOW

     The simplest pattern of this configuration is the plane,
two-dimensional problem represented on Figure 2: two parallel
flows, one of fuel and one of oxidiser, separated by a wall; at
the end of the wall the two flows mix and, after initiation, the
combustion develops along the flow. Equivalence ratio ϕ passes
from zero, pure oxidiser, to infinity, pure fuel, the flame position
being located near the stoichiometric ratio ($\phi = 1$). The luminous

part of the flame appears near the stoichiometric line, in the
fuel-rich part ($\phi > 1$). Temperature is maximum close to $\phi = 1$.
The burned products diffuse on either side of the flame, while fuel
and oxidiser diffuse toward the flame.

Mass and energy transport take place perpendicular to the flow.
The zone of chemical reaction, located around $\phi = 1$, is usually
thin relative to the thickness concerning mass and heat transport
phenomena , and that is why the diffusion phenomenon is the main
phenomenon governing the flame. Thus, in a general way, three
main phenomena intervene in this type of flame:
  i. convection, corresponding to the gas injection velocity u,
     with a characteristic time $t_c \simeq (\ell^*/u)$
 ii. species diffusion, with a characteristic time $t_d \simeq (d^{*2}/D)$
     with a diffusion coefficient D;
iii. chemical reaction, with a characteristic time $t_h \simeq (\rho/\dot{\omega})$,
     where ρ is the density of the mixture and $\dot{\omega}$ the mass production
     of the important chemical species; ℓ^* and d^* are characteristic
     lengths along the direction of the evolutions.

The comparison of these times (Damköhler numbers) makes it
possible to assess the pilot processes; in this type of flame, it
is usually small relative to t_c and t_d, which in particular
expresses the importance of diffusion and aerodynamics.

This plane configuration is the most classical, but we find
in practice cylindrical configurations; the burner used in them
is similar to that represented on Figure 3, due to Mitchell,
Sarofim and Clomburg,[2] which is well studied from the aerodynamic
viewpoint and allows a simple theoretical model, also given on
Figure 3.

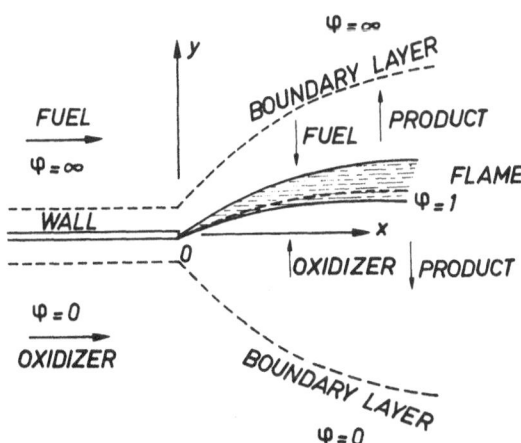

Fig. 2.   Structure of a diffusion flame for a bidimensional
          configuration.

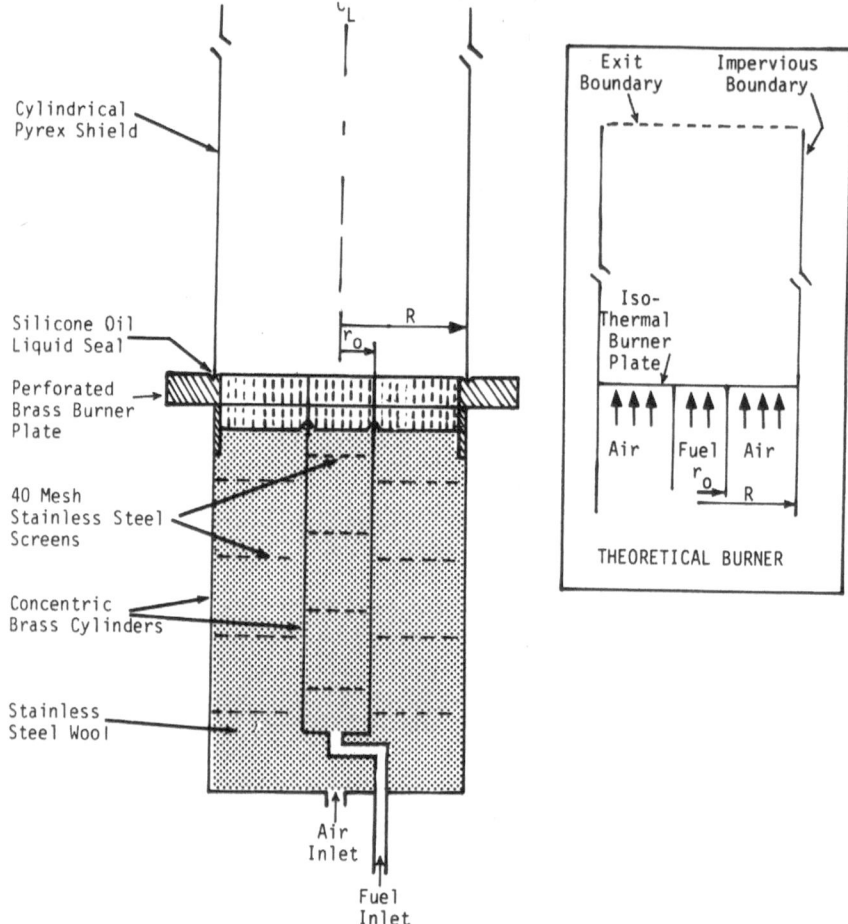

Fig. 3. Schematic of laboratory burner.[2]

The theory of flames of this type is very similar to the
analysis of two-dimensional boundary layers. In an $|x,y|$ plane
such as that defined on Figure 2, the balance equations in
permanent regime write:

Mass balance

$$\frac{\partial \rho u}{\partial x} + \frac{1}{y^k}\frac{\partial (\rho v y^k)}{\partial y}$$

u and v being the velocities along x and y, k depending on the
symmetry (k = 0 for the plane, k = 1 in a cylindrical configuration).

Species balance

$$\rho u \frac{\partial Y_i}{\partial x} + \rho v \frac{\partial Y_i}{\partial x} = \frac{1}{y^k} \frac{\partial}{\partial y} (y^k \cdot g) + \dot{\omega}_i$$

        convection   diffusion   chemical
                                             production

Y_i is the mass fraction of species i, g the diffusion flux along y, i.e. into the gradient zone, and $\dot{\omega}_i$ the chemical production of species i.

Momentum balance

$$\rho u \frac{\partial u}{\partial x} + \rho v \frac{\partial u}{\partial y} = \frac{1}{y^k} \frac{\partial}{\partial y} (y^k \cdot \tau)$$

τ is the component of the viscosity tensor; here again, we admit a component along y, in a direction of strong gradient.

Energy balance

$$\rho u \frac{\partial h}{\partial x} + \rho v \frac{\partial h}{\partial y} = \frac{1}{y^k} \frac{\partial}{\partial y} y^k q - \sum_{i=1}^{i=N} h_i^o \dot{\omega}_i$$

q corresponds to the heat flux, $h = \int_{T_o}^{T} C_p \, dT$ is the sensitive enthalpy, h_i^o the heat of formation of species i, $\dot{\omega}_i$ the production of species i.

As a conclusion, we take into account in these balance equations a two-dimensional convection, and mass, momentum and energy transfers along y, which is the direction of important gradients, neglecting the diffusion along x.

The boundary conditions are as follows:
- at y = ∞ the velocity component u tends towards zero as well as the mass fractions, and
- at ordinate y = 0 the gradients are zero $\partial u/\partial y = 0$ and u = 0 if the fuel is solid.

Composition should be defined by two variables, if we admit a simple chemical reaction of the form

$$F + SO \rightarrow (1 + S)P$$

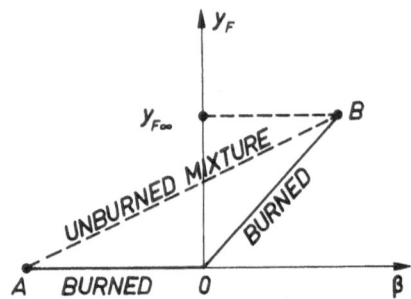

Fig. 4. Definition of the composition.

We may take Y_F and Y_O for instance, i.e. the mass fractions
of fuel and oxidiser; S is the stoichiometric proportion.

In the case of fast combustion, i.e. of a thin flame, we
obtain the variation of compositions in the $[Y_F, \beta]$ plane, with
$\beta = Y_F - (Y_B/S)$ as indicated in Figure 4. Let us place the
flame in $O\{Y_F = 0, Y_O = 0,$ hence $\beta = 0\}$. On the oxidiser side
$\beta \le 0$, as $Y_F = 0$ and $\beta = - (Y_O/S)$; in particular, in A there is
only pure oxidiser; on the fuel side $\beta \ge 0$, $Y_O = 0$ and $\beta = Y_F$;
in particular, point B corresponds to pure fuel $Y_{F\infty}$. Segments
AO and OB correspond to burned products, and AB is a mixture
of unburned fuel and oxidiser. The local composition is located
within the AOB triangle. The variables chosen to give the
composition vary with authors and with the analysis methods used,
but can be expressed from those defined above, (Y_F, Y_O) or (Y_f, β).

For instance Yarin[3] uses an integral method similar to that
of the boundary layer, defined by the composition of $Y = \overline{Y_F} - \overline{Y_O}$
and $\Delta Y = Y + 1$ with $\overline{Y_F} = (Y_F/Y_O)S$ and $\overline{y}_O = (Y_O/Y_{O\infty})$. Using the
self similar constants a, b, c, and the constants A, B, C determined
by particular integrals, the velocity profile is given, in the
case of a fuel jet, by:

$$\frac{u}{u_m} = F'(\xi)$$

with $u_m = Ax^a$ and $\xi = Byx^b$. The concentration profile outside
the flame is given by:

$$\frac{\Delta Y}{\Delta Y_m} = \pi(\xi)$$

with $\Delta Y_m = Cx^c$.

For an axisymmetric flame we obtain:

$$F'(\xi) = 1 - th^2\xi$$

$$\pi(\xi) = (1 - tg^2\xi)^{Sc}$$

Sc being the Schmidt number,

Within the jet, the oxygen concentration is nil, $Y_0 = 0$, and

$$\frac{\overline{Y}_F + 1}{\overline{Y}_{Fm} + 1} = (\xi)$$

The flame length corresponds to $x = \ell_f$ and $\overline{Y}_F = 0$; hence, $C\ell_f = 1$ and $\overline{x}_f = x_f/\ell_f$ is given by:

$$\overline{x}_f^{-c} \pi(\xi_f) = 1$$

If the Lewis number is equal to 1, the temperature and concentration profiles are identical, hence:

$$\frac{T - T_f}{T_m - T_f} = \frac{\overline{Y}_F}{\overline{Y}_{F_m}} = \frac{\pi(\xi) - \pi(\xi_f)}{1 - \pi(\xi_f)}$$

and

$$\overline{Y}_0 = 1 - \frac{\pi(\xi)}{\pi(\xi_f)}$$

This is the classical method to approach this problem but there exist others. Mitchell[2], for instance, utilizes a finite element numerical method: the hypotheses are as follows. The chemistry is still as simple and the production of each species $\dot{\omega}_F$, $\dot{\omega}_0$, $\dot{\omega}_P$ is related by stoichiometry:

$$\dot{\omega}_F = \frac{\dot{\omega}_0}{S} = -\frac{1}{1 + S} \dot{\omega}_P$$

so that species conservation becomes

$$\underline{\nabla} \cdot (\underline{G}\beta - \rho D\underline{\nabla}\beta) = 0$$

with still $\beta = Y_F - (Y_0/S)$; vector $\underline{G}$ corresponds to the global mass balance equation $\underline{\nabla} \cdot \underline{G} = 0$, hence to unit flowrate, D is a mean diffusion coefficient: $D_{F,mixture} = D_{0,mixture} = D$ if we

assume that the coefficient of diffusion of F with the mixture is
equal to the one of the oxidiser with mixture.

The combustion is that of methane with oxygen according to
the reaction:

$$CH_4 + iO_2 = aCO + (1 - a)CO_2 + (2i - 2 + a)H_2O$$

$$+ (4 - 2i - a)H_2$$

The momentum and energy balances become:

$$\begin{cases} \underline{\nabla} \cdot (G\underline{v}_r + \underline{\tau}_r) + \dfrac{\partial p}{\partial r} = 0 \\[2ex] \underline{\nabla} \cdot (G\underline{v}_z + \underline{\tau}_z) + \dfrac{\partial p}{\partial z} = 0 \end{cases}$$

and

$$C_p \, \underline{\nabla} \cdot (\underline{G}\, T) - \underline{\nabla} \cdot (\lambda \underline{\nabla} T) - \rho \sum_{L=1}^{N} D_{i,mel} \, C_{p_i} (\underline{\nabla} T \cdot \underline{\nabla} Y i) = - \sum_{L=1}^{N} h_i^o \dot{\omega}_i$$

$\underline{\tau}_r$ and $\underline{\tau}_z$ being the components of the viscosity tensor, and h_i^o
the heat of formation of species i.

The solution is more easily obtained if we introduce the
rotational ω and the stream function ψ:

$$\omega = \frac{\partial}{\partial z} \left(\frac{Gr}{p}\right) - \frac{\partial}{\partial r} \left(\frac{Gz}{r}\right)$$

$$rGr = - \frac{\partial \psi}{\partial z} \; , \; rGz = \frac{\partial \psi}{\partial r}$$

The reaction is assumed stoichiometric at the fuel-oxidiser
interface, and departs from stoichiometry on either side.

The comparison between theory and experiment is satisfactory;
this comparison will anyway allow us to describe the structure of
a laminar diffusion flame. Figure 5 gives the temperature, velocity
and concentration profiles [CH_4, H_2, CO, CO_2, H_2O, N_2] from the
axis of symmetry of the burner, in a plane perpendicular to
the axis, at 2.4 cm above the exhaust section. We can see that
the flame is wider than predicted by the theory: this is due to
the assumptions made; the velocity is almost correctly restituted;
the luminous zone is rather close to the maximum of temperature.
We also notice that the fuel-rich zone is at a high temperature,
sufficient to promote soot formation, oxygen is practically
consumed near the luminous zone. The flame geometry and the flame

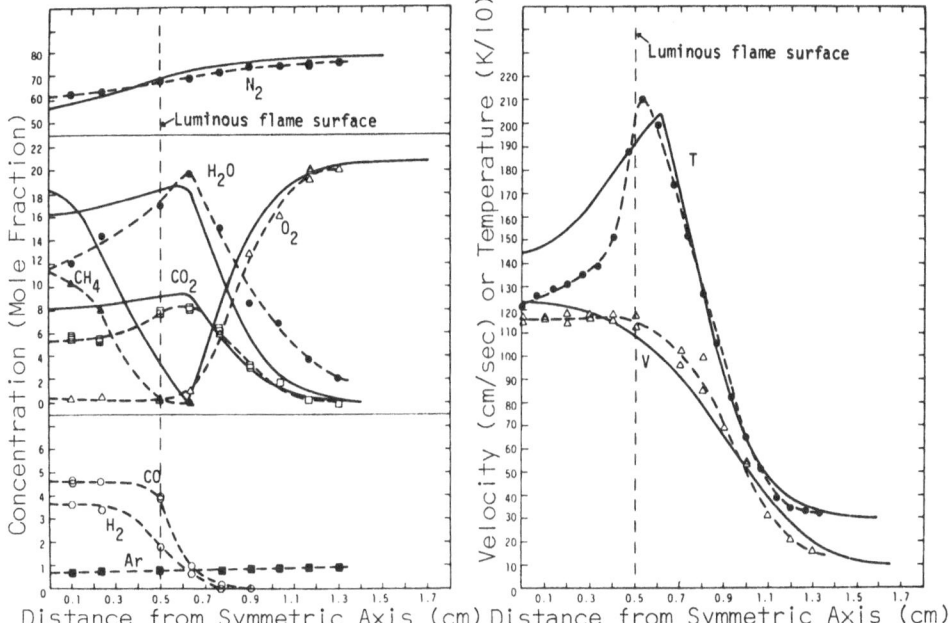

Fig. 5.   Concentration, temperature, and velocity profiles established 2.4 cm above the burner plate. Solid lines denote
theoretical profiles. Dashed lines denote curve drawn
through data points. (■ Ar; ▲ CH$_4$,○CO; □ CO$_2$; ◇ H$_2$;●H$_2$O;
◆N$_2$, △O$_2$).[2]

height are correctly calculated if we take a stoichiometric
coefficient i = 1.76 instead of 2 of the complete reaction, the
flame position being difficult to determine by optical methods
(Fig. 6). This height depends on the incoming velocity of the
fuel, hence, on the corresponding Reynolds number, and the
height, as shown on Figure 7, is proportional to the corresponding
Reynolds number.

    Two figures seem to be of interest: the variation of
temperature and composition as a function of equivalence ratio
(ϕ = ∞ corresponds to fuel, ϕ = 1 to stoichiometric ratio, and
ϕ = 0 to pure oxidiser), drawn on Figure 8, and the spatial
distribution of equivalence ratio (Fig. 9). These two figures
giving the distribution of fuel-rich zones and temperatures,
are at the basis of any determination of the formation of carbon
particles, by superposing to the energetic problem that of the
kinetics of particle formation.

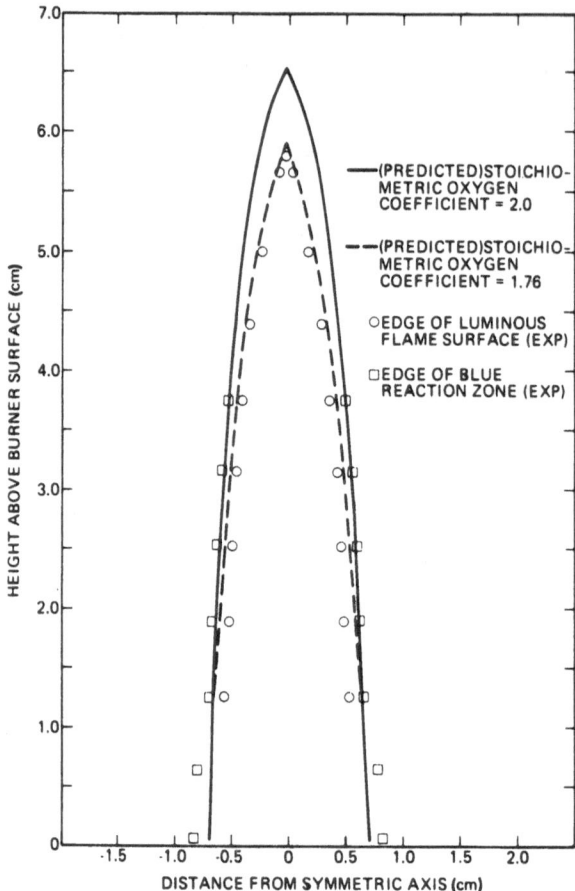

Fig. 6. Flame profile and height of the flame.[2]

We would like, at the occasion of this analysis, to note the precautions taken to design this burner.[2] (Fig. 3)

- the care taken to obtain in the ducts an unperturbed flow of fuel and oxidiser (presence of screens);
- the devices placed at the end of the burner to obtain uniform velocities and to reduce the importance of the boundary layers;
- the cylinder placed around the flame to guide the flow and obtain a confined flame with a well defined boundary conditions.

A major drawback in the study of this type of flame, but which intervenes less in the problem here investigated, is derived from the Lewis number, which may be different from unity.

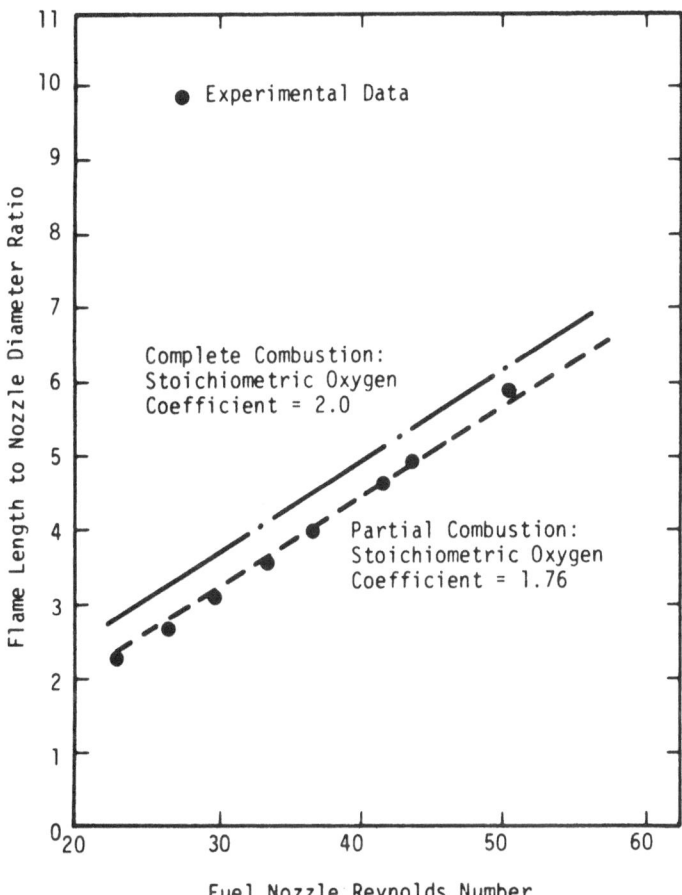

Fig. 7.   Influence of nozzle-flow rate on the length of laminar
          methane-air diffusion flame.[2]

        Allison and Clarke[4] studied for instance the combustion by
diffusion of the hydrogen-nitrogen mixture with the oxygen-
nitrogen mixture. The theoretical part rests on an asymptotic
method, the chemistry being represented by 6 reactions and the
diffusion coefficients by:

$$D_i = \sum_{i=1}^{N} \frac{Y_i}{m_i} \Big/ \sum_{L=1}^{N} \frac{Y_i}{m_i D_{ij}}$$

where m_i is the molar mass of species i and D_{ij} a binary
diffusion coefficient. The asymptotic development concerns
the parameter ε, inverse of the Damköhler number D:

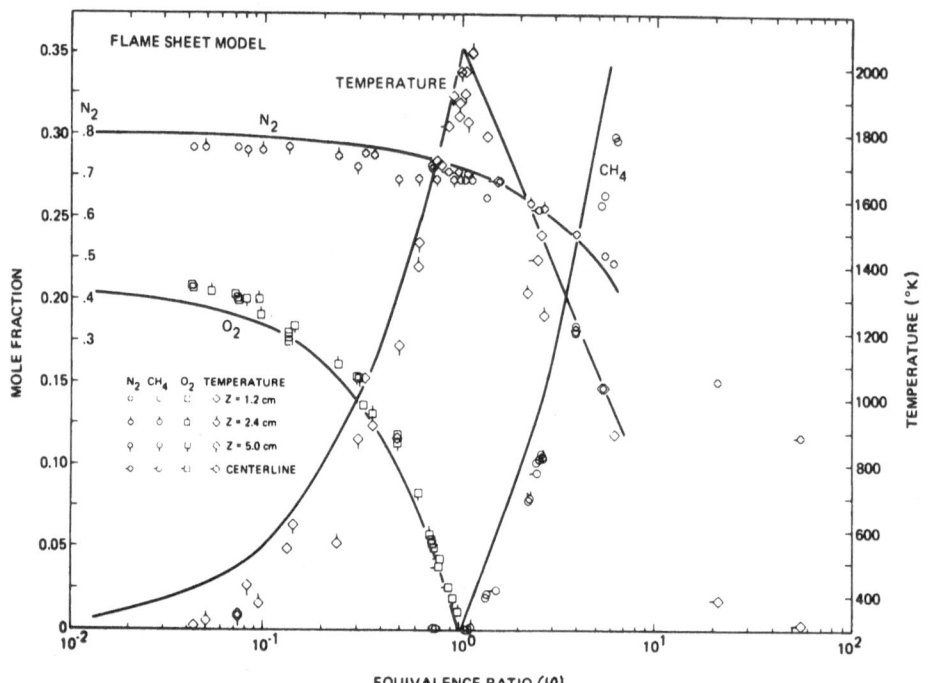

Fig. 8. Comparison of predicted and experimentally-obtained
         concentrations and temperatures as a function of the
         local-equivalence ratio.[2]

$$\varepsilon = D^{-1} = \frac{u_c}{S_c L \Omega_c}$$

being the velocity in the combustion zone, S_c the Schmidt
number, L a length and Ω_c a frequency characteristic of combustion, inverse of a chemical time.

This complex analysis shows in this case a disagreement
between theory and experiment, as revealed by Figures 10 and 11,
but constitutes a first approach to the general problem.

LAMINAR DIFFUSION FLAME WITH SPHERICAL SYMMETRY

The results obtained in spherical symmetry are rather similar
to those just described for a plane configuration. This symmetry
is often encountered in practice in the combustion of liquid fuels
(drops) or of solids with spherical symmetry. Combustion schematics
are given on Figure 12 for the combustion of a fuel drop. The
diffusion flame surrounds the drop and the heat transfer vaporizes
the drop. The chemical reaction of combustion, always assumed to
be fast, takes place over a very thin sphere and the burned products
diffuse on either side of this sphere.

We only meet in the gas part a mixture of burned products and
fuel between drop and flame, and of burned products and oxidiser
outside the flame; burned products and diluting gases (nitrogen)
may diffuse up to the drop surface. Figure 13 gives the experimental
composition on either side of the flame, for a porous sphere fed
with heptane; the oxygen is practically burnt within the flame and
does not diffuse into the fuel zone.

There exist many theories of this complex problem, but the
simplest analyses lead to realistic results. Chemistry is still
as simple: reaction of the second order of the type:

A + O → P

with a production:

$$\dot{\omega}_A = - k_A \rho^2 Y_A Y_o \exp(- \frac{T_A}{T})$$

The balance equations (mass, species, energy) are of the
same type, with a complication arising from the spherical
coordinates, but the flow velocity is coaxial with the diffusion
velocity.

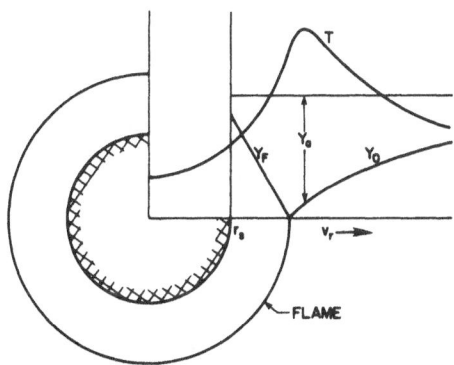

Fig. 12. Sketch of the diffusion flame model for drops.

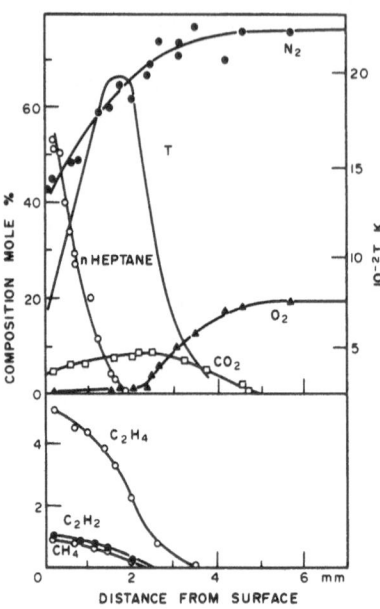

Fig. 13. Temperature and concentration profiles at the lower
          stagnation point of a 9.2 mm diameter porous sphere,
          burning n-heptane in still air.

- We usually admit that the phenomenon is permanent (constant
  drop radius), as the time characteristic of drop regression
  is great relative to those characteristic of transport
  (mass, heat) and chemical phenomena.
- The Lewis number is equal to unity.
- Natural convection is neglected and we maintain the
  spherical symmetry.
- Forced convection intervenes in the regression laws obtained
  through the Nusselt number $N_u = h_c D/\lambda$, where h_c is the
  convection coefficient, D the drop diameter, and λ the mean
  conduction coefficient within the gas, e.g.

$$N_u = 2\left[1 + f(R_e, P_r)\right]$$

where the Reynolds number is relative to the drop and P_r
is the Prandtl number.

The mass and energy balance equations involve global chemical
production $\dot{\omega}$ such that

$$\dot{\omega}_j = \pm Y_j^* \dot{\omega}$$

the sign + corresponding to the products formed and the sign −
to the reactants.

We may define a reduced temperature T such that:

$$T^* = - \frac{1}{Cp} \sum_j h_j^o \dot{\omega}_j$$

being a mean specific heat.

The thin flame hypothesis leads to the following temperature evolution betweeen the drop radius r_s and the flame radius r_f:

$$T = T_s + \frac{\Delta l}{Cp} \{exp \frac{\mathring{m}_F Cp}{4\pi\lambda} (\frac{1}{r_s} - \frac{1}{r}) - 1\}$$

being the temperature on the drop surface and $\mathring{m}_F$ the fuel mass flow rate, Δl the heat necessary to vaporize the liquid fuel, with:

$$\mathring{m}_F = \frac{4\pi\lambda r_s}{Cp} \ln \left\{ 1 + \frac{CpT^*}{\Delta l} \left[\frac{T_\infty - T_S}{T^*} + \frac{Y_{o,\infty}}{Y_o^*} \right] \right\}$$

The fuel evolution is:

$$Y_F = 1 - Y_p = 1 - (1 - Y_{F,S}) \; exp \frac{\mathring{m}_F Cp}{4\pi\lambda} (\frac{1}{r_s} - \frac{1}{r})$$

and for r between r_f and infinity:

$$T = T_f - \left[T^*(\frac{1 + Y_o^*}{Y_F^*}) - (T_f - T_s) - \frac{\Delta l}{Cp} \right] \; exp \left\{ \left[\frac{\mathring{m}_F Cp}{4\pi} (\frac{1}{r_f} - \frac{1}{r}) \right] - 1 \right\}$$

and

$$Y_o = 1 - Y_p = \frac{Y_o^*}{Y_F^*} \left\{ exp \left[\frac{\mathring{m}_F Cp}{4\pi} (\frac{1}{r_f} - \frac{1}{r}) \right] - 1 \right\}$$

We may also determine the fuel concentration on the drop surface:

$$Y_{F,S} = 1 - \frac{1 + Y_{o,\infty} \frac{Y_F^*}{Y_o^*}}{1 + \frac{CpT^*}{\Delta l} \left[\frac{T_\infty - T_s}{T^*} + \frac{Y_{o,\infty}}{Y_o^*} \right]}$$

where r_f is the flame radius and T_f the flame temperature.

From this theory we bring to light a number of interesting results:

i. The drop regression during combustion is of the form $D^2 = D_0^2 - Kt$, D_0 being the initial diameter and the constant K equal to:

$$K = \frac{8\lambda}{\rho_1 Cp} \ln \left[1 + \frac{Cp(T_\infty - T_s) + CpT^* \frac{Y_{o,\infty}}{Y_O^*}}{\Delta l} \right]$$

This law, as shown on Figure 14, is rather well verified experimentally

ii. We may take into account the forced convection by taking a value of K accounting for this convection, K being replaced by $KN_u/2$, a classical law being:

$$N_u = 2 + 0.555 \, R_e^{1/2} Pr^{1/3}$$

iii. We may introduce a more complex and slower chemistry, and the regression law becomes:

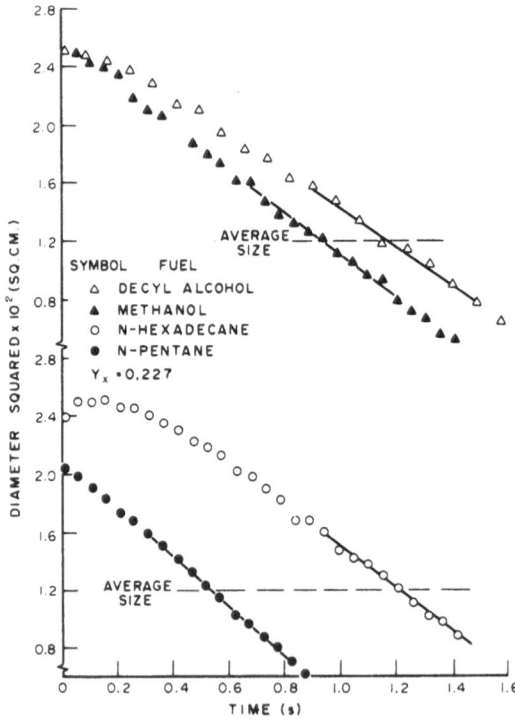

Fig. 14. Drop diameter variation during burning in a combustion gas environment; V = 0.625 m/s. T = 2530 K.

$$D^n = D_0^n - Kt$$

with $n \leq 2$, $n = 2$ when diffusion is the principal phenomenon and $n = 1$ if the action of chemistry is predominant (for instance combustion of a drop of monopropellant).

We may improve this theory, but at the price of great complications, by taking into account:

a) the unsteady character of the phenomenon,
b) a more complex chemistry,
c) a Lewis number different from unity in some cases,
d) a more realistic effect of the flow, in particular free convection,
e) a convection effect within the drop,
f) critical conditions in some cases,
g) a complex liquid composition (mixture).

The problem treated during these meetings being soot formation, the drop diameter is an important problem as it determines the residence time in the reaction zone: if the drop diameter is smaller than 1 mm for instance the soot has not enough time to form; on the contrary, if we simulate with porous spheres a spherical symmetry with a radius of several millimetres we observe very bright flames, with formation of soot. If the mixture ratio is important, as well as the temperature, the residence time of the products in this zone is also an important parameter this time

$$\frac{(r_f - r_s)^2}{\text{Diff}} \simeq \frac{r_s^r(\frac{r_f}{r_s} - 1)^2}{\text{Diff}}$$

where Diff is a mean diffusion coefficient; this residence time is thus very sensitive to drop radius r_s.

TURBULENT DIFFUSION FLAMES

If we increase the injection velocity of the fluids or the Reynolds number ρ_{vd}/μ, where d is the duct diameter, the flow becomes turbulent and the flame structure then depends on the nature of the flow: we observe that the flame has a more complex aspect, the diffusion and reaction zone is thicker, as shown on Figure 15.

The two problems arising then are the following: on the one hand, concerning the nature of the flow, its repercussion on the flame structure and, on the other hand, the incidence of combustion on the flow structure.

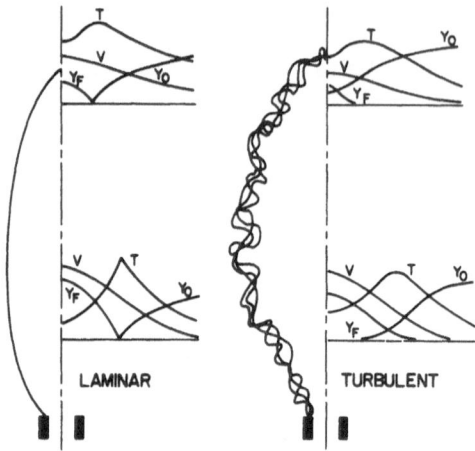

Fig. 15.   Schematic diagram of overventilated laminar and
turbulent diffusion flames.

Before examining these problems we shall give some experimental results.[6] Let us first emphasize that the flame diagnostic
techniques display an ever higher performance through the use of
lasers for determining instantaneous flow velocities as well as,
in certain cases, temperature and composition fluctuations; this
field is in full evolution.

Figure 16 is an example of signals obtained from a hot wire
and a laser velocimeter and concerning velocity u(t) and the
use of these records. First, we process the signal, determining
the mean values $\bar{u}$, equal to:

$$\bar{u} = \frac{1}{t_s} \int_{t_o}^{t_o+t_s} u(t)\ dt$$

during a time interval t_s. The fluctuating part of the signal
is:

$$u' = u(t) - \bar{u}$$

then we apply to the signal a Fourier transform:

$$X(N) = \int_o^{t_s} u(t)\ \exp(-i2\pi Nt)d\,N$$

where N is the frequency. The spectral power E(N) is calculated
from X(N)

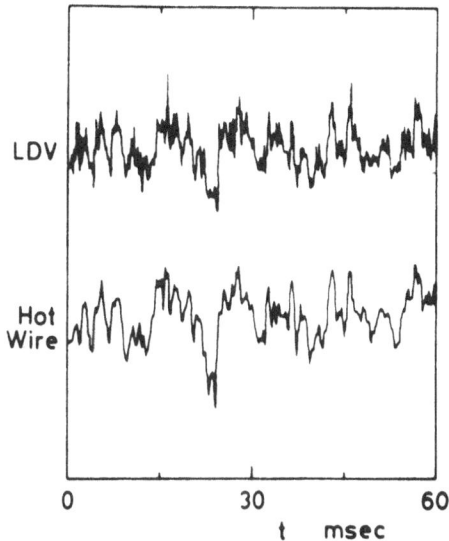

Fig. 16. Simultaneous recording of LDV and HWA signals.
          ($\bar{U}$ = 4.5 m/s).

$$E(N) = \frac{[X(N)]^2}{t_s}$$

The square of turbulence intensity is obtained from:

$$u'^2 = \int_0^\infty E(N) \; dN$$

and the autocorrelation coefficient:

$$R(\tau) = \frac{\overline{u(t) \; u(t + \tau)}}{u'^2} = \frac{1}{u'^2} \int_0^{N_{max}} E(N) \; \cos 2\pi N \tau \, dN$$

which makes it possible to determine the integral scale:

$$\ell = \bar{u} \int_0^{\tau_0} R(\tau) \; d\tau$$

τ_0 is the minimum value of τ when $r(\tau)$ reaches zero. Let us
emphasize that ℓ corresponds to a macroscale of turbulence;
we may also define a microscale of turbulence λ, for instance
by considering the curvature of $r(\tau)$ at $\tau = 0$:

$$\lambda^2 = \frac{2u'^2 \bar{u}^2}{(\frac{\partial u}{\partial t})^2}$$

For example the recording of Figure 16 at a mean velocity $\bar{u}$ = 4.5 m/s gives with the laser $u'/\bar{u}$ = 0.5, ℓ = 6.2 mm and λ = 0.15 mm, and with the hot wire $u'/\bar{u}$ = 0.23, ℓ = 5.2 mm and λ = 0.48 mm.[6]

Turbulent diffusion flames have been analyzed with these techniques. Let us mention for instance the works of T. Tagaki et al.[6] concerning flames of hydrogen-nitrogen mixtures with oxygen-nitrogen mixtures. An example of the results obtained is given on Figures 17A and B at a Reynolds number of 11,000 along the axis, without and with combustion. We may notice the following facts:

- increase of mean velocity, hence lengthening of mixing length with combustion due to the volume increase,
- velocity fluctuations u' and v' less intense with combustion, the ratio u'/v' is approximately the same with and without combustion,
- the nitrogen may be considered as an inert tracer, and expresses the diffusion evolution and the influence of the combustion on the turbulence.

When we examine radial sections, we notice:
- an increase of u', v', w' in the combustion zone,
- a very different distribution of species, e.g. evolution of nitrogen along the radius faster, at least at the beginning of the combustion zone.

Spectral power may be expressed as a function of wave number $K = 2\pi N/\bar{u}$, so that $E(K) = E(N) \bar{u}/2\pi$. Figure 18 gives $E(K)/u'^2$ as a function of K at various points along the axis with and without flame; small structures correspond to high values of K. With combustion we observe at each point of the axis a rather great dispersion for all structure dimensions, while without flame the small structures have about the same dimension; the tendency due to the existence of the flame is a widening of the turbulence scale, and the presence of combustion accentuates the turbulence non-isotropy.

On the theoretical viewpoint, an excellent review has been made by R. W. Bilger.[7]

If we examine the balance equations for a property f, which may be the mass fraction Y_i of species i, the velocity $\underline{v}$, the internal energy e or the enthalpy h, we obtain:

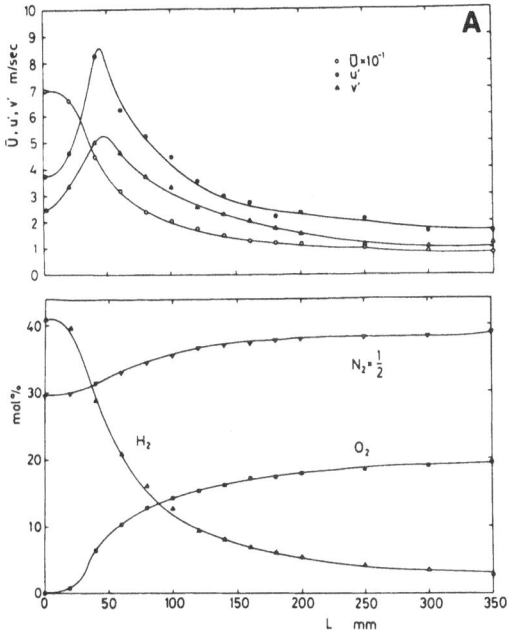

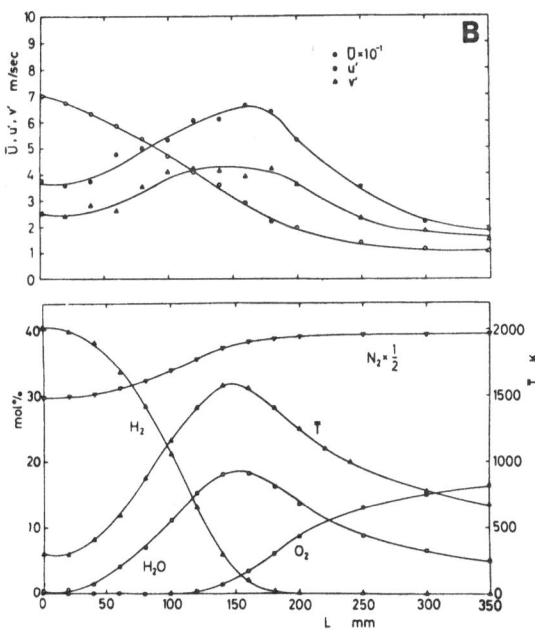

Fig. 17. Axial profiles of $\bar{U}$, u', v', $\bar{T}$ and gas species concentration on the jet axis in the case: A- without flame, B-
with flame (R_e = 11,000).[6]

$$\frac{\partial(\rho f)}{\partial t} + \underline{\nabla} \cdot \underline{J}_F = \dot{\omega}_F$$

(vectors are underlined once, and tensors twice) where we find again the terms of accumulation, of flux and of production; thus, turbulence intervenes in a more complex manner in the last two terms:

- in flux, $\underline{J}_F$ is the sum of two terms: a convective term and a transport term $\underline{j}_F$:

$$\underline{J}_F = \rho \underline{v} f + \underline{j}_F$$

A classical approach supposes $f = \bar{f} + f'$ (i.e. sum of a mean ter $\bar{f}$ and a fluctuating term f'); we shall remark that the mean convective term, for instance :

$$\overline{(\rho v)f} = \overline{(\rho v)}\,\bar{f} + \overline{(\rho v)'f'}$$

introduces a complementary turbulent transfer $\overline{(\rho v)'f'}$, more important than the molecular transfer $\underline{j}_F$. This way of doing things introduces new parameters which should be determined by various processes: it is the problem of closure of the system as we introduced new parameters.

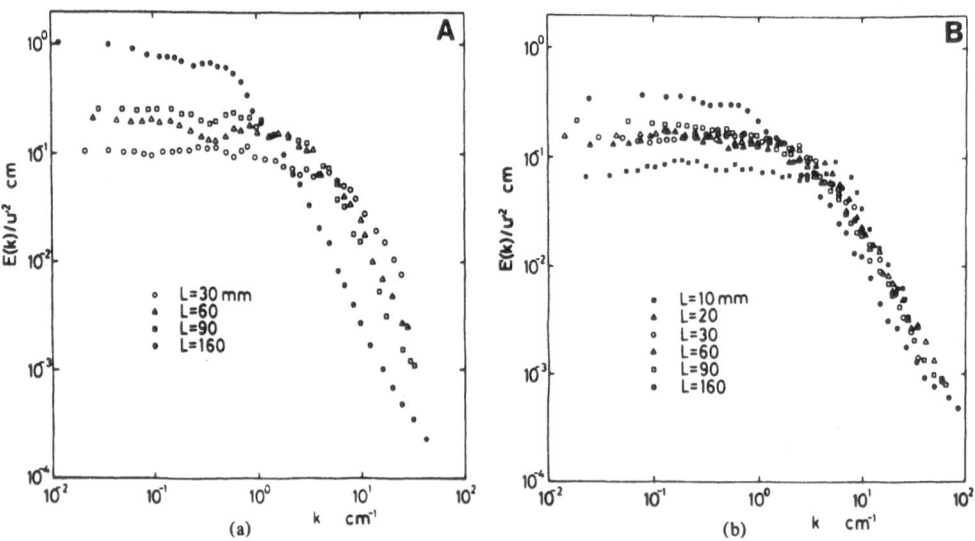

Fig. 18. Power spectrum of axial velocity fluctuation at points on the jet axis. (a) the case with flame, (b) the case without flame.

For instance, we admit that $\overline{(\rho u')f'}$ is proportional to a gradient of a parameter $\bar{f}$ exactly as in the definition of $\underline{j}_F$, the proportionality coefficient being a coefficient of turbulent transport depending on a characteristic length or Prandtl mixing length; it is the simplest closure method.

-The chemical production $\dot{\omega}$ is studied in the same way, for a second order reaction $i + j \rightarrow p$, e.g.:

$$\dot{\omega} = K(T)\ Y_i Y_j$$

If, for instance, temperature plays no part:

$$\overline{K(T)} = K(\bar{T})$$

--which remains to be proved-- and

$$\bar{\dot{\omega}} = \bar{K}\ (\overline{Y}_i \overline{Y}_j + \epsilon \overline{Y_i' Y_j'})$$

where ϵ = + 1 if reactants i and j are premixed
        ϵ = - 1 if they enter separately into the reaction volume.
It is the case of a diffusion flame; in this case, turbulence may reduce the production $\dot{\omega}$.

Another way of calculating mean values, useful in case of turbulent reactive medium, is to introduce a probability density function (pdf)P so that the mean value is given by:

$$\bar{T} = \int_0^\infty \int_0^1 \int_0^1 TP(T,Y_1,Y_2;\underline{x})\,dY_1\,dY_2\,dT$$

for a composition defined by Y_1 and Y_2 at a point $\underline{x}$ in space. For the product of two variables we obtain the mean value from the same function P:

$$\overline{T'Y_1'} = \int_0^\infty \int_0^1 \int_0^1 Y_1 TP(T,Y_1,Y_2;\underline{x})\,dY_1\,dY_2\,dT - \bar{T}\bar{Y}_1$$

This function P is usually determined experimentally but is calculated in some simple cases. It is characteristic of the flow structure, and not always Gaussian, as shown on Figure 19 in the case of a mixing layer, a jet or a reactor.

We may also calculate the mean value of chemical productions, at constant pressure for instance:

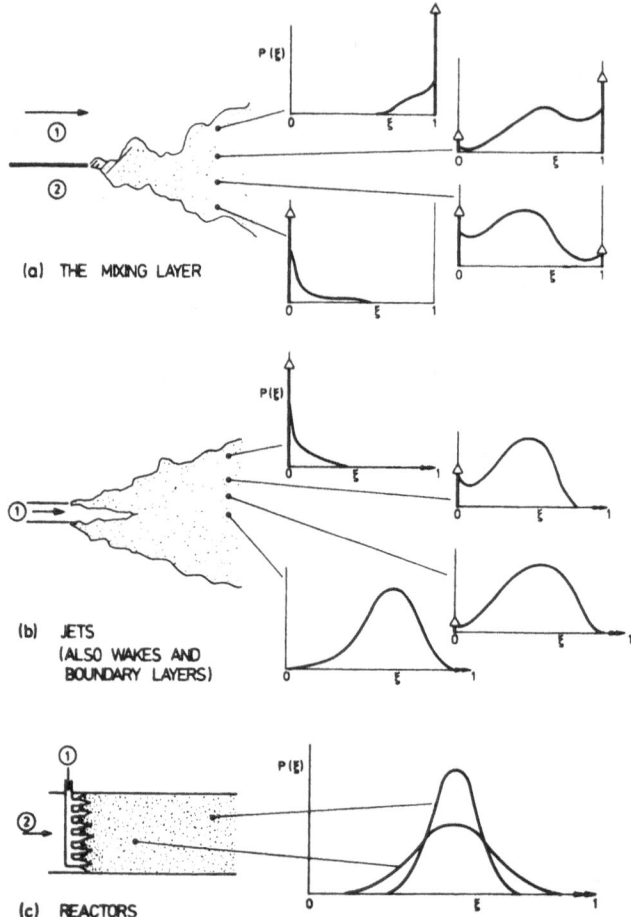

Fig. 19.   Probability density function forms for a conserved scalar
           in various types of flow (schematic only).

$$\bar{\omega} = \int_0^\infty \int_0^1 \int_0^1 KY_1Y_2 P(T,Y_1,Y_2;\underline{x})\ dY_1 dY_2 dT$$

These few examples show that the problem gets more complicated
when the number of variables increases, in particular when we
introduce the chemical variables, which should be reduced as far
as possible.

Still with the hypothesis of a single chemical reaction of
the type:

F + SO → (S + 1) P

We had seen, in the laminar case, that the production was:

$$\dot{\omega}_F = \frac{\dot{\omega}_O}{s} = -\frac{\dot{\omega}_P}{1 + S}$$

which makes it possible to find linear combinations in order to eliminate the chemistry; we had set, to characterize the composition:

$$\beta = Y_F - \frac{Y_O}{S}$$

which we shall now call β_{FO}, as it is now possible to consider also:

$$\beta_{FP} = Y_F + \frac{Y_P}{S + 1}$$

$$\beta_{OP} = Y_O + \frac{SY_P}{S + 1}$$

Let us consider a flow constituted by two fluxes 1 and 2 getting mixed; we may defined a new mixing variable β or , normalizing β:

$$\xi = \frac{\beta - \beta_2}{\beta_1 - \beta_2}$$

The stoichiometric value is given by:

$$\xi_S = \frac{Y_{O_2}}{SY_{F1} + Y_{O_2}}$$

We then have a single composition variable ξ , which may fluctuate, and compare it with the stoichiometric ratio, hence, to ξ_S which is constant (with a rich mixture $\xi > \xi_S$ and with a lean mixture $\xi < \xi_S$). In case of a fast reaction we obtain:

$$\beta_{Fo} \leq 0 \quad \xi \leq \xi_S \quad Y_F = 0 \quad Y_O = S\beta_{Fo} = SY_{F1} \frac{\xi - \xi_S}{1 - \xi_S}$$

$$Y_P = (S + 1)Y_{F1}\xi$$

$$\beta_{Fo} \geq 0 \quad \xi \geq \xi_S \quad Y_F = Y_{F1} \frac{\xi - \xi_S}{1 - \xi_S} = \beta_{Fo}, \quad Y_O = 0$$

$$Y_P = (S + 1) \, Y_{F1} \, \xi_S \, \frac{1 - \xi}{1 - \xi_S}$$

So that, at a given point of the flow and at a given time, the composition is known, e.g.:

$$\bar{Y}_F = \frac{Y_{F1} S}{1 - \xi_S} \int_{\xi_S}^{1} (\xi - \xi_S) \, P(\xi; \underline{x}) \, d\xi$$

if we introduce the pdf $P(\xi; \underline{x})$. The mean values introduced in these problems are averages in the sense of Favre.

There exit other closure methods, which are perhaps less well adapted to the study of turbulent combustion.

Although the problem of a turbulent diffusion flame is not yet solved, it is of interest to describe the flame structure.

Let us first examine that of the flow: there exist different scales of turbulence which, in decreasing order and for the main ones, are:
- the length L defining the size of the coherent structures corresponding to parts of the flow which are organized – vortices, eddies – and which, carried away by the flow, get deformed in time;
- the integral scale ℓ , which is a macroscale of turbulence, but which corresponds to random motions;
- the microscale λ, given by the evolution of the auto-correlation function, near the starting-point;
- the microscale η, which is the Kolmogorov length, depending on the turbulent Reynolds number $Re_T = (K^{1/2}\ell)/\nu$ where $K = 1/2(u'^2 + v'^2 + w'^2)$ corresponds to the kinetic energy of turbulence; ν is the viscosity. We can also write:

$$\eta/\ell = (Re_T)^{-3/4}$$

Then in the flow there exist many scales, but in reality there is not a line spectrum, but a continuous evolution of scales, as shown by real turbulence spectra.

To this convective structure due to the flow should be super-imposed a transport phenomenon and a chemical reaction phenomenon. It is thus a complex system which, as in the laminar case, depends on the characteristic times and the characteristic thicknesses of each of the processes.

Let us take a few examples to show the difficulty of the problem:

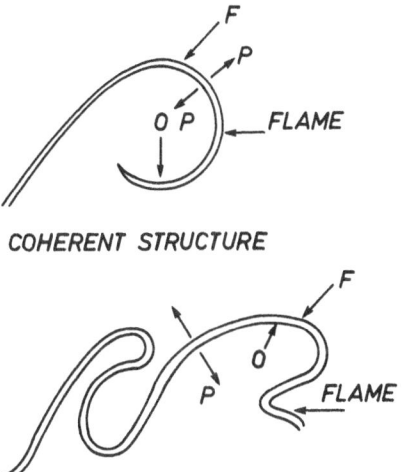

COHERENT STRUCTURE

LARGE SCALE OF THE TURBULENCE

Fig. 20.   Large scale of the turbulence and thin flames.

- Vortex zones and large scale turbulence

These two cases are presented on Figure 20, where scales L
and ℓ are large relative to the diffusion zone and to the chemical
reaction thickness. This case is similar to that of laminar diffusion flames, but may also depend on radiation, radius of curvature
and stretch of the flame, and also on some conditions of reaction
time, function of T_A/T (T_A being the activation temperature).

- Small scale (η) turbulence

In most of the models considered these phenomena are ignored,
although they may be important: for instance, an increase of
turbulence, hence of the turbulent Reynolds number,
i) reduces the scale $\eta \sim Re_T^{-3/4}$ so that turbulence penetrates
more easily into the reaction zone e,
ii) increases the effective transfer coefficient ν^*,
iii) increases the effective reaction time τ_c, as $\tau_c = \rho/\bar{\omega}$
and $\bar{\omega} = K(T)|\bar{Y}_O\bar{Y}_F - \overline{Y_O'Y_F'}|$: in a diffusion flame the mean value
of the product of concentration fluctuations $\overline{Y_O'Y_F'}$ increases,
which reduces the chemical production $\bar{\omega}$, hence increases τ_c. There
results a thickening of the reaction zone in a premixed flame, as
$e \sim \sqrt{\eta^* \tau_c}$.

This thickening of the reaction zone may play a role in these phenomena, and should be taken into account in the modelling of turbulent flames.

So it is necessary, in each case, to ascertain the relative times of every process and the evolution of the zones concerned.

STABILITY OF DIFFUSION FLAMES

There exist few works concerning the study of the stability of diffusion flames; this problem is, however, important as it may, for instance, modify the turbulence structure and, in particular, generate coherent structures. We shall mention some results of a work by Baev and Yasakov [8] to emphasize the interest of this problem.

Let us consider (Fig. 21) a duct of diameter d fed with fuel (velocity U_F) surrounded by a flow of air (velocity U_A); a diffusion flame is initiated above the potential core, at a distance H from the exhaust section of the nozzle. This distance is a function of the jet velocity, of the mixing of fuel and air and of a characteristic combustion time lag.

In different conditions this detachment of the flame may be stable or unstable. This instability is characterized by two frequency levels (high and low) and an amplitude function of H, rather of $\bar{H} = H/d$. The frequency is especially related to τ_c. High frequencies are of the order of 6 times low frequencies, and are given by the relationship: $N = 1/8\tau_c$.

DIFFUSION FLAMES IN MULTIPHASE FLOWS

This situation arises in the case of combustion of a fuel injected in the form of drops and when there appear, in given conditions, solid soot particles in the flame. In this case three phases - solid (soot), liquid (drops), and gas - coexist in the flow (Fig. 22).

It would be too long to write and discuss the balance equations, and we shall only make a few remarks:

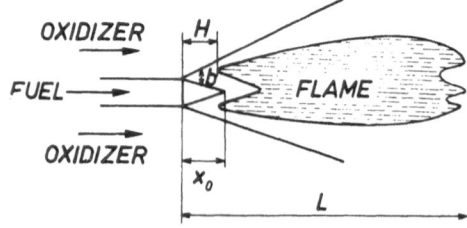

Fig. 21. Detached diffusion flame.

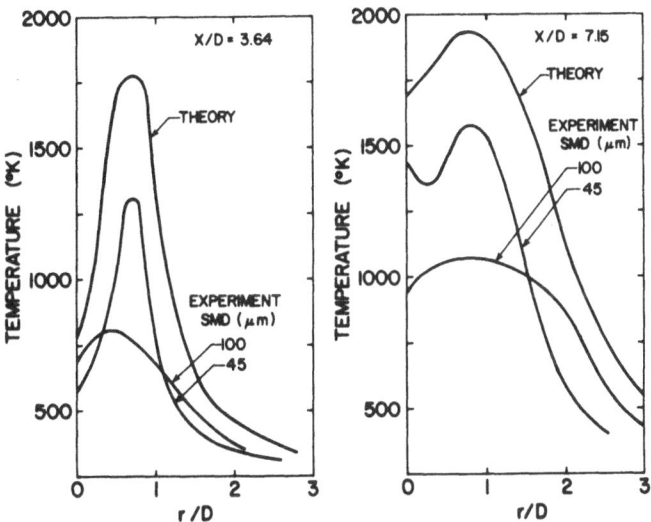

Fig. 22. Two-stage process: formation of nuclei, growth of
particles, soot formation.

- We consider that these particles (solid or liquid) are of
various sizes, and we should know the distribution function
$f_K(r, \underline{x}, \underline{v}, t)$ dr d$\underline{x}$ d$\underline{v}$ giving the number of particles of
category $\bar{K}$ whose radius is comprized between r and r + dr within
the space between $\underline{x}$ and $\underline{x}$ + d$\underline{x}$, and having a velocity between
$\underline{v}$ and $\underline{v}$ + d$\underline{v}$ at time t. This function should satisfy the
balance equation:

$$\frac{\partial f_K}{\partial t} + \frac{\partial}{\partial r}(R_K f_K) + \underline{\nabla}_{\underline{x}}(\underline{v}f_K) + \underline{\nabla}_{\underline{v}}(\underline{\underline{F}}_K f_K) = \psi_K + \phi_K$$

where $\underline{\nabla}_{\underline{x}}$, $\underline{\nabla}_{\underline{v}}$ represent gradients in the space and velocity
fields;
$R_K = (df/dt)_K$ represents the law of variation of radius with
time; for a vaporization or combustion of the
drops, $R_K = - a_K/r^{n-1}$. As the law utilized is

$r^n = r_o^n - Kt$, $F_{=K}$ is given by the drag law used,

ψ_K is a function modifying f_K by particle collision; ϕ_K is a function expressing the modification of f_K by a process of nucleation or pulverization of the particles (formation or bursting of particles).

This assumes that the characteristics of the injector are known, i.e. the initial distribution function of the drops; the most commonly used law is that of Rosin-Rammler, of the form:

$$f_{Ko} = ar^p \exp(-br^n)$$

r being the radius and a, b, p, n constants.

This law of balance giving the evolution of the distribution function may be utilized for soots if we define a new distribution law, but we must know the conditions of formation of the particles and the laws giving the evolution of the particle diameter within the flame.

A possible scheme of formation is given in Figure 22, the formation being preceded by that of a critical nucleus, then by an evolution of the nucleus with time.

The critical nucleus r may be determined by the thermo-dynamic equilibrium, $dF = 0$, from the free energy F, and the increase by a study of collisions Z; for instance dZ/dt is propor-tional to the number of particles per unit volume ($\sim 1/r^3$), to the particle velocity ($\sim 1/r^{3/2}$) and to the collision section area ($\sim r^2$), which leads to a law in $r^{5/2}$, proportional to time. This mode of reasoning may provide the function R_j, ψ_j and ϕ_j.

- The description of the evolution of the gas and particles rests on the solution of the equations of mass, momentum and energy balance, establisehd first for the (ballistic) particles, then for the gas-particle flow, which eliminates the interaction between gas and particles. It is a problem very complex in its generality, as it concerns an unsteady, sometimes three-dimensional flow, with two classes of particles, evoluting and perhaps interacting; more-over, this flow is the seat of chemical reactions, some of them highly exothermic and acting on the temperature, pressure, velocity and composition. There is thus a whole set of parameters to be determined, related by more or less known phenomena, which makes it mandatory to make important simplifications such as considering the multiphase flow as a quasi gas.

To illustrate this an example is given on Figure 24, showing the differences between theory and experiment.

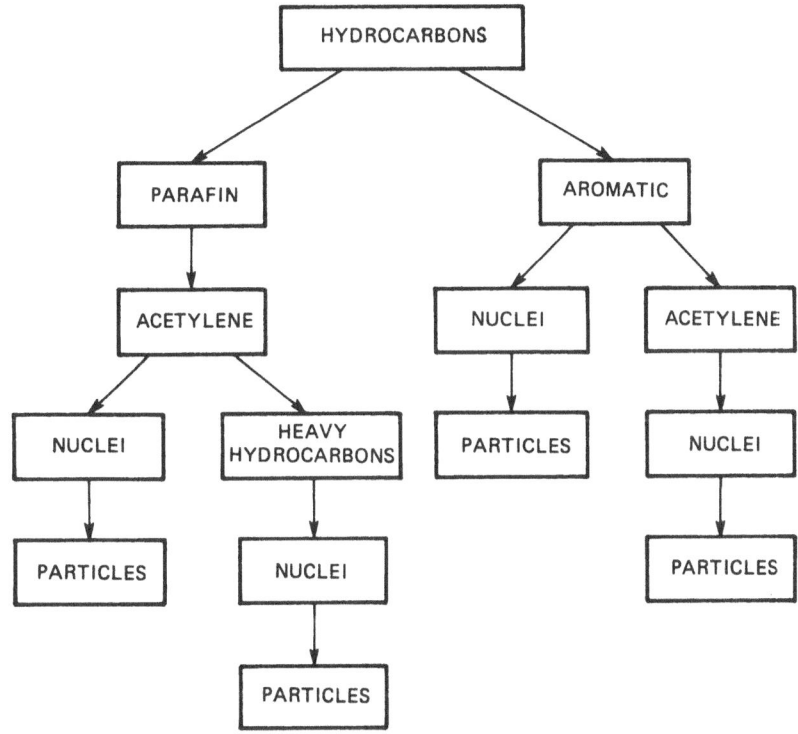

Fig. 22. Two stage process: formation of nuclei, growth of particles, soot formation.

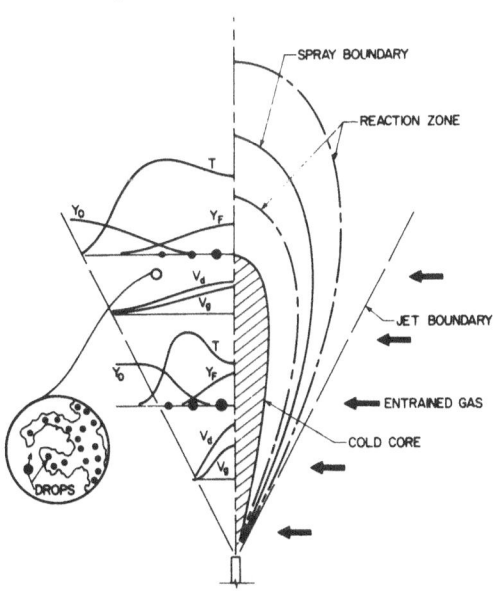

Fig. 23. Schematic representation of a coaxial spray diffusion flame.

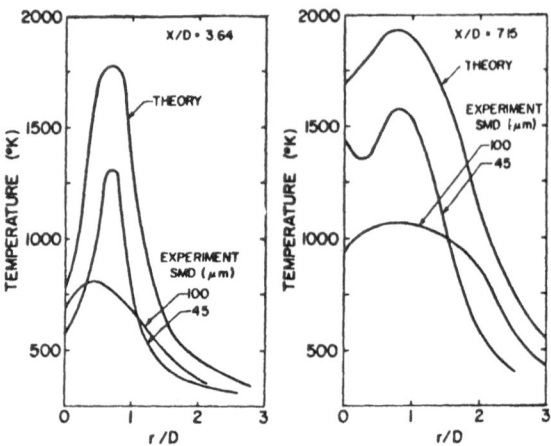

Fig. 24.   Comparison between measured and predicted temperatures
in a hollow-cone spray flame.[5]

CONCLUSION

We intended in this paper to emphasize the complexity of the
problem of soot formation in regards of physical parameters. The
simplifications used should be justified as, considering the
important number of parameters and the many constants involved,
it is easy to adjust the constants afterwards to harmonize theory
with experiment.

So one should be very prudent. However, we may indicate here
that:

- the laminar diffusion flame is rather well understood and
  may be correctly modelled when the Lewis number is close
  to unity;
- complications appear when the Lewis number is different
  from unity (hydrogen flames) or when chemistry becomes
  complicated;
- turbulent diffusion flames are getting better understood in
  some cases, but a great effort remains to be done to obtain
  a good modelling;
-spherical symmetry may provide research themes important
  for the understanding of the phenomena, for instance by
  varying the flame radius (hollow sphere);
- the study of multiphase flows is in progress, but is still
  insufficient to model correctly the soot formation in a
  combustor, as many important physical phenomena are still
  little known and hence neglected.

These few words give a measure of the amount of work still to be accomplished.

REFERENCES

1. F. H. Holderness and J. J. Macfarlane, Soot Formation in Rich Kerosene Flames at High Pressure, AGARD Conf. Proc. 125 (1973)

2. R. E. Mitchell, A. F. Sarofim and L. A. Clomburg, Experimental and Numerical Investigation of Confined Laminar Diffusion Flames, Combustion and Flames 37:227 (1980)

3. L. P. Yarin, Some Problems in the Aerodynamics of Gas Flames, Combustion, Explosion and Shock Waves (USSR) 5(2):157 (1969)

4. R. A. Allison and J. F. Clarke, Theory of a Hydrogen-Oxygen Diffusion Flame, Combustion Science and Technology 23(3-4):113 (1980)

5. G. M. Faeth, Current Status of Droplet and Liquid Combustion, in"Progress in Energy in Combustion Science" Pergamon Press 3:4 (1977)

6. T. Tagaki et al., Properties of Turbulence in Turbulent Diffusion Flames, Combustion and Flame 40:121 (1981)

7. R. W. Bilger, Turbulent Flows with Nonpremixed Reactants, Turbulent Reacting Flows, Topics in Applied Physics, 44 Springer Verlag (1980)

8. V. K. Baer and V. A. Yasakov, Stability of a Diffusion Flame in Single and Mixed Jets, Combustion, Explosion and Shock Waves, (USSR) , 12(2):163 (1975)

9. S. G. Graham, The Modelling of the Growth of Soot Particles, Proceedings of the Royal Society of London, 377(1769): 119 (1981)

SOOT FORMATION AND BURN-OUT IN TURBULENT COAL LIQUID FUEL FLAMES

J. M. Beér, M. Toqan, W. Farmayan, M. T. Jacques[*]
and G. Prado[**]

Department of Chemical Engineering and
the Energy Laboratory
Massachusetts Institute of Technology
Cambridge, Massachusetts, USA
[*]Present address, British Petroleum Britannic House
London, England
[**]Present address, CNRS, Mulhouse, France

INTRODUCTION

Combustion scientists and engineers have for a long time been interested in understanding the processes leading to the formation and disappearance of soot in flames. The practical reasons for developing a capability of predicting soot concentration were mainly connected with the effects of soot upon the radiative emission from furnace flames, and the clean combustion of hydrocarbon fuels in furnaces and gas turbines. More recently new interest has been generated by the concern over health effects of soot and polycyclic aromatic hydrocarbon species, the latter of which condense in the flue gas to form submicron-size aerosols. A major research program on the formation of particulates in flames, their emission and effects on human health, is in progress at M.I.T. under the sponsorship of the National Institute of Environmental Health. Investigations under this program include flat flame and stirred reactor studies of hydrocarbon vapor pyrolysis and the kinetics of formation of PAH and soot in pyrolyzing liquid fuel droplet streams. Parallel with these bench scale experiments detailed studies are carried out on turbulent diffusion flames using the M.I.T. Combustion Research Facility. In these flames the thermal and chemical environment of industrial flames is closely simulated.

There are a number of comprehensive reviews of soot formation and combustion[1,2,3] which understandably concentrate their discussion on the thermodynamics and chemical kinetics of the problem. There have been only few attempts to apply the chemical kinetic information to turbulent flames.[4,5] Magnussen et al[5] have used an "eddy dissipation combustion model" and a simplified empirical relationship of soot formation to predict soot concentration in a turbulent-free jet acetylene-air diffusion flame. They show good agreement between predicted and experimental data along the flame axis but it appears from their results that when changes in fuel concentration in the turbulent-free jet were made by N_2 dilution, the agreement between prediction and experiment was poor, especially towards the tail end of the flame, the significant flame region from the point of view of soot emission. Experiments reported by Prado et al.[4] emphasize the importance of chemical kinetic information in the formation of soot and in particular the relationship between soot and polycyclic aromatic hydrocarbon species which appear to be precursors of soot formation. They use a simple aerodynamic analysis based on a model by Appleton and Heywood[6] but apply this model only downstream of the region in which soot has been formed, in their burners.

More recently, Eickhoff and Grethe[7] have developed a flame zone model for hydrocarbon combustion assuming that the instantaneous gas composition in a turbulent flame is that corresponding to local chemical equilibrium over a wide range of mixture ratios around the stoichiometric, and that the shifting equilibrium reactions freeze at a unique fuel rich mixture fraction. Their simplified model is based on Libby's idea who suggested subdividing a laminar flame into regions of equilibrium and frozen flow. They have then combined the model with a pdf representation of the fuel mass fraction in a turbulent flame. The application of this reaction model together with a k-ε turbulent transport model seems to have a potential for predicting species concentration distributions in turbulent hydrocarbon-air flames provided that a concise chemical kinetic scheme can be found which will permit satisfactory accuracy without requiring excessive computational effort.

In the pioneering work carried out at the International Flame Research Foundation, Hubbard[8] reported the effects of fuel type (C/H ratio) upon the soot concentration distribution in the flame and interpreted the data in terms of the jet momentum at source, and the fuel input rate. Strong correlation between C/H ratio of the fuel and soot concentration in the flame also was obtained by Holliday[9] from measurements in laboratory size turbulent diffusion flames. Some of Holliday's results are shown in Figure 1.

Figure 2 illustrates the effects of the main variables in Hubbard's experiments.[8] The steam atomized fuel oil issued through a nozzle of 10 mm diameter along the axis of the 2 x 2 meter

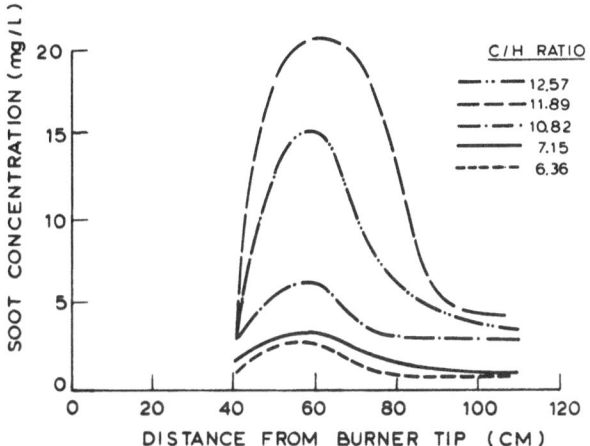

Fig. 1. Effect of C/H ratio in fuel upon the soot concentration along the flame axis (After Holliday).

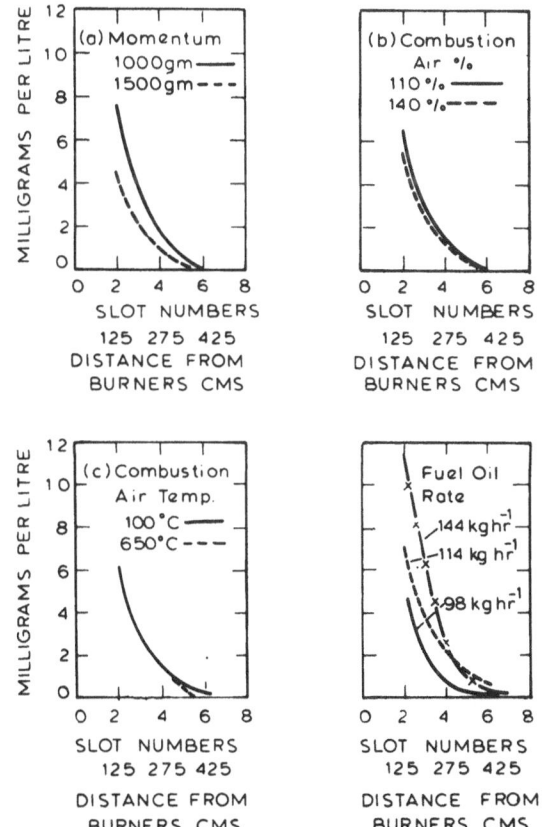

Fig. 2. Effect of main variables on concentration of soot particles on flame axis (After Hubbard).

cross-section, 6.3 meter long furnace. Peak values of soot
concentration occurred very close to the burner within a corresponding residence time of less than 125 msec, and thus it can be said
that Hubbard's data show the burn-out rather than the formation of
soot. The strong effect of the jet momentum upon the decay of soot
concentration along the flame implies that the rate of soot burn-out
in the flame is mixing controlled. It is expected, however, that in
the tail end of the flame, particularly for cases of strong heat-
extraction from the flame, the rate of soot burn-out will be controlled by chemical kinetics rather than mixing.

Holden and Thring [10] analyzed Hubbard's data to show the
effects of velocity and mixing rate upon soot formation. They
assumed that the amount of soot in the flame at any axial station
was proportional to the product of soot concentration and velocity
at the axis and the cross-sectional area of the jet at the section
considered. The values thus obtained were then rationalized in
terms of fraction of fuel which goes to form soot. These have been
plotted against time taken to reach a given mixing ratio. The
authors found that within the range considered the correlation
between the fraction of fuel converted to soot and time, for a
given mixture ratio in the turbulent jet was linear. The regression
lines so obtained are shown in Fig. 3. While this approach was
attractive because of its simplicity, its application was limited
because no account was taken of the differences in temperature and
concentration history of fuel-air pockets in the flame.

For jet flames the development of which are not solely dominated
by the jet momentum at source, the soot formation near the burner
depends upon additional parameters such as fuel atomization quality,
the spatial distribution of the spray and the near-field flow pattern; the latter refers to coherent structures of flow such as wakes
of bluff body flame holders, or toroidal recirculation zones produced
by swirling jet flow. [11]

It appears that models for predicting the formation and oxidation
of soot in turbulent flames whether based on relatively simple jet
theory or detailed fluid dynamic computations are in an initial
stage of their development. Significantly more information on the
chemical kinetics of the formation of soot and its burn-out at the
end of the flame is needed for improved confidence in prediction
procedures. Also more detailed experimental data on the spatial
distribution of soot and other hydrocarbon species concentration in
turbulent flames is necessary.

In the present contribution experimental data obtained by measurements of soot and polycyclic aromatic hydrocarbon concentrations
in coal derived liquid fuel flames are reported. There is special
interest in these fuels due to their strongly aromatic nature, high
nitrogen content and high C/H ratio of their elemental composition.

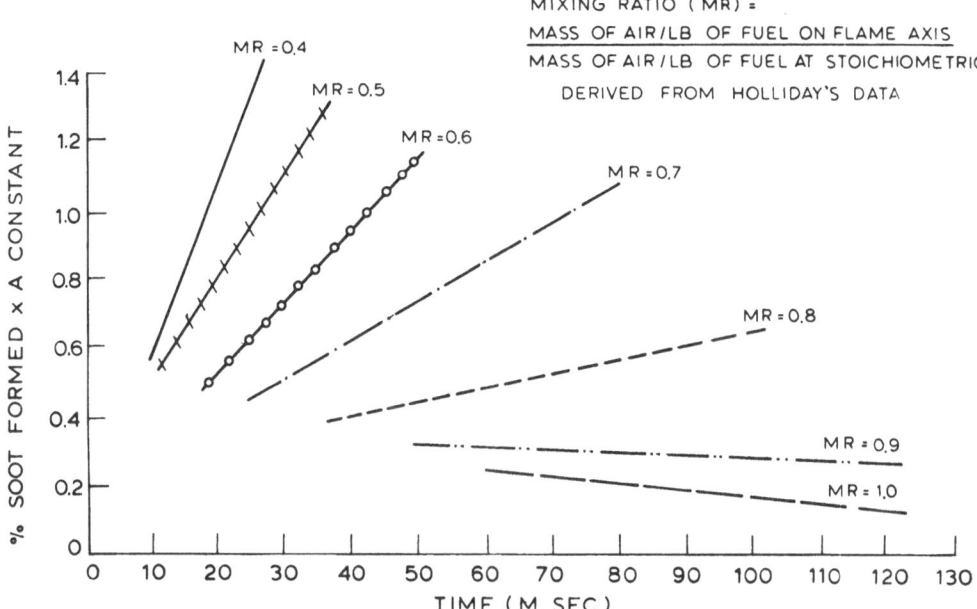

Fig. 3. Effect of velocity and mixing rate on soot formation
         (IJmuiden) (After Holden and Thring[10]).

Also, some of the polycyclic aromatic hydrocarbons (PAH) in these
flames are known to be biologically active and it is important
therefore to follow the fate of these compounds in turbulent flames
as they form and decay to their emission levels. The thermal and
chemical environment in these flames and the residence times were
characteristic of industrial type and scale turbulent diffusion
flames.

Experimental apparatus

     The experimental program involved the use of the M.I.T. Combus-
tion Research Facility (CRF) which was designed especially to
facilitate detailed experimental investigations of large turbulent
diffusion flames. The CRF is a 1.2 x 1.2 m cross-section, 10 m
long combustion tunnel equipped with a single burner of up to 3MW
thermal multi-fuel firing capability. The combustion tunnel com-
prises a number of individual 30 cm wide sections all of which are
water cooled with some sections having a refractory lining on the
fireside while the rest have bare-metal fireside walls. The sections are interchangeable, an arrangement which permits variable
heat sink for control of heat extraction along the length of the
flame (Fig. 4).

The burner used in these experiments is equipped with a
variable-swirl generator (movable block type IFRF burner); the
combustion air enters in the form of a swirling annular jet, the
annulus being formed around the 60 mm diameter oil spray gun. At
the burner exit a 35° half angle water cooled quarl assists in
the formation of a toroidal recirculation zone in the central
region of the jet.

Measurements

Spatial distributions of time average values of velocity,
gas temperature and of gaseous and solids species concentrations
were determined by probe measurements, sampling, and chemical
analyses. The incident radiation at the wall and the flame
emissivity variation along the flames was also determined but is
reported elsewhere.[12]

Sampling

Most of the probe measurement techniques are those developed
at the IFRF [13] except a novel type soot sampling probe which uses
water injection at the probe tip for quenching and for transport-
ing the soot sample along the length of the probe. The design is

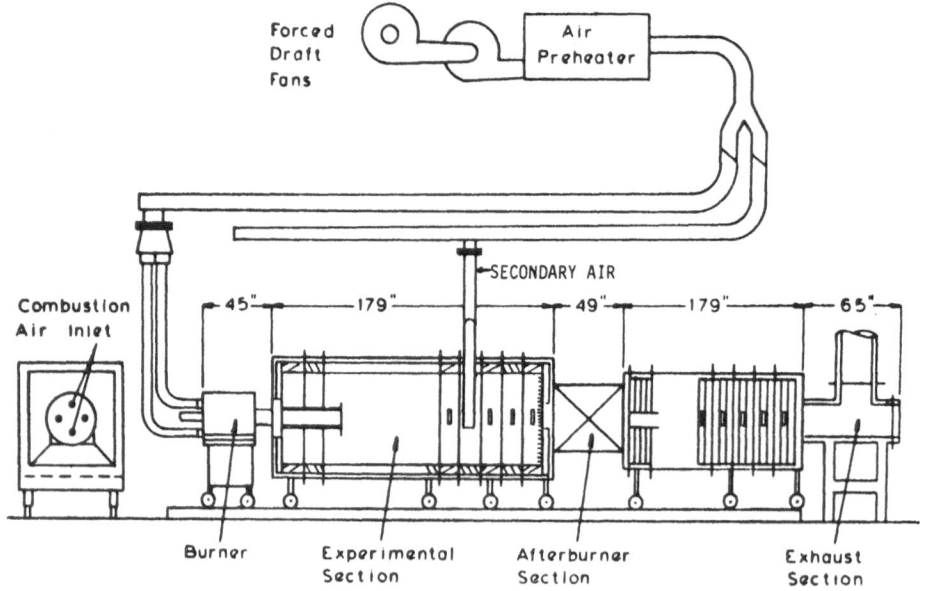

Fig. 4.   Furnace assembly and air staging system.

based on a similar probe developed by Prado et al. (4). The sample
stream comprised of water, gases and particulates, is passed through
a glass-fiber filter where part of the PAH and the soot are retained.
The filter is backed by a hydrocarbon resin trap (XAD-2) to ensure
quantitative recovery of PAH. The filter and soot deposit are
extracted with methylene chloride. The XAD-2 resin is soxhlet
extracted with methylene chloride.

Analytical procedure

The methylene chloride extract is concentrated to a volume from
5 to 100 mℓ, depending upon the concentration of the organics, and
analyzed by gas chromatography. A fused silicone stationary phase
capillary column is used to separate PAH compounds including many
isomers. The system is programmed from 45°C to 280°C at 10°C/min
after sample injection. A standard solution consisting of 25 compounds representative of 3 to 5 rings PAH is used to calibrate the
flame ionization detector, which is used in conjunction with the
above capillary column.

The experimental fuels

Experiments were performed with two SRCII fuel types; a blend
of 2.9 parts of middle distillate to one part of heavy distillate
(2.9/1) and an unblended heavy distillate. Elemental analyses of
these fuels are given in Table 1.

Experimental conditions, input variables

The experiments reported in this paper ran parallel with an
extensive study on the low NO_x combustion of the high nitrogen
content SRCII fuels under the sponsorship of the Electric Power
Research Institute (EPRI).[12]   Because of the general tendency of
these fuels to form large amounts of NO_x the combustion process had
to be modified by staging of the combustion air.

The experimental conditions selected for the soot and PAH
investigation included two flame types, a single stage flame and
a flame in which the fuel rich stage equivalence ratio was $\phi_b = 1.3$
and the rest of the combustion air was injected by high velocity
opposed jets 2.6 m downstream of the burner. In both cases the
heat input was 1MW and the O_2 in the flue gas was maintained at
2 percent (∿ 10% excess air).

Details of the input conditions are given in Table 2.

Experimental results - discussion

Soot concentrations (g/Nm^3) along the axis of SRCII fuel flames
plotted for the effect of fuel type are shown in Fig. 5. As these

Table 1.   SRCII Fuels Analyses.

	SRCII (middle)	SRCII (2.9/1)	SRCII (heavy)
API Gravity	13.4	10.4	1.3
Viscosity SSU at 100°F	-	-	-
SSU at 140°F	36.4	36.8	67.2
SSU at 210°F	31.8	33.0	41.3
Flash Point °F	147	150	265
Water and Sediment %	0.10	0.04	1.5
Pour Point °F	<-20	<-20	+8
Heat of Combustion			
Gross BTU/lb	17000	17000	17120
Net BTU/lb	16190	16340	16420
Elemental Analysis			
Carbon %	83.79	86.28	88.37
Hydrogen %	9.04	8.83	7.38
Nitrogen %	0.94	0.96	1.17
Sulfur %	0.23	0.28	0.43
Ash %	<0.01	<0.01	<0.07
Oxygen %	4.84	3.68	1.17

Table 2.   Input Conditions.

Fuel type: SRCII 2.9/1 and SRCII heavy distillate

Heat input: 1 MW

Flame type: unstaged and staged

Atomizing nozzle: Sonicore steam atomized

Combustion air:

   Exit velocity at the burner: U_o = 30 m/s

   Swirl number: S = 0.5

   Secondary air jets: horizontal, opposed jets U_s = 50 m/s

flames are produced with a combination of swirl (S = 0.5) and a
35° half angle diffuser at the exit of the annular air flow from
the burner, a recirculation zone is established in the central
region of the jet which extends to about 15 cm (X/D ≃ 3) downstream
of the burner. There is intense soot formation in this region of
"coherent structures" of the jet with significantly higher peak concentration in the SRCII heavy distillate flame. This can be explained
qualitatively by the higher concentration of high molecular weight
polycyclic aromatic hydrocarbons (PAH) some of which act as soot
precursors. Also, the C/H ratio in the SRCII heavy distillate is
12, and in the lighter 2.9/1 blend, 9.8.

As can be seen in Fig. 5, the decay of soot concentration along
the jet is similar for the two fuels leading to nearly identical
values of about 0.05 g/Nm^3 at positions beyond 3 m (X/D > 13) from
the burner. It appears that the small amount of soot that is emitted
at the end of the flame originates from flamelets in the high shear
zone of the jet which might have become extinct due to extensive
stretch. The chemical kinetics of soot oxidation reaction is fast
enough at temperatures of about 1550 K and 2% O_2 - conditions at
the exit from the combustion chamber - for mixing to control the
burn-out of soot at the end of the flame.

Radial profiles of soot concentration in a SRCII heavy distillate
fuel flame are shown in Fig. 6. For the profile nearest to the
nozzle, samples were taken at a distance of 0.15 m (X/D≈1) from the
burner. This traverse is well within the toroidal recirculation
zone formed in the central region of the swirling annular jet. When
the radial distribution profiles are normalized, (Fig. 7), it can
be seen that, close to the burner, where in the recirculation zone
coherent structures dominate, the soot concentration attains its
peak value, and the radial distribution shows slow decay with radial
distance. Further downstream, in the region of 2 < X/D < 9 the
profiles show a steeper radial decay of soot concentration but level
off at the edge of the jet approaching a value corresponding to the
soot concentration of the combustion products recirculated on the
outside of the jet; the external recirculation typical in enclosed
jets causes the profiles taken at increasing axial stations, X/D,
to tail off at decreasing values of the r/x ratio. Finally the
concentration distributions at large axial distances, (X/D > 9),
have characteristics similar to the distribution of nozzle fluid
properties in turbulent jets. At this point the temperature and O_2
concentration is rather uniform across the jet and the chemistry of
soot burn-out is fast relative to mixing.

The effect of staged combustion of SRCII heavy distillate fuel -
partial oxidation of the fuel in a fuel rich zone with fuel/air
equivalence ration of ϕ = 1.3 and a residence time of about 2 seconds
followed by the injection of secondary air in the form of transverse
opposed jets - is represented in Fig. 8. Peak soot concentrations

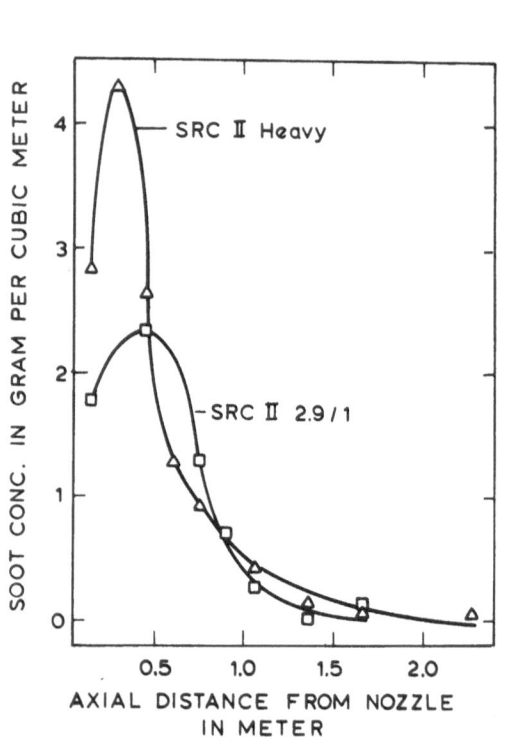

Fig. 5. The effect of fuel type upon the soot concentration along the flame axis (Unstaged combution of SRC II 2.9/1 and SRC II heavy distillate fuels).

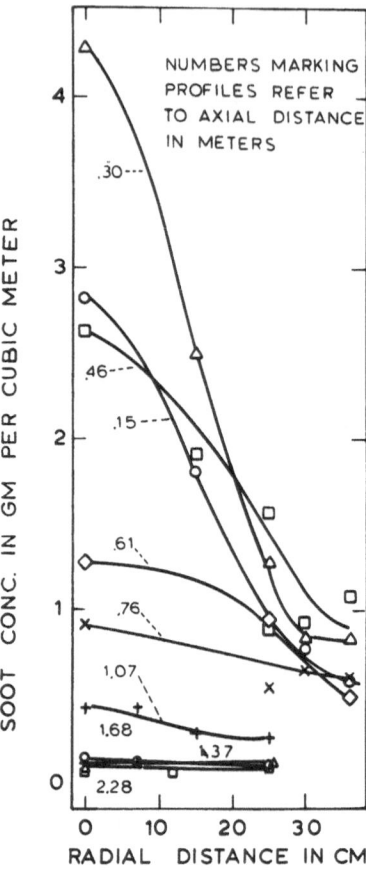

Fig. 6. Radial soot profiles in an unstaged SRC II heavy fuel flame.

in the stage flame are about twice as high as in the single stage flames. Time average measurements of O_2 concentrations in the fuel rich zone begin to show a rise approximately 1 meter upstream of the secondary air injection point as a result of some backmixing of the secondary air; at a distance of 0.6 meters upstream of the secondary air injection point the O_2 concentration is 0.5%. This follows the initial entrainment of the burner air into the jet and explains the observed decay of soot concentration in the fuel rich zone of the flame.

Downstream of the secondary air jets the soot concentration decay rapidly to a level close to that in the unstaged flame of

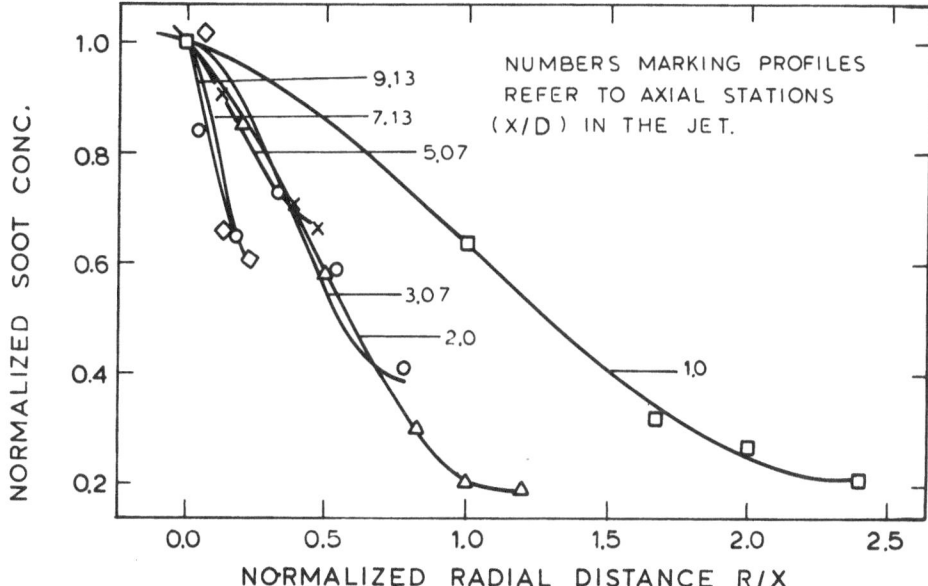

Fig. 7. Normalized soot concentration along radial distance. Unstaged SRC II heavy distillate.

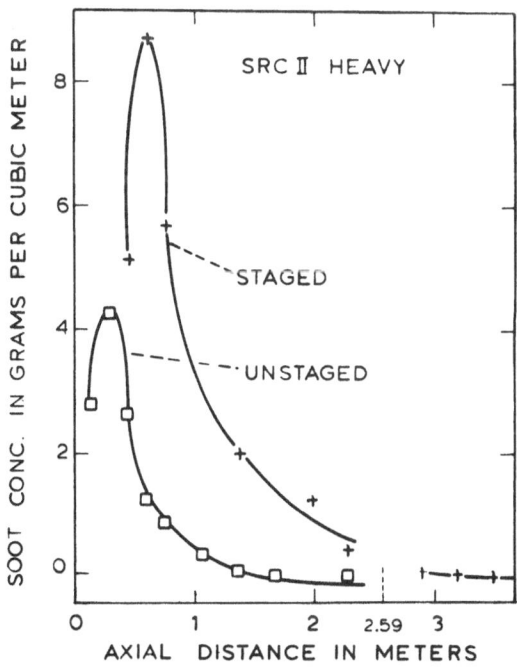

Fig. 8. Effect of air staging upon soot concentration along the flame axis (SRC II heavy distillate).

about .02 - .05 g/Nm³. The high velocity transverse secondary air
jets are designed to entrain the total amount of the fuel rich
pyrolysis products before reaching the axis of the furnace. The
soot burns in these secondary air jets to a low level of concentration. It is thought that the emission of low concentrations of
soot can be attributed to the same mechanism of excessively stretched
flamelets in the high shear region of the jet as mentioned earlier.

The staged combustion study was extended to include experiments
with reduced residence times by doubling the fuel input and maintaining the fuel rich equivalence ratio (ϕ = 1.3), the overall
excess air (10%) and the burner-secondary jet nozzle geometry.
Figure 9 shows the effect of reduced residence time upon soot burn-
out. As can be seen the soot burns to a low concentration in the
transverse air jets which now have their Reynolds number, based on
the jet nozzle diameter, increased from Re_{1MW} = 80.000 to Re_{2MW} =
160.000. There is a slight increase in the soot concentration of
the flue gas for the case of 2MW; but the increase is too small to
conclusively state that limits of chemical kinetic control have been
reached.

SOOT AXIAL PROFILES

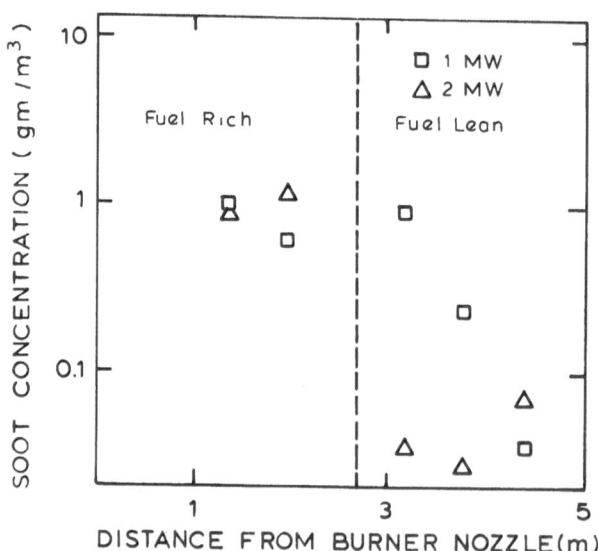

Fig. 9. Effect of fuel input upon soot concentration
         along the flame axis (Staged combustion of SRC II
         2.9/1 fuel at 1 MW and 2 MW thermal inputs).

Polycyclic Aromatic Hydrocarbons (PAH)

Samples were collected from the flame and analyzed for PAH concentration. The majority of PAH compounds in the samples have been individually identified and quantified using capillary gas chromatography. Concentration of PAH compounds was also determined in the SRCII fuels. The detailed data of fuel and sample analyses have been reported elsewhere.[14] Because of the limited scope of this paper the discussion is restricted to a few of the compounds which dominate the concentration distribution or are of interest because of their known biological activity.

When the fuel composition of PAH compounds is compared with those in the samples extracted from the flames, it is generally found that long chain alkylated PAH in the fuel have either disappeared, or are present in the flame samples in very low concentrations. Other compounds such as acenaphthalene of napthalene can be found in the flame samples as a result of pyrosynthesis; these are not present in the fuel. Further, some longer non alkylated PAH such as phenanthrene, anthracene, fluoranthene and pyrene seem to be rather resilient to burn-out below a certain low limit of concentration and they are chosen therefore for illustrating the general trends of PAH concentration variation in the flames. In Fig. 10 the axial concentrations of three PAH compounds, pyrene, fluoranthene and naphthalene, along an unstaged SRCII heavy distillate fuel flame are plotted. Axial soot concentrations are

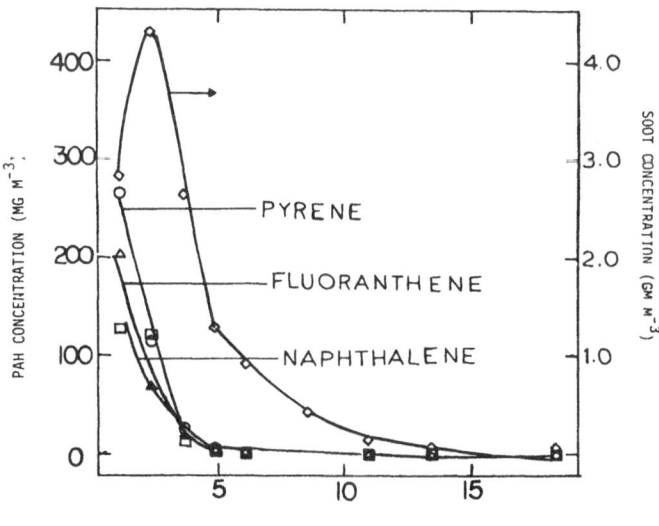

AXIAL DISTANCE FROM NOZZLE (X/D)

Fig. 10. PAH and soot concentration along axial distance of the flame (SRC II heavy distillate).

also shown to illustrate the point that the PAH compounds may well
be considered to be soot precursors. Close to the burner - X/D
< 5 - the mixture is strongly fuel rich and pyrolysis of the fuel
produces high concentrations of PAH. As soon as soot begins to
form the PAH concentration reduces to a low level and then continues
to decay further but at a slower rate. Figure 11 illustrates the
burn-out of PAH and soot for axial distances X/D > 5. It is thought,
that while the four and five ring compounds are relatively stable
their survival in low concentrations along the flame is not due to
chemical kinetic limitation but is caused by unmixedness due to the
extinction of reaction in stretched flamelets emanating from the
turbulent flame brush. The effect of staging of the combustion air
upon the axial concentration distributions of pyrene and naphthalene
is shown in Fig. 12. Both PAH compounds have significantly higher
peak concentrations in the fuel rich zone of the staged than in the
unstaged flame, the ratio of concentrations being especially high
for the pyrosynthetized napthalene. It is interesting to note that
the decay of the PAH concentrations along the axis occurs in the
fuel rich stage similarly to that in the unstaged flame. As there
is little or no oxygen in this region ($\phi = 1.3$), the fast decay of
PAH and the increase in soot concentration (Fig. 13) lend further
support to the suggestion that the PAH is consumed by the soot
formation process leaving only the most stable 3-5 ring compounds.
Some other PAH compounds, not present in the fuel are formed in
this rich flame zone. For the case of air staging when the zone
of fuel rich composition ($\phi = 1.3$) extends to a distance of X/D $\sim$ 20

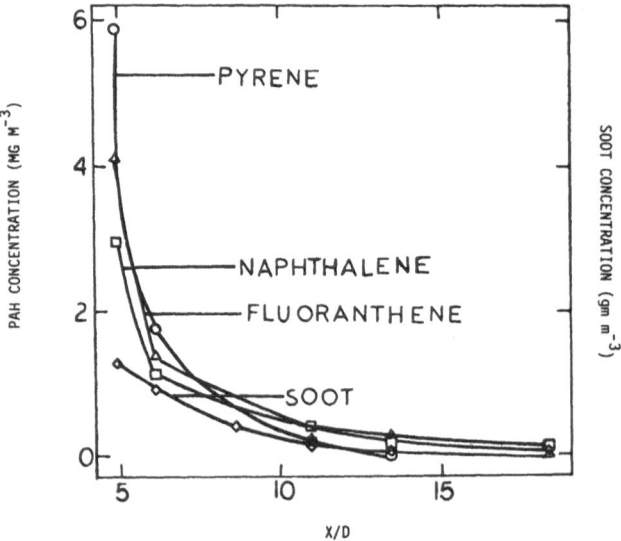

Fig. 11.   PAH and soot concentrations in the burn-out region
           (SRC II heavy distillate).

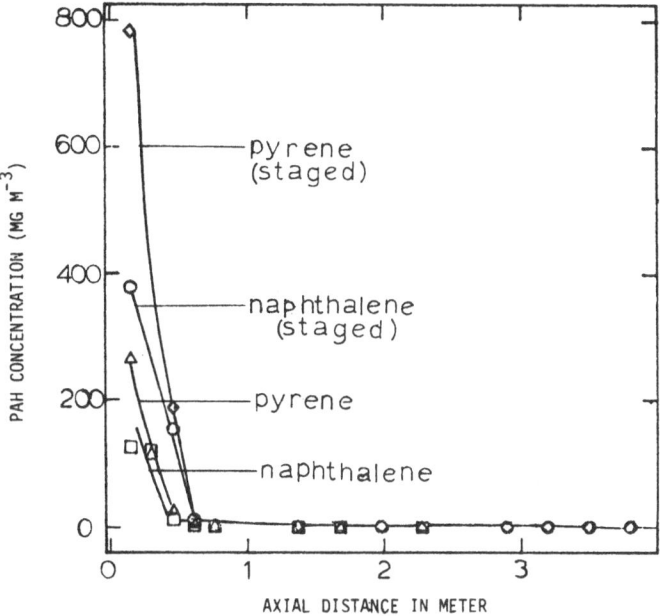

Fig. 12.   PAH concentration along axial distance for staged and
           unstaged combustion (SRC II heavy distillate).

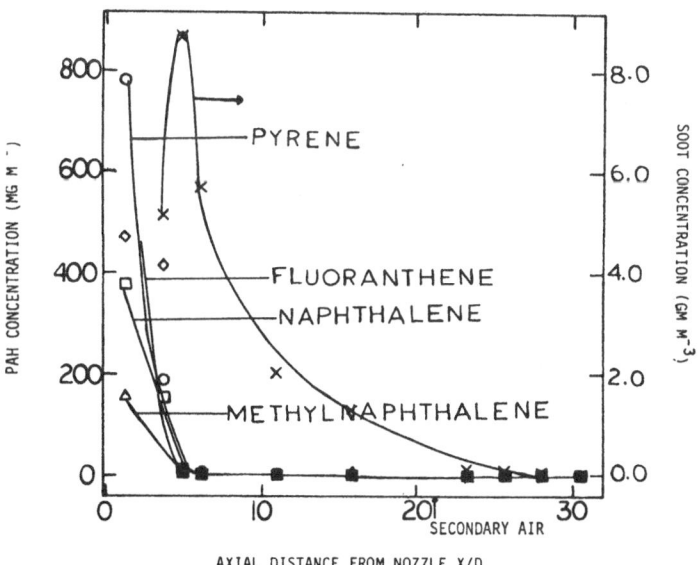

Fig. 13.   PAH concentration along axial distance for staged
           combustion (SRC II heavy distillate).

in the combustor, higher peak concentrations of PAH species are
found.

CONCLUSIONS

o   Soot peak concentrations were higher in the SRCII heavy distillate flame than in the flame produced with the lighter distillate,
SRCII 2.9/1. The emission levels of soot were also somewhat higher
(by a factor of 2) both in the unstaged and staged SRCII heavy distillate flames.

o   In both the unstaged and staged flames PAH concentrations measured along the flame axis sharply decay as soot is being formed.
This behavior underlines the suggestion that PAH compounds act as
soot precursors in flames.

o   In the burn-out region of the flame, where most of the oxidant
has been entrained into the jet, the soot concentration decays fast
at first but then approaches asymptotically a low level which is
thought to be determined by the freezing of the reaction due to
stretching of small eddies in high shear zones of the turbulent jet.

o   Of the PAH compounds attention was paid to some stable compounds
such as pyrene, phenanthrene and fluoranthene, which tend to maintain relatively high concentrations as PAH decays. The burn-out of
fluoranthene was followed with special interest because of its known
biological activity.

REFERENCES

1.   H. B. Palmer and C. F. Cullis, The formation of carbon from
         gases, in: "Chemistry and Physics of Carbon," P. L. Walker,
         ed., Marcel Dekker, N. Y., Vol. 1, p265 (1965).
2.   J. D. Bittner and J. B. Howard, Role of aromatics in soot
         formation, in: "Alternative Hydrocarbon Fuels: Combustion
         and Chemical Kinetics," C. T. Bowman and J. Birkeland, eds.,
         Progress in Aeronautics and Astronautics, American Institute
         of Aeronautics and Astronautics, New York, Vol. 62, p335
         (1978).
3.   B. S. Haynes and H. G. Wagner, "Soot Formation," (unpublished
         manuscript), Institut fur Physikalische Chemie der
         Universität Göttingen, 3400 Göttingen, West Germany.
4.   G. P. Prado, M. L. Lee, R. A. Hites, D. P. Hoult, and J. B.
         Howard, Soot and hydrocarbon formation in a turbulent dif-
         fusion flame, in: Sixteenth Symposium (International) on
         Combustion, The Combustion Institute, Pittsburgh, p649
         (1977).

5. B. F. Magnussen, B. H. Hjertager, J. G. Olsen and D. Bhaduri,
    Effects of turbulent structure and heat concentrations on
    soot formation and combustion in C_2H_2 diffusion flames, in:
    Seventeenth Symposium (International) on COmbustion, The
    Combustion Institute, Pittsburgh, p. 1383 (1979).
6. J. P. Appleton and J. B. Heywood, The effects of imperfect
    fuel-air mixing in a burner on NO formation from nitrogen
    in the air and the fuel, in: Fourteenth Symposium (Inter-
    national) on Combustion, The Combustion Institute, Pittsburgh,
    p777 (1973).
7. H. E. Eickhoff and K. Grethe, A flame-zone model for turbulent
    hydrocarbon diffusion flames, Combustion and Flame, 35:267
    (1979).
8. E. H. Hubbard, Comparison between turbulent jet diffusion
    flames of gaseous, liquid and pulverized coal, University
    of Sheffield Fuel Society Journal (1958).
9. D. K. Holliday, "The Radiation from Turbulent Jet Diffusion
    Flames of Liquid Hydrocarbons," Ph.D. Thesis, Department of
    Fuel Technology and Chemical Engineering, University of
    Sheffield, (Nov. 1955).
10. C. Holden and M. W. Thring, Emissivity and radiation of steam
    and air-atomized liquid-fuel flames, in: "Residential Con-
    ference on Major Developments in Liquid Fuel Firing,"
    (proceedings), The Institute of Tuel, London, pA-60 (1959).
11. J. M. Beér and G. Monnot, "Resultats des Etudes sur les Bruleurs
    à Pulverisation Mécanique," 5ème Journée d'Etudes sur les
    Flammes, French National Committee of the International
    Flame Research Foundation, Paris, 1963.
12. J. M. Beér, M. T. Jacques, W. F. Farmayan, J. D. Teare, "Design
    Strategy for the Combustion of Coal-Derived Liquid Fuels,"
    prepared by the Mass. Inst. of Techn. Energy Laboratory, for
    the Electric Power Research Institute (RP1412-6), Palo Alto,
    California (1981).
13. J. Chedaille and Y. Braud, "Measurements in Flames," Vol. 1 of
    industrial Flames, J. M. Beér and M. W. Thring, ed., Inter-
    national Flame Research Foundation, Crane Russak & Company,
    Inc., New York, N. Y., (1972).
14. Center for Health Effects of Fossil Fuels Utilization, Second
    Annual Report of Progress to the National Institute of
    Environmental Health Sciences (Center Grant No. 5P30 ESO2109-
    02), Mass. Inst. of Technology, Harvard University-MIT Divi-
    sion of Health Sciences and Technology, and MIT Energy Labor-
    atory, Cambridge, Mass. (June 30, 1980).

ACKNOWLEDGEMENTS

    The research work reported in this paper was supported by a
grant from the National Institute of Environmental Health of the
U.S.A., and in part by a contract from the Electric Power Research

Institute under which the SRCII fuels where made available. Special
thanks are due to Dr. W. C. Rovesti of EPRI for valuable discussions
on the characteristics of SRCII fuels. The valuable assistance was
given by the technical team of the MIT Combustion Research Facility,
Mr. Rolf Steendal, Mr. Don Bash and Mr. Joseph Gartland, graduate
students Mr. Calvin Galbriel and Mr. Vince Paul, and undergraduate
students Miss Tina Bahadori and Miss Hou Yee.

THEORETICAL MODELS FOR THE INTERPRETATION OF LIGHT SCATTERING

BY PARTICLES PRESENT IN COMBUSTION SYSTEMS

Antonio D'Alessio, Antonio Cavaliere and Pietro Menna

Istituto di Ricerche sulla Combustione, C.N.R.
Istituto di Chimica Industriale e Impianti Chimici
Università
Piazzale Vincenzo Tecchio - 80125 NAPLES (Italy)

INTRODUCTION

Particles are almost always present in practical combustion systems in form of input fuel and/or combustion products. Submicronic soot particles are exclusively present only in fuel rich flames produced by gaseous fuels, whereas other classes of particulates have to be considered simultaneously in the combustion of liquid fuels and pulverized coals. These are the fuel droplets or the coal particles in the micronic range, the cenospheres produced by incomplete oxidation of heavy hydrocarbons fuels or char, the micronic and submicronic inorganic components of the ashes.

A primary objective of the optical diagnostics is to obtain, through light scattering and extinction measurements, a time and space resolved characterization of the different classes of particulates in terms of shape, size distribution and optical properties. Similar problems are present in meteorology and astrophysics, and, therefore, some methodologies or results obtained in those studies could be transferred to combustion investigations. However, the high temperature, the fast kinetics and the space and temporal gradients typical of combustion systems require that much of the basic optical research has to be specifically forwarded to this field. The need of a systematic analysis of the optical properties of the combustion particulates has been only recently focused by the combustion scientific community.

In this paper we intend to make a preliminary exploration, on the basis of the theoretical predictions, of some potentialities of different elastic light scattering effects for the problem

327

outlined above. The formalisms of the complete scattering matrix
are presented and the optical properties of soot particles in the
visible and u.v. are shortly discussed. Properties of different
matrix elements are illustrated for spherical particles from
the Rayleigh limit (D << λ) up to geometrical optics formulation
(D >> λ), using the exact Lorenz-Mie theory in the size/wavelength
intermediate region. Characteristics of the scattering due to
spheroidal particles in the Rayleigh and intermediate size region
are also discussed as well as the properties of particles with
irregular shapes. The scattering properties of clouds where
small soot particles and larger oil droplets are simultaneously
present are also computed, as an example.

THE SCATTERING MATRIX

The properties of the scattered beam are fully characterized
by relating its four Stokes parameters with those of the incident
beam through a 16 element matrix.[1,2,3] When the first two Stokes
parameters of the incident beam are substituted by the light
intensities polarized in the vertical and horizontal plane I_{oV}
and I_{oH} and the corresponding ones of the scattered beam by the
energy fluxes analyzed in the vertical and horizontal plane P_V
and P_H, the matrix can be expressed as:[4,5,6]

$$
\begin{vmatrix} P_V \\ P_H \\ U \\ V \end{vmatrix} = \Delta V \, \Delta\Omega \begin{vmatrix} Q_{VV} & V_{HV} & 0 & 0 \\ Q_{VH} & Q_{HH} & 0 & 0 \\ 0 & 0 & F_{33} & F_{34} \\ 0 & 0 & F_{43} & F_{44} \end{vmatrix} \begin{vmatrix} I_{oV} \\ I_{oH} \\ U_o \\ V_o \end{vmatrix} \tag{1}
$$

The scatterers are assumed to be randomly oriented in
space and consequently the matrix independent elements to be
reduced to six [7,8] because $F_{34} = - F_{43}$ and $Q_{VH} = Q_{HV}$. $Q_{ij}(\theta)$ are
the scattering coefficients and they represent the energy flux
scattered by a scattering volume $\Delta V = 1$ under a solid angle
$\Delta\Omega = 1$.

The last two Stokes parameters U and V are defined by the
equations

$$
U = \frac{< 2E_{ox} E_{oy} \cos \delta >}{< E_{ox}^2 > + < E_{oy}^2 >}
$$

$$
V = \frac{< 2E_{ox} E_{oy} \sin \delta >}{< E_{ox}^2 > + < E_{oy}^2 >}
\tag{2}
$$

It is worth noting that the two adimensional elements of the matrix F_{33} and F_{43} are simultaneously determined when the incident light is plane polarized at 45° with respect to the scattering plane (U_o = 1 and V_o = 0); in this case:

$$U = F_{33} \quad \text{and} \quad V = F_{43} \tag{3}$$

The other two elements F_{34} and F_{44} are determined from the measurements of the U and V vectors of the scattered light when the incident light is completely circularly polarized (U_o = 0 and V_o = 1); in this case :

$$U = F_{34} \quad \text{and} \quad V = F_{44} \tag{4}$$

The angular scattering coefficients Q_{HH}, Q_{HV}, Q_{VH} can be expressed in nondimensional terms as polarization ratio $\gamma = Q_{HH}/Q_{VV}$ and depolarization ratios $\rho_V = Q_{HV}/Q_{VV}$ and $\rho_H = Q_{VH}/Q_{HH}$.

The Q_{ij} coefficients depend on the scattering angle θ, the scatterers number concentration N, their size, shape and refractive index. In the single scattering regime they are directly proportional to N so that, for a monodisperse system:

$$Q_{ij}(\theta) = N \, C_{ij}(\theta) \tag{5}$$

where $C_{ij}(\theta)$ is the angular cross section ($cm^2 sr^{-1}$); more particularly C_{VV} may be expressed in terms of incident wavelength λ and total scattering cross section C_{scatt} as:

$$C_{VV}(\theta) = \frac{\lambda^2}{4\pi^2} \, i_\perp(\theta) = \frac{C_{scatt}}{4\pi} \, p_{VV}(\theta) \tag{6}$$

where $i_\perp$ and p_{VV} are the adimensional scattering and phase functions.

OPTICAL AND PHYSICAL PROPERTIES OF SOOT AND OTHER PARTICULATES

The classical interaction between e.m. waves and a material medium is characterized by a complex refractive index m = n - ik, which generally varies with wavelength. Experimental determinations of these properties for carbonaceous materials like graphite, carbon, soot and coal have been carried out by different authors.[9,14]

Previous evaluations were mostly based on spectral reflectivity measurements on polished samples at room temperature. Only more recently this quantity has been estimated by spectral transmission

measurements on particle suspensions either at room temperature
or in flame.[13,14]

Data obtained from reflectivity are criticized because of the
difficulties in obtaining a true specular surface with a low void
fraction from sample made with compressed particles.[13,15,16]

The shape and size distributions of the particles have to be
known independently when the refractive index is derived from
transmissivity measurements. Therefore, this method is also open
to criticism.

Recently the refractive index of single micron sized carbon
particle has been measured by Pluchino et al.[17] from angular
scattering measurements on electrostatically suspended isolated
particles.

Figures 1 and 2 report the values of n and k obtained by
Taft and Philipp[10] for graphite and glassy carbon[18], Carter
et al.[11] for graphite, and finally those predicted by the
dispersion equations proposed by Dalzell and Sarofim[12] and Lee
and Tien[14] for soot particles; the range of the determination
of Pluchino et al. at λ = 488 nm is also reported.

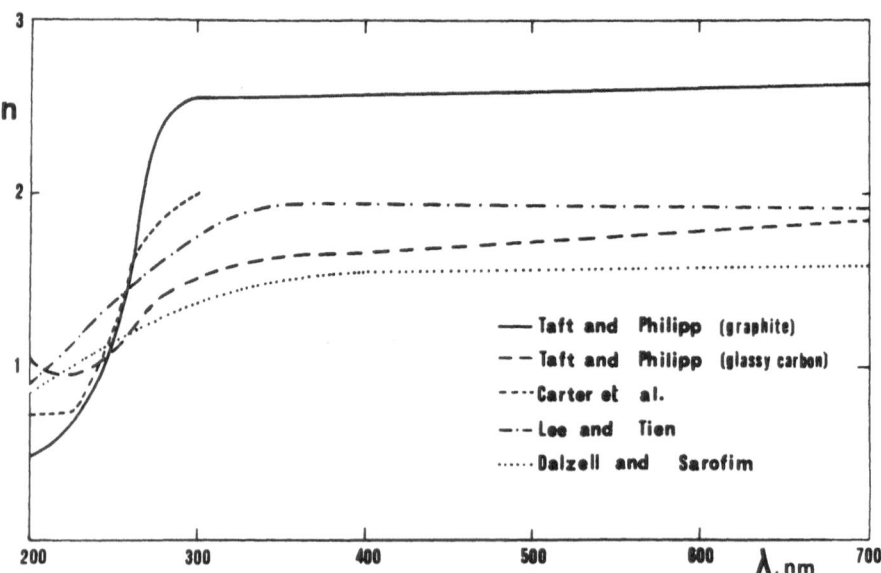

Fig. 1.   Real components of refractive index n in the spectral
range 200-700 nm for carbonaceous materials.

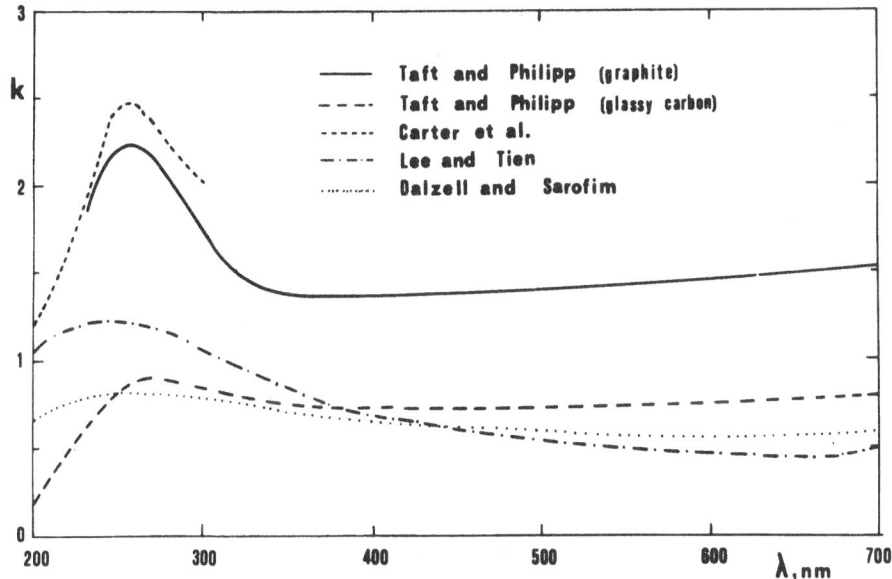

Fig. 2. Imaginary components of refractive index k in the spectral range 200-700 nm.

It is worth noting that the experimental results for graphite evidence a strong resonance in the u.v. which is not predicted by the soot dispersion equations and, furthermore, that the results of Dalzell and Sarofim are consistently lower than the other published data.

Futher implications of these data will be discussed in a subsequent section on the scattering and extinction properties of small soot particles in the Rayleigh limit. From a physical point of view, among the carbonaceous particles, are to be distinguished the small submicronic soot particles generated by the gas phase pyrolytic processes and the cenospheres produced by the heterogeneous pyrolytic processes in liquid or solid phases during the combustion of heavy hydrocarbons fuel and coal. Soot particles in the first nucleation and surface growth period are roughly spherical with diameter ranging from 2 to 30 nm; they become irregularly shaped clusters of spheres in the later coagulation regime and the typical size and compactness of the aggregates depend heavily on the residence time and physico-chemical conditions. conditions.

Cenospheres are the residual skeleton of hydrocarbon droplets and coal particles, and, therefore, they have the same size and shape as the parent particle between 5 and 300 nm but with a much higher porosity.

The partially burned coal particles fall also in this category: they have normally a more regular shape than the input particles since they loose their angularity while traversing the flame front.[19]

Liquid fuel droplets are constituted by a mixture of saturated, unsaturated and aromatic hydrocarbons, mainly substituted monocyclic compounds.

Figure 3 shows the trend of the real part of the refractive index, for paraffines, substituted monoring compounds and some two rings compounds as function of the number of carbon atoms.

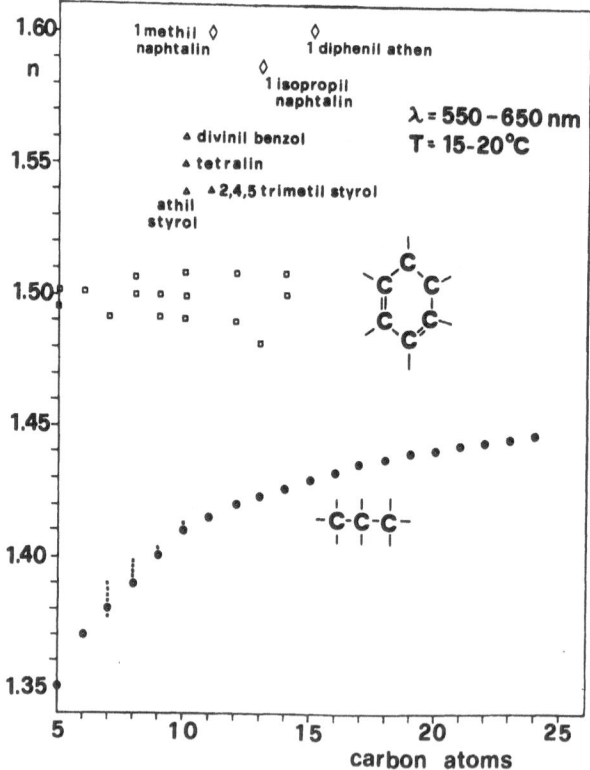

Fig. 3.   Real component of refractive index n for different liquid
          hydrocarbons as function of the number of carbon atoms.

The linear saturated compounds have refractive index values between 1.35 and 1.45 whereas aromatics range between 1.48 and 1.52 and polycyclic compounds exhibit values higher than 1.55.

Saturated and unsaturated chain compounds and the lighter aromatics do not exhibit absorption in the visible and only heavier polycyclic hydrocarbons absorb also in the visible, whereas banded absorption in the u.v. are normally shown by aromatics and unsaturated compounds.[20] The mineral matter and ashes present in coal combustion systems include a large variety of metal oxides, silicates and sulphates with different sizes.[21] Recent determination of the physical and optical properties of single electrostatically suspended fly ash particles have shown that most of them had spherical shapes with a refractive index spanning in the range 1.48-$1.57 \pm i$ $(0$-$0.01)$, at $\lambda = 632.8$ nm.[22]

The hypothesis that ash particles have a relevant absorption index has been advanced by Lowe et al.[23] on the basis of their radiation measurements in large pulverized coal flames, although different interpretations may be given to their results.[24]

RAYLEIGH SCATTERING

The scattering and extinction properties of particles much smaller than the wavelength are well described by the Rayleigh dipolar expression.

Soot particles in their early formation processes and metal oxides formed from vapour phase[25] fall in this category; also larger particles may be considered as Rayleigh scatter if the ratio $\alpha = \pi D/\lambda$ is reduced to values smaller than 0.3 by increasing the wavelength of the incident light source, using for instance a CO_2 laser.[26]

The matrix elements for spherical particles are given by the expressions:

$$C_{VV}(\theta) = \frac{\lambda^2}{4\pi^2} \left| \frac{m^2 - 1}{m^2 + 2} \right| \cdot \alpha^6$$

$$C_{HH}(\theta) = C_{VV}(\theta) \cos^2\theta$$

$$F_{33}(\theta) = \frac{2 \cos \theta}{1 + \cos^2 \theta}$$ \hfill (7)

$$F_{34}(\theta) = 0$$

We note that for spheres of any size:

$$C_{HV} = C_{VH} = 0$$

$$F_{33} = F_{44}$$

(8)

Consequently, measurements of angular distribution of the scattered light in any polarization conditions and of phase shift do not give information of the size of the scatterers.

The absorption coefficient is given by the expression:

$$C_{abs} = \frac{\lambda^2}{\pi} \alpha^3 I_m \left| \frac{m^2 - 1}{m^2 + 2} \right|$$

(9)

and the extinction coefficient by

$$C_{ext} = C_{abs} + C_{scatt}$$

$$= \frac{\lambda^2}{\pi} \left\{ \alpha^3 I_m \left| \frac{m^2 - 1}{m^2 + 2} \right| + \frac{2}{3} \cdot \alpha^6 \left| \frac{m^2 - 1}{m^2 + 2} \right| \right\}$$

(10)

The scattering contribution to the extinction is definitely negligible for the refractive index of the carbonaceous particles whereas it becomes more relevant for larger particles with a very low imaginary part of the refractive index. However, it is to note that in this case also the absorption prevails over the scattering; for instance, a 30 nm ash particle, with $m = 1.5 - i\ 0.01$, presents in the visible ($\lambda = 500$ nm) a contribution of less than 10% of the scattering to the total extinction.

Therefore, the ratio between scattering and extinction is proportional to the volume of the particle. This is the only method which gives the particle size in the Rayleigh regime. However, in practical flame conditions, it is very difficult to obtain the local values of the extinction coefficient. The other interesting application of the Rayleigh scattering is related to the wavelength variation of the scattering and extinction coefficient; in fact, they vary with the fourth and first inverse power of λ, respectively, only when the optical properties of the particles do not change with the wavelength.

Consequently, the difference between the experimental results and the theoretical predictions are to be attributed to changes in n and k, the values of which might be determined in principle with this method.

An example of such considerations is given in Fig. 4 where
the Rayleigh angular scattering at θ = 20° and the extinction cross
sections in the range 200-350 nm are reported for the complex
refractive index determinations illustrated in Figs. 1 and 2.

Scattering and extinction present maxima, at coincident
wavelengths, when the experimental data for graphite or carbons
are used, whereas they evidence a monotonic decrease where the
dispersion equations proposed by Dalzell and Sarofim and Lee and
Tien are employed.

The parameters of the oscillation of the bounded electrons,
with natural frequency near 250 nm, were treated as adjustable
parameters in both papers in order to obtain the best fit for the
experimental data in the visible and i.r. Therefore, their

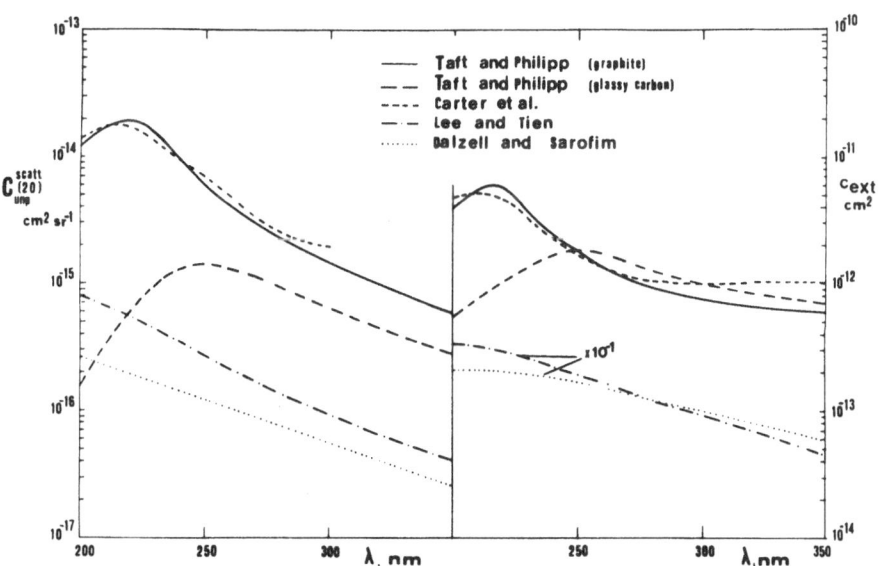

Fig. 4.   Spectral profiles of scattering cross section at θ = 20°
for unpolarized incident light (on the left) and of
extinction cross section (on the right) in the range
200-350 nm.

extrapolation in the u.v. is doubtful and, consequently, they fail
to predict maxima in the spectral scattering and extinction data.

Furthermore, it is interesting to note that the scattering
maxima is shifted toward larger wavelength for glassy carbon
with respect to that predicted for graphite.

From these preliminary observations, it is to be expected
that the position and shape of the u.v. scattering peaks might
give useful informations on the local and high temperature
optical properties of soot particles and, consequently, on their
chemical nature.

The properties of the Rayleigh absorbing spheroid allow a
first approximation discussion on the scattering and extinction
properties of irregularly shaped particles and explain some
particular features of the soot particles themselves.[27,28]

The theoretical expressions of the relevant cross sections
have been reported by Kerker [2], and a more detailed analysis on
their dependence on the complex refractive index and semiaxis ratio
has been done by Ferrara.[29]

The ratio between the scattering and extinction cross section,
with respect to those presented by the equal volume sphere, is
always slightly higher than 1; this ratio reaches limit values
of 1.2 for prolate spheroids and 1.6 for oblate spheroids with
semiaxis ratios higher than 30, for m = 1.56=i0.56.

The ratio C_{scatt}/C_{ext} is independent of the particle shapes
in the Rayleigh regime, as pointed out by Van De Hulst, and is
proportional to the volume of the particle.

The anisotropy of the scatterer introduces the depolarized
element in the scattering matrix Q_{HV} which does not depend upon
the scattering angle. The depolarization ratio ρ_V is again a
function of the complex refractive index and the semiaxis ratio[30]
and its measurements might give informations on these parameters.

VERY LARGE PARTICLES

When the size parameter α is much larger than one, the
interaction between the e.m. waves and the particle can be followed
using the classical theory of the ray optics, and the scattered
light can be computed as the sum of contributions due to diffraction, refraction, internal and external reflections.

The diffraction contribution is described by the Fraunhofer
expression through the first-order Bessel function:

$$P_{VV}(\theta) = 2 \cdot \alpha^2 \frac{J_1(\alpha \cdot \sin \theta)}{\sin \theta} \qquad (11)$$

Diffraction takes into account half of the total scattered light and it is confined in a very narrow solid angle in the forward region; for instance, a particle with $\alpha = 100$ scatters 90% of the diffracted light in a scattering angle range 0°-1°.[31]

The refraction and reflection contributions are computed through the Fresnel equations and their angular patterns on the vertical and horizontal planes of polarization are reported on Fig. 5 for a refractive index m = 1.5 $\pm$ i0.

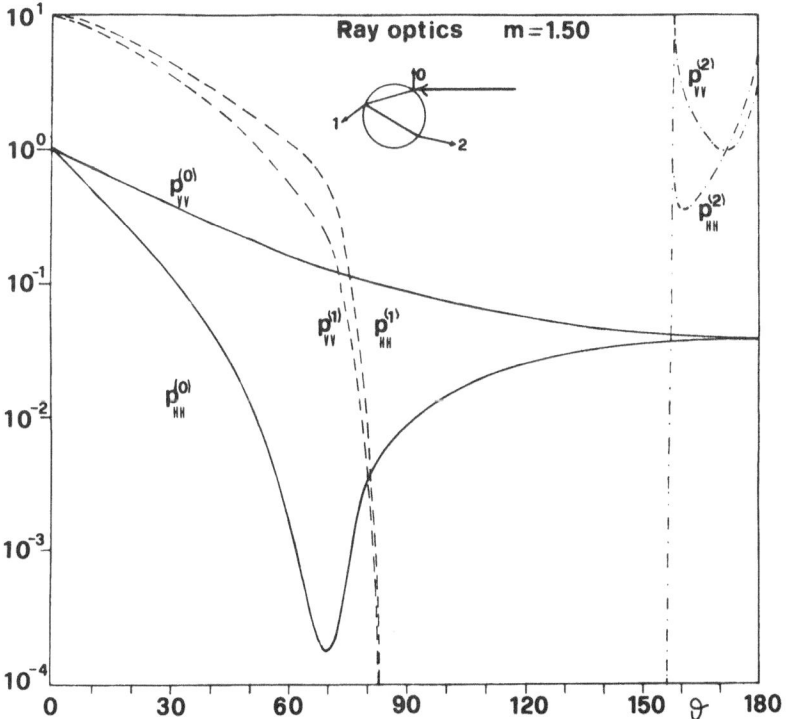

Fig. 5.   Angular patterns of components of phase function p_{VV} and p_{HH}. The superscript (o), (1), (2) stand for external reflection, simple refraction and internal reflection components.

The phase function due to the external reflection $(p_{VV}^{(o)}, p_{HH}^{(o)})$ are the only contribution presented by a sphere with high imaginary part of refractive index since, in this case, the light traversing the particle is completely absorbed. The vertically polarized light is slightly dissymmetric and it becomes more isotropic when the real and imaginary parts of refractive index increase. The limit tendency of this behaviour is that of the perfect reflector $(m \to \infty)$, where the scatter is completely isotropic. The horizontally polarized light shows a sharp minimum, in correspondence with the Brewster angle which shifts toward lower scattering angles, for material with higher reflectivity.

The components $(p_{VV}^{(1)}, p_{HH}^{(1)})$ of the scattered light due to simple refraction are prevailing over the reflected ones in the forward region $(0° < \theta < 180°)$. The $\gamma^{(1)} = p_{HH}^{(1)}/p_{VV}^{(1)}$ polarization ratio has values near one with a slight prevalence of the horizontal component, particularly near the Brewster angle $(\gamma^{max} \sim 1.4)$. Again, the presence of an absorption contribution to the refractive index decreases dramatically the relative amount of the refracted contribution to the total scattering, and consequently the overall pattern in the forward zone tends toward that produced by the external reflection effects alone; therefore, the γ ratio near the Brewster angle is a very sensitive indicator of even low imaginary parts of the refractive index.

Single and multiple reflections do not contribute significantly to the total angular scattering except for very localized angles where high intensity rainbows occur. The very marked primary rainbow exhibit a sharp peak at $\theta = 157°$ for $m = 1.5$ and it is strongly polarized in the vertical plane $(p_{VV}^{(2)} > p_{HH}^{(2)})$. The secondary rainbow is due to light emerging from the sphere after two internal reflections; it takes place near $\theta = 90°$ and, since it is weaker than the primary one, it is easily masked when the refractive index is complex.

The ray optics theory is not strictly correct since it does not take into account the interference effects due to change of phase of the incident radiation inside the particle. In fact, the intensity of the rainbows, and the presence of glory in the backscattering regions are not satisfactorily predicted by this approximation; furthermore, it does not define the minimum size of the sphere for which the angular patterns on both polarization planes are adequately described. The exact solution of the scattering of spheres of any size is given by the Lorenz-Mie theory in terms of the i parameters defined by eq. 6. [1,2]

A number of terms in the series expansion of the order of α has to be considered in order to ensure the convergence [1,32], and furthermore the angular patterns present a sequence of oscillations, due to resonant interference effects also proportional

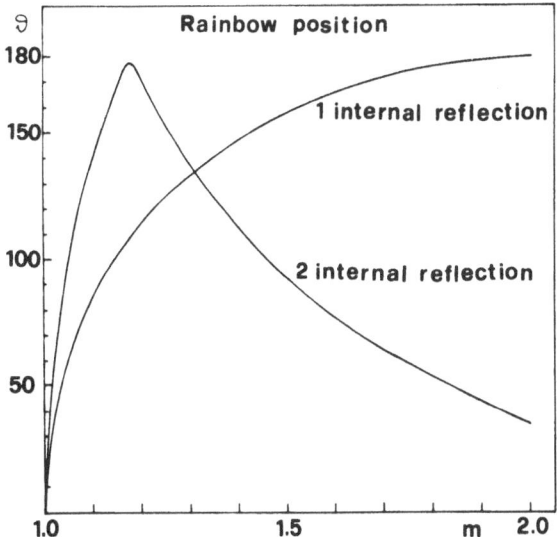

Fig. 6.   Angular position of primary rainbow (1 int. reflection)
and secondary rainbow (2 int. reflection) as function
of real part of refractive index m.

to α. Obviously, a polydispersion of sizes reduces drastically the
oscillatory behaviour also for a small standard deviation.[33]

Fig. 7 reports a comparison between the vertical components of
the scattered light computed with the ray optics, and the Lorenz-
Mie solution for α = 300, and in this case the geometric optics
follow the average values of the exact theory. The same agreement
is shown by the γ ratio in Fig. 8, although the oscillations in
the θ = 100°-150° range are particularly relevant.

A numerical analysis shows that, for m = 1.5, the agreement
between the ray optics and the Mie theory is acceptable for the
horizontal scattered component as low as α = 100. For the

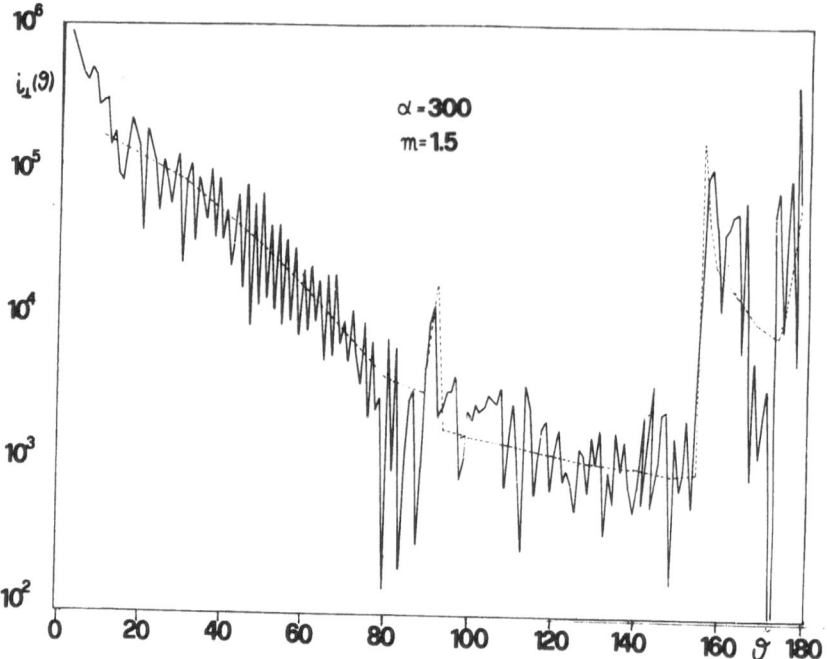

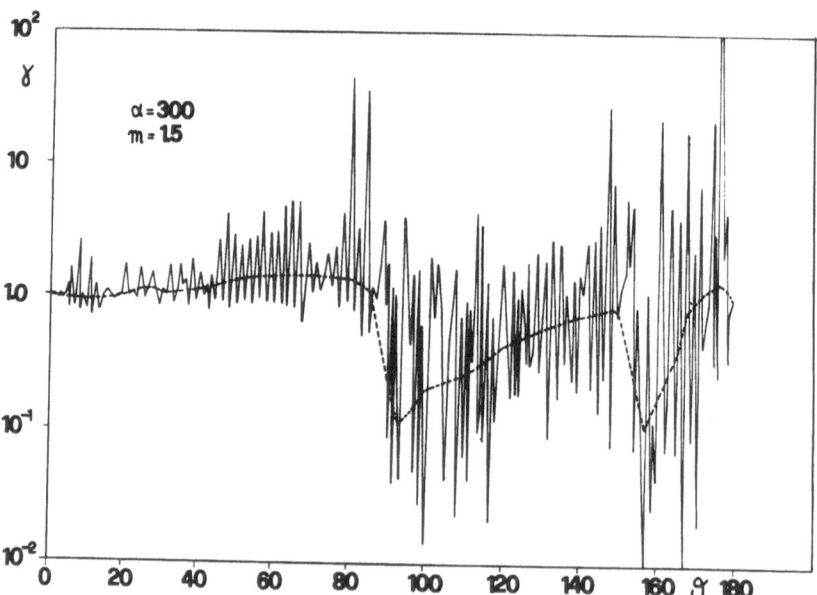

Figs. 7-8.   Angular patterns of scattering function $i_\perp(\theta)$ and
              $\gamma = i_{//}/i_\perp$ polarization ratio for size parameter $\alpha = 300$.

vertical one it extends to $\alpha = 80.$[34]

SPHERES OF INTERMEDIATE SIZE ($\alpha < 20$)

The calculations for large particles of the previous section are also representative of the scattering of micronic fuel droplets, pulverised coal and the larger components of ashes. However, the α values might be easily decreased by a factor of 20 by using a CO_2 infra-red laser source ($\lambda = 10.6$ μm) instead of a visible light source ($\lambda = 0.5$ μm). Conversely, the α value of a submicronic scatterer, like a primary soot particle or a metal oxide nucleated from vapour phase, might be increased by employing an u.v. incident source. Therefore, the scattering and extinction properties of spheres between $\alpha = 1$ and $\alpha = 20$ have to be explored for dielectric and absorbing cases.

Figure 9 reports the vertically polarized light intensities presented by spheres with $\alpha = 17$ for $m = 1.5$ and $m = 1.57-0.56i$.

Although the differentiation in reflection, refraction and diffraction should be properly used only in the limits of ray optics, this terminology may be useful also for smaller particles. For instance, it is evident, in the case of $m = 1.5$, that the presence of high intensity scattered light in the $150°-180°$ angular range is due to one internal reflection and the continuous angular decrease in the forward region, characteristic of a refraction dominated zone. Finally, the forward lobe restricted in the $\theta = 0°-10°$ range is clearly due to diffraction effects, and is not dependent on refractive index; namely the first minimum at $\theta = 10°$ is clearly evidenced also for $m = 1.57-0.56i$.

The oscillations in this case are damped, as compared to the case of $m = 1.5$, by the absorption of the light traversing the particle; they are detectable only in the forward region and some of them, in the range $\theta = 30°-80°$, are in phase with the angular oscillations for both refractive indices.

The γ polarization ratio, reported in Fig. 10, for the same size and refractive indices, presents a simple external refraction behaviour for the absorptive particle with an averaged minimum in correspondence with the Brewster angle; for $m = 1.5$, their values are greater than on in the forward region, thus showing the preeminent contribution of refraction, as in the ray optics limit.

All these features tend to disappear when decreasing the particle size: only four large oscillations are shown for a dielectric particle with size parameter $\alpha = 4$, as illustrated in Figure 11. The forward lobe is clearly enlarged reaching $40°$

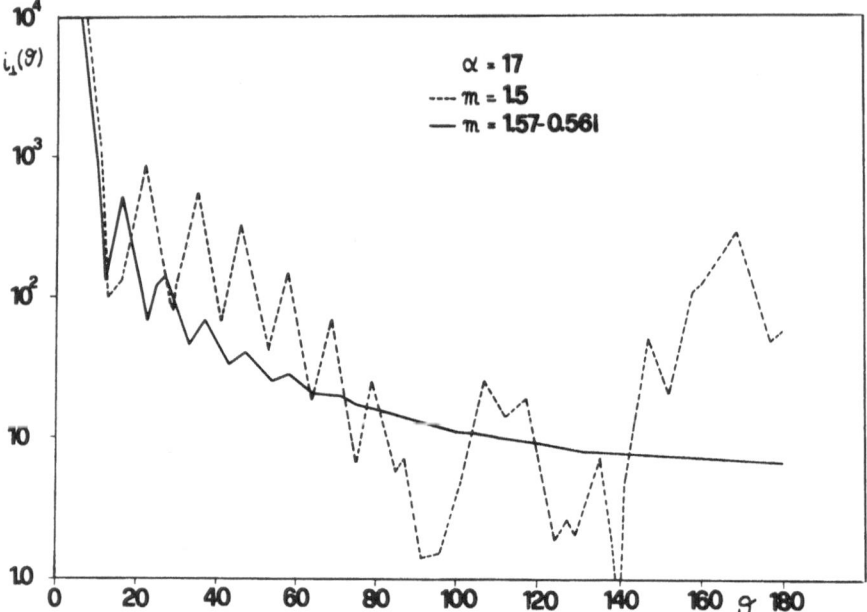

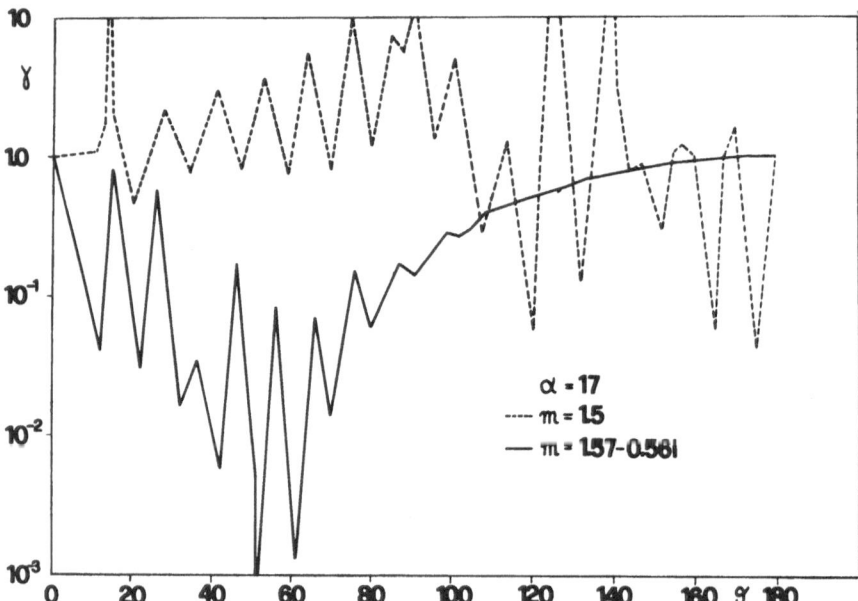

Figs. 9-10. Angular patterns of scattering function $i_\perp(\theta)$ and
             $\gamma = i_\parallel / i_\perp$ polarization ratio for a size parameter
             $\alpha = 17$ and for refractive indices = 1.5 and
             $m = 1.57-0.56i$.

and the strong peak, in the backscattering region, evidences that
it is still relevant to the contribution due to two internal
passages of light.

A more smoothed behaviour is exhibited by the angular pattern
when the refractive index is complex, some oscillations are evident
in the forward region, thus following the behaviour of larger
particles. Finally the γ polarization ratio of Fig. 12 loses all
relation with the large particle trends, showing for both indices
similar oscillations, the average values of which decrease with
the presence of absorption effects.

The properties of absorbing spheres in the regime of transition
between the Rayleigh limit and the Lorenz–Mie treatment ($0.4 < \alpha
< 1.5$) are of interest as a first approximation of the behaviour
of soot aggregates. Many characteristics of polydispersed clouds
of such particles have been examined in a previous paper:[30]
the dissymetry, polarization and extinction/scattering ratios
have been computed for different parameters of a log-normal
distribution and the sensitivity of those methods for diagnostic
purposes has been discussed too.

Finally, it is useful to comment on the physical meaning
of the F_{33} and F_{34} scattering matrix elements on the ground of
their predicted trends.

Since U and V represent the degree of diagonality and
circularity of the light, F_{33} and F_{34} are respectively a measure
of conservation of the diagonally plane polarized light and
of transformation of a circularly polarized light into diagonally
plane polarized light. The scattering is dominated in this
region by a double refraction, which does not change the polarization state of the radiation traversing the particle and consequently
F_{33} is near one and F_{34} is near zero.

In the backscattering region F_{33} decreases and eventually
becomes negative for very large particles, up to the regions
where the primary rainbow and glory enhance F_{33} again to positive
values.

Liou and Hansen [35] suggested for the computation of F_{33} the
simplified expression:

$$F_{33} = \frac{2 \cdot \sqrt{P_{VV} \cdot P_{HH}}}{P_{HH} + P_{VV}}$$

The comparison with the Mie solution shows a reasonable
agreement of the two profiles in the forward region; F_{34} has

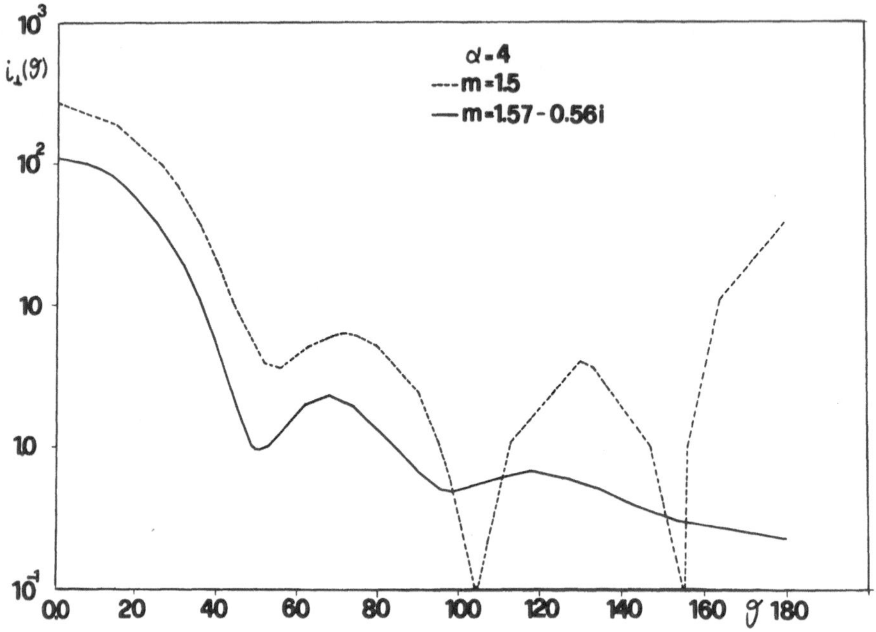

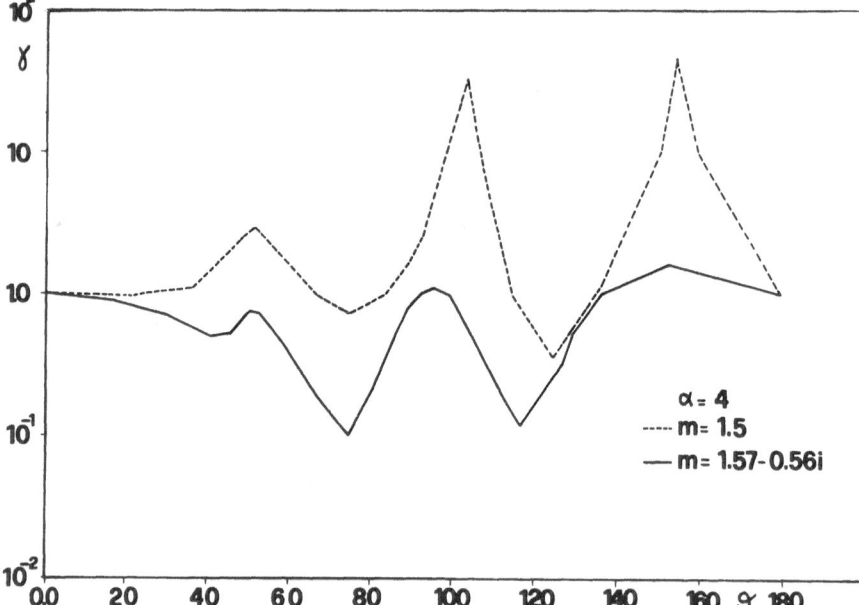

Figs. 11-12. Angular patterns of scattering function $i_\perp$ and
$\gamma = i_{//}/i_\perp$ polarization ratio for a size parameter
$\alpha = 4$ and for refractive indices m = 1.5 and
m = 1.57-0.56i.

been computed only for real refractive index and it shows a
region of significant values in the side scattering.[35,36,37]

PARTICLES WITH SPHEROIDAL AND IRREGULAR SHAPES

Some of the particulates present inside combustion systems
do not have spherical shapes: the more evident case is presented
by the fully agglomerated soot particles, but coal and ashes
fragments may also fall in this category. Oblate and prolate
spheroids, the limit shapes of which are plates or needles
respectively, are often a good approximation for irregular particles which do not exhibit concavities.

Some properties of the Rayleigh spheroid have been briefly
reported in a previous paragraph; scattering and extinction
properties in the transition regime between Rayleigh and Mie may
be computed with the so-called third order expansion of Stevenson[1]
which can be applied to spheroids with a larger axis up to 1/3rd
of the wavelength ($\alpha < 0.95$) , with error no bigger than 5%.

Heller and Nakagaki [38] have applied this theory to randomly
oriented spheroids with real refractive index up to m = 1.4 and
the computation has been extended, more recently, to absorbing
spheroids.[29,30] The results predict that the i_{VV} angular pattern
follows quite closely the one presented by the isovolume sphere
and that the volume does not influence the depolarization ratio ρ_V,
which is function only of the semiaxis ratio and the refractive
index, C_{HV} cross section has small angular variation and C_{HH} has
still a minimum at $\theta = 90°C$.

A more complete characterization of the properties of randomly
oriented spheroids is given in a remarkable paper by Asano and
Sato.[39] The authors have presented the exact solutions for all
matrix elements of spheroids with α up to 20 and with real refractive index equal to 1.33, in comparison to the corresponding values
for spheres.

Their results evidence that the scattering cross sections
values are equal to those for the volume equivalent sphere up
to a size parameter $\alpha = 5$, and this equality extends to $\alpha = 15$
for the absorption cross section. This last result is in agreement
with the general rule which states that the absorption is proportional to the volume of the scatterer for $2\alpha K < 1$ and for any kind
of spheroids.

Their computed angular patterns show that the unpolarized
scattered light exhibits quite reasonable agreement in the forward
region with the area equivalent spheres, whereas in the backward
region the characteristic resonance effects due to rainbow and
glory disappear.

The γ polarization ratio (corresponding to their degree of linear polarization) is less than one in the side scattering region and this tendency increases for very elongated prolate and very thin oblate spheroids. The ρ_V depolarized ratio is surprisingly more intense for spheroids with smaller semiaxis ratios, while ρ_V decreases for very elongated shapes. For randomly oriented spheroids F_{44} is in general larger than F_{33} and the difference between these two elements is related to the depolarized scattered light through the inequality relationship $\rho_V \leq F_{44} - F_{33}$, as it is also pointed out by Perry et al. [8]

The authors have also noted that the prolate spheroids present generally scattering patterns more similar to the sphere than the oblate ones. They suggest too that the angular analysis of the depolarization and polarization ratios appear to be particularly promising for evidencing shape effects.

Non absorbing irregular particles with relatively compact shapes have scattering and extinction characteristics very similar to those predicted for spheroids. The extinction cross sections and the angular pattern follow the prediction of the Mie theory for small values of the ratio of a characteristic dimension to the wavelength. The similarity of the phase function with the sphere is restricted to the forward lobe ($\theta < 60°$) when the particles become larger; the observed patterns are much higher in the side scattering region ($80° < \theta < 140°$) than those predicted by equal volume spheres but conversely they are much lower in the backscattering. [8,40]

Pollack and Cuzzi [41] tried to model the scattering and extinction of this class of particles with a combination of Mie theory, physical optics, geometrical optics, and parametrization. The diffracted and reflected contributions were easily assimilated to those due to spheres, whereas the transmitted component is primarily responsible for the deviation in the scattering behaviour of irregular particles from that of their spherical counterparts. [42]

Randomly oriented irregular absorbing particles with convex shapes have their scattering diagram completely determined by diffraction and Fresnel reflection only, when the absorption is strong enough to ensure that the refracted part of radiation is totally absorbed. [6] Aggregates of partially fused smaller particles present many concavities and have "fluffy" surfaces; the interpretation of the scattering properties of this important class of particulate is not possible with reference to spheroids or spheres.

Zerull et al. [6] have measured the angular pattern of absorbing aggregates with $\alpha = 31$ and $m = 1.65-0.25i$ in the microwave region and found that the scattering in the side and backward regions is more isotropically distributed than it is expected from a Fresnel analysis showing an enhancement near 180°.

Loose aggregates of soot particles might be treated as chains or clusters of smaller spherical elementary particles with diameter around 200-300 Å. Approximate solutions for linear chains and clusters formed by absorbing Rayleigh spheres have been presented by Ravey and Jones.[43,44] The latter author calculates the angular patterns of polarized and depolarized components as functions of the diameter of the elementary Rayleigh particles, of their number, for complex refractive indices near those expected for soot particles.

SCATTERING BY DIFFERENT CLASSES OF PARTICLES.

It happens often that in practical combustion systems different types of particles are present, on time averaged basis, in the scattering volume. Therefore, it should be necessary to apply the considerations, outlined in the previous paragraphs, on the scattering properties of particles of different sizes, shapes and refractive index to this situation.

It is clear that the possibilities of combining concentrations and types of particulates are enormous but it is not the aim of the present paper to examine systematically this matter. Only some calculations will be presented and discussed on a particular system composed by small submicronic absorbing particles with $\alpha = 0.3$ and $m = 1.56 - i.0.57$, and large non-absorbing particles with $\alpha = 300$ and $m = 1.5$.

This mixture composition refers to the formation of soot particles in presence of oil droplets on the spray combustion of liquid hydrocarbons fuel. However, it might be considered representative also of the late stages of coal combustion where soot particles coexist with non absorbing ashes. The angular properties of the Q_{VV} scattering coefficient and of the ratio $\gamma = Q_{HH}/Q_{VV}$ have been investigated by keeping constant the total volume fraction ϕ_{tot} occupied by the particles, and changing the relative amount of the different particles. In this case

$$\phi_{tot} = X \frac{N_s \pi D_s^3}{6} + \frac{N_d \pi D_d^3}{6} (1 - X) \tag{13}$$

with

$$X = \frac{\phi_s}{\phi_{tot}} \tag{14}$$

where the subscripts s refer to soot particles and d to the fuel droplets, X = 0 is representative of an exclusive presence of droplets in the scattering volume. Figure 13 reports the angular behaviour of the quantity Q_{vv}/ϕ_{tot} for X ranging from $1 \cdot 10^{-3}$ to 0.8. Curves with very small X values follow obviously the typical pattern of the large spheres with relevant primary and secondary rainbows and a large amount of light scattered in the Fraunhofer forward lobe. The levelling effect due to small particles begins to be seen at X = 0.04 where the secondary rainbow is much less evident; for X = 0.4, also, the primary rainbow is barely evident. It is worth noting that when the large particles are in relatively small amount (X = 0.8) their contribution is appreciable only in the forward lobe.

The   ratio is reported in Figure 14 for the same X values: it is always near one when few soot particles are present (X = $1 \cdot 10^{-3}$) except for a marked vertically polarized region in correspondence to the primary rainbow. The polarization ratio declines at all the angles when the amount of soot is increased and at X = 0.8 its pattern follows quite closely the Rayleigh law.

Therefore, it is evident that a dissymmetry ratio which includes the primary rainbow angle and a polarization ratio near θ = 90° are the most sensitive measurement techniques, in order to distinguish between the contribution to the total scattering due to small and large particles. The last technique has been recently used by the authors in their experiments on unconfined diesel oil spray flames.[45,46]

CONCLUSIONS

The purpose of this paper is to give a broader prospect to the application of light scattering techniques to the diagnostics of soot forming flames. Therefore, topics as wavelength variaitons of the scattered and absorbed light in presence of resonance of the optical properties, properties of large and intermediate size spheres, semiquantitative characteristics of spheroids and other irregular shapes were only briefly touched. A systematic analysis of all these subjects and of their diagnostic application to a multicomponent scattering system will require a much longer and detailed discussion.

The final comment is that in a near future it will be possible to follow with optical techniques the chemical and physical evolution of many particulates, including soot, in practical combustion systems, if the progresses in the theoretical knowledge on the optical properties and on the tunable laser source are properly employed by the combustion scientists.

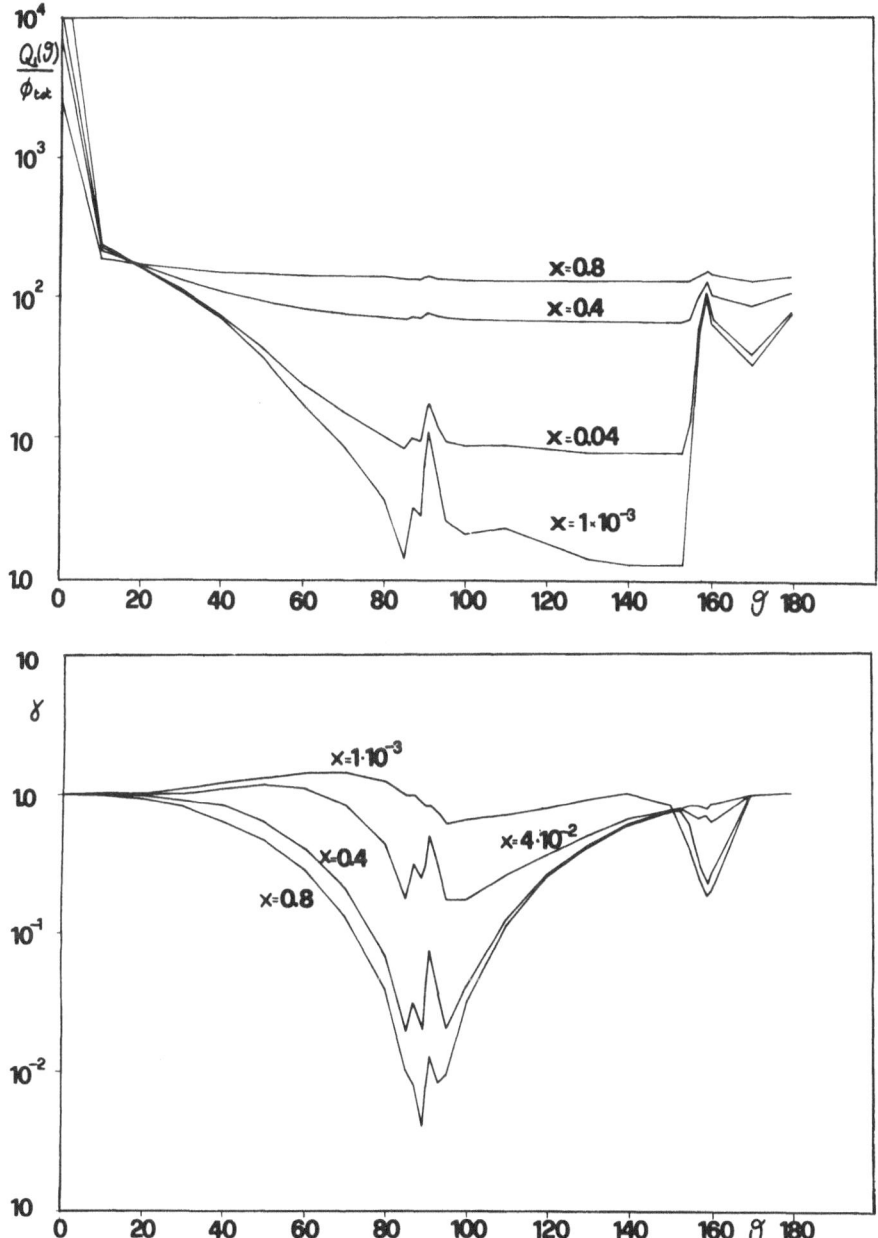

Figs. 13-14.   Angular patterns of the scattering coefficient Q_{VV}
divided by the total volume fraction ϕ_{tot} and the
$\gamma = Q_{HH}/Q_{VV}$ polarization ratio for a mixture of monodisperse soot particles with size parameters $\alpha = 0.3$
and of monodisperse oil droplet with $\alpha = 300$.

REFERENCES

1. H. C. Van de Hulst, "Light Scattering by Small Particles",
   J. Wiley, New York (1957)
2. M. Kerker, "The Scattering of Light and Other Electro-
   magnetic Radiations", Academic Press, New York (1969)
3. A. R. Jones, Scattering of electromagnetic radiations
   in particulate laden fluids, Progr. En. Comb. Sci.
   5:73 (1979)
4. F. Beretta, A. Cavaliere and A. D'Alessio, Experimental
   and theoretical analysis of the angular pattern
   distribution and polarization state of the light scattered by isothermal sprays and oil flames. ASME Winter
   Meeting, Two Phase Combustion, November (1981)
5. R. G. Pinnick, D. E. Carroll and D. J. Hoffmann, Polarized
   light scattered from monodisperse randomly oriented
   nonspherical aerosol particles measurements, Appl.
   Opt. 15:384 (1976)
6. R. H. Zerull, Scattering measurements of dielectric and
   absorbing nonspherical particles, Beiträge zu
   Physik der Atmosphäre 49:168 (1976)
7. F. Perrin, Polarization of light scattered by isotropic
   opalescent media, J. Chem. Phys. 10:414 (1942)
8. R. J. Perry, A. J. Hunt and D. R. Huffman, Experimental
   determinations of Mueller scattering matrices for
   nonspherical particles, Appl. Opt. 17:2700 (1978)
9. H. Senftleben and E. Benedict, Uber die Optischen
   Konstanten und die Strahlungsgesetze der Kohle, Annalen
   der Physik 54:65 (1918)
10. E. A. Taft and H. R. Philipp, Optical properties of
    graphite, Phys. Rev. 138:A197 (1965)
11. J. G. Carter, R. H. Huebner, R. N. Hamm and R. D. Birkhoff,
    Optical properties of graphite in the region 1100 to
    3000 A, Phys. Rev. 137:A639 (1965)
12. W. H. Dalzell and A. F. Sarofim, Optical constants of
    soot and their application to heat-flux calculations,
    J. Heat Transfer (Trans. ASME Ser. C) 91:100 (1969)
13. J. Jänzen, The refractive index of colloidal carbon, J.
    Coll. Interf. Science 69:436 (1979)
14. S. C. Lee and C. L. Tien, Optical constants of soot in
    hydrocarbon flames, 18th Symp. (Int'l) on Combustion,
    The Combustion Institute, Pittsburgh (1981)
15. S. C. Graham, The refractive indices of isolated and of
    aggregated soot particles, Comb. Sci. Tech. 9:159 (1974)
16. W. G. Egan and T. Hilgeman, Anomalous refractive index
    of submicron-sized particulates, Appl. Opt. 19:3724
    (1980)
17. A. B. Pluchino, S. S. Goldberg, J. M. Dowling and C. M.
    Randall, Refractive-index measurements of single micron-
    sized carbon particles, Appl. Opt. 19:3370 (1980)

18. E. A. Taft, Personal communication (1969)

19. L. Smoot, M. D. Horton and G. A. Williams, Propagation of Laminar pulverized coal size flames, 17th Symp. (Int'l) on Comb., The Combustion Institute, Pittsburgh, p.375 (1979)

20. H. N. Jaffè and M. Orchin, "Theory and application of ultra-violet spectroscopy", J. Wiley, New York (1962)

21. T. F. Wall, A. Lowe, L. J. Wibberley and I. McC. Stewart, Mineral matter in coal and the thermal performance of large boilers, Prog. Energy Combust. Sci. 5:1 (1979)

22. P. J. Wyatt, Some chemical, physical and optical properties of fly ash particles, Appl. Opt. 19:975 (1980)

23. A. Lowe, I. McC. Stewart and T. F. Wall, The measurement and interpretation of radiation from fly ash particles in large pulverised coal flames, 17th Symp. (Int'l) on Comb., The Combustion Institute, Pittsburgh, p.105 (1979)

24. A. F. Sarofim, Comment to ref. 23, p.113

25. R. C. Flagan, Submicron particles from coal combustion, 17th Symp. (Int'l) on Comb., The Combustion Institute, Pittsburgh, p.97 (1979)

26. K. Sassen, Infrared (10.6 m) scattering and extinction in laboratory water and ice clouds, Appl. Opt. 20:185 (1981)

27. J. Embury, Absorption by small non-spherical particles in the Rayleigh region, in: "Light Scattering by Irregularly Shaped Particles", D. W. Scherman , ed., Plenum Press, New York, p.97 (1980)

28. D. R. Huffman and C. F. Bohren, Infrared absorption spectra of non-spherical particles treated in the Rayleigh-ellipsoid approximation, in: "Light Scattering by Irregularly Shaped Particles", D. A. Scherman, ed., Plenum Press, New York, p.97 (1980)

29. P. Ferrara, Analisis teorico-sperimentale delle forme e polidispersione di particelle carboniose in fiamme di metano ed ossigeno premiscelati con misure ottiche "in situ" , Tesi di Laurea in Ingegneria Chimica, Napoli (1977)

30. A. D'Alessio, Laser light scattering and fluoresence diagnostics of rich flames produced by gaseous and liquid fuels, in: "Particulate Carbon: Formation During Combustion", D. C. Siegla and G. W. Smith, eds., Plenum Press, New York, in press

31. J. R. Hodkinson and I. Greenleaves, Computations of light-scattering and extinction by spheres according to diffraction and geometrical optics, and some comparisons with the Mie theory, J. Opt. Soc. Am. 53:577 (1963)

32. H. M. Nussenzveigh, Complex angular momentum theory of the rainbow and the glory, J. Opt. Soc. Am. 69:1068 (1979)

33. J. V. Dave, Effects of coarseness of the integration increment on the calculation of the radiation scattered by polydispersed aerosols, Appl. Opt. 8:1161 (1969)

34. G. Viola, Analisi teorica delle proprietà di diffusione
     della luce da particelle di gasolio e di fuliggine,
     nell'ambito della teoria di Lorenz-Mie, Tesi di Laurea
     in Ingegneria Chimica, Napoli (1981)

35. K. Liou and J. E. Hansen, Intensity an polarization for
     single scattering by polydisperse spheres: a compari-
     son of ray optics and Mie theory, J. Atm. Sci. 28:995
     (1971)

36. A. J. Hunt and D. R. Huffman, A polarization modulated
     light scattering instrument for determining liquid
     aerosol properties, Japan J. Appl. Phys. 14(Suppl):
     14-1 (1975)

37. R. Eiden, Determination of the complex index of refraction
     of spherical aerosol particles, Appl. Opt. 19:962
     (1980)

38. W. Heller and M. Nakagaki, Light scattering of spheroids.
     Depolarization of the scattered light, J. Chem. Phys.
     61:3619 (1974)

39. S. Asano and M. Sato, Light scattering by randomly
     oriented spheroidal particles, Appl. Opt. 19:962
     (1980)

40. A. C. Holland and G. Gagne, The scattering of polarized
     light by polydisperse systems of irregular particles,
     Appl. Opt. 9:1113 (1970)

41. J. B. Pollack and J. N. Cuzzi, Scattering by nonspherical
     particles of size comparable to a wavelength: a new
     semi-empirical theory and its application to tropo-
     spheric aerosols, J. Atm. Sci. 37:868 (1980)

42. R. H. Giese, K. Weiss, R. H. Zerull and T. Ono, Large
     fluffy particles: a possible explanation of the
     optical properties of interplanetary dust,
     Astron. Astrophys. 65:265 (1978)

43. J. C. Ravey, Light scattering by aggregates of small
     dielectric or absorbing spheres, J. Coll. Interf. Sci.
     46:139 (1974)

44. A. R. Jones, Electromagnetic wave scattering by assemblies
     of particles in the Rayleigh approximation, Proc. R.
     Soc. Lond. A366:111 (1979)

45. F. Beretta, A. Cavaliere, A. Ciajolo, A. Di Lorenzo, C.
     Langella and C. Noviello, Laser light scattering,
     emission/extinction spectroscopy and thermogravimetric
     analysis in the study of soot behaviour in oil spray
     flames, 18th Symp. (Int'l) on Comb., The Combustion
     Institute, Pittsburgh (1981)

46. F. Beretta, A. Cavaliere, A. D'Alessio, C. Noviello and
     C. Scodellaro, Investigation on oil spray flame with
     laser light scattering and extinction techniques, La
     Rivista dei Combustibili 34:383 (1980).

DISCUSSION

H. Gg. Wagner (Universität Göttingen)

Did you find an indication for glory scattering?

D'Alessio

The scattered light due to glory is confined in a very narrow scattering angular region between 175-180° for large particles ($\alpha > 100$).

We have evidenced by theoretical computation that for smaller sizes the glory can be embedded in the rainbow region and the same should happen for real refractive index higher than 1.7 since the rainbow position shifts backward with increasing refractive index.

We did not measure scattered intensity for angle greater than 170°, but this angular region has been widely used by meteorologists and astrophysicists.

LASER LIGHT SCATTERING AND FLUORESCENCE IN FUEL RICH FLAMES:

TECHNIQUES AND SELECTED RESULTS

A. D'Alessio, F. Beretta, A. Cavaliere and P. Menna

Istituto di Ricerche sulla Combustione C.N.R.
Istituto di Chimica Industriale e Impianti Chimici
Università
P. le Vincenzo Tecchio 80125 Naples, Italy

INTRODUCTION

Practical flames are normally produced starting from hydro-carbons mixtures often poorly defined from a chemical point of view; they are introduced into the system in the form of liquid and/or solid particles. The presence of regions where the fuel pyrolysis takes place are almost unavoidable given the intrisic times of the vaporization and devolatilization processes. Conse-quently lighter compounds like C_2H_2, C_2H_4 are generated together with heavier ones like Polynuclear Aromatic Hydrocarbons (PAH) and eventually soot particles. In such systems, the development of optical diagnostics for simultaneous presence of different classes of particles and for the greater variety of the hydrocarbon species presents problems different from those in the lean or stoichiometrc combustion of gaseous fuels. Furthermore the flames show intense luminosity, space inhomogeneity and time fluctuations, due to turbulence or massive recirculation.

The theoretical background for the interpretation of elastic light scattering effects has been exposed in another contribution to this volume.[1] The present paper deals with experimental potentialities and limitations and treats of interference effects among different classes of scatterers. A selection of problems and results obtained by the authors, rather than a systematic framework, is presented, with an emphasis on the measurements on spray oil flames, gaseous diffusion and premixed flames.

SCATTERING AND EXTINCTION EXPERIMENTAL SYSTEM

A schematic view of a typical set up for the scattering /
extinction measurements is reported in Fig. 1. A laser is normally
used as light source since it presents the advantage of a high
intensity monochromatic and almost collimated incident light beam.
The traditional xenon and mercury high pressure lamps are still
useful particularly in the spectral region near 200 nm, where
tunable dye lasers have low efficiency.

Any polarization state of the incident beam is easily obtained
by a suitable combination of polarizers and retarders. The scattering angle and the polarization state are therefore considered as
independent variables. When the scattering angle $\theta = 0°$, the
experiment detects the extinction on the optical pathway whereas,
for $\theta \neq 0°$, the scattering from the volume element ΔV is analyzed.

The shape and size of the scattering volume, determined by
the intersection of the incident and scattered beams depend on
the focusing of both beams and on the scattering angle. Incident
laser beam can be easily focused down to 200 μm; very small
scattering volume ($\sim 10^{-6}$ cm^3) is consequently acquired. Local
measurements are achieved when the attenuation of the in-out

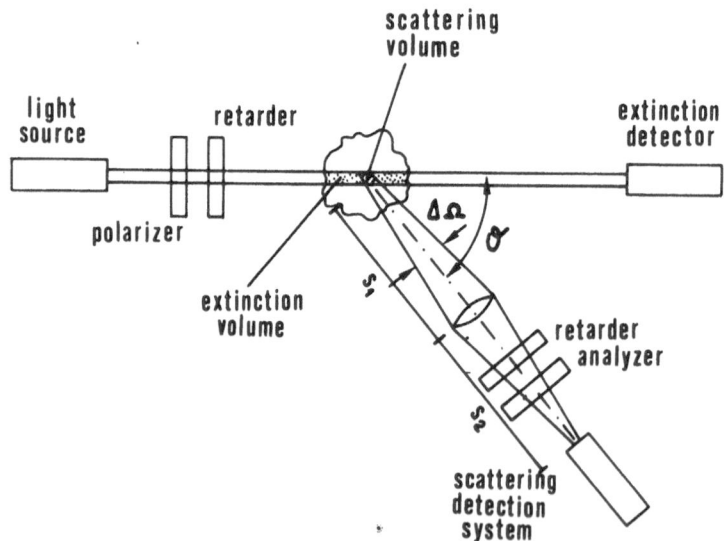

Fig. 1. Sketch of the optical apparatus.

radiation is negligible or is taken into account.

The state of polarization of the scattered beam is analyzed
with the same components of the incident one and, therefore,
the scattering matrix elements are obtained.[2,3] The spectral
analysis of scattering allows the distinction between elastic
effects detected at the same wavelengths of the incident beam
$(\lambda = \lambda_o)$, and inelastic ones $(\lambda \neq \lambda_o)$.

The intensity of the scattered beam depends linearly on the
energy flux of the source $I_{o\lambda_o}$, on the scattering volume, and on
the solid angle of the collection optics $\Delta\Omega$, as it is evidenced
in Fig. 1. These quantities are determined by the detector slit
width W_s, the optics linear magnification ratio S_1/S_2, the
aperture stop diameter D, the laser beam diameter D_L, through
the equations:

$$\Delta V = \frac{\pi}{4} D_L^2 \cdot W_s \cdot \frac{S_1}{S_2} \tag{1}$$

$$\Delta\Omega = \frac{\pi}{4} D^2 \cdot \frac{1}{S_1^2} \tag{2}$$

The scattering intensity signal S is also proportional to
an angular scattering coefficient $Q(\theta)$, a function of the polarization and spectral properties of both beams, which takes into
account the medium and e.m. field interaction. Then:

$$S = k\, I_o\, \Delta V\, \Delta\Omega\, Q(\theta) \tag{3}$$

where k is an optical and electronic efficiency factor.

The detectability of one species scattering signal depends
on its absolute intensity in respect with the luminosity and
fluctuation of the background and on the interference with other
elastic or inelastic effects.

In an extinction measurement, the relevant quantity is the
extinction coefficient k_{ext} which is defined by:

$$k_{ext}(x) = -\frac{dI(x)}{I(x)} \cdot \frac{1}{dx} \tag{4}$$

Experimental attenuations are normally measured along in-
homogeneous pathlengths and, therefore, the local values of k_{ext}
appear as a variable of an integral equation:

$$\ln \frac{I_o(x=0)}{I(x=L)} = \int_0^L k_{ext}(x) \; dx \qquad (5)$$

ELASTIC SCATTERING AND EXTINCTION COEFFICIENTS

The elastic scattering coefficients are directly proportional to the scatterers number density N in the single scattering regime which prevails in dilute systems:

$$Q = N \cdot C \qquad (6)$$

C is the angular cross section of the effect and it is predictable by the theory.[4] Rayleigh cross sections of gas phase compounds range between 10^{-28} and 10^{-25} cm^2sr^{-1}, and their number density ranges between 10^{17} and 10^{19} cm^{-3}. Soot primary particles have a characteristic size in the order of 2-20 nm and the maximum number density in the nucleation zone is 10^{15} particles/cm^3. Soot aggregates produced by gaseous fuels have typical sizes between 100 and 200 nm and their number density in the fully coagulated regime is around 10^8 cm^{-3} at atmospheric pressure. The vertically polarized cross section C_{VV} is in the range 10^{-25}-10^{-12} cm^2 sr^{-1} and, thus, the scattering coefficient is in the range 10^{-10}-10^{-4} cm^2 sr^{-1}. Fuel droplets might exhibit diameters from 1 μm to 10^3 μm and the corresponding cross sections vary from 10^{-10} to 10^{-4} cm^2 sr^{-1}; their maximum number density, outside the spray core, does not exceed 10^5 cm^{-3}.

Table 1 synthesizes the number concentration, cross section of gas phase compounds, soot particles and fuel droplets. It is evident that the lowest values for soot, 10^{10} cm^{-1} sr^{-1}, overlap with those for the gas phase background; therefore, it is difficult to detect the smallest soot nuclei, in the first formation region, with scattering measurements. Emission or absorption are more promising techniques since they avoid the interference effects with low molecular mass gaseous compounds.

Significant overlap exists also between the scattering due to soot and condensed phase fuel, and it is particularly relevant in the fuel rich regions near the droplet introduction zone. However, it is possible to distinguish between these two classes of scatterers using measurements in different polarization planes, as it will be discussed later.

The extinction cross section C_{ext} is defined by analogy with equation $k_{ext} = N \cdot C_{ext}$; it ranges, for the submicronic soot particles, between 10^{-20} and 10^{-9} and , for the micronic fuel droplets, between 10^{-8} and 10^{-4} cm^2.

Table 1.

$C_{VV}(90°)$, $Q_{VV}(90°)$		Number concentration cm^{-3}		Cross Section $cm^2 sr^{-1}$		Scattering coefficient $cm^{-1} sr^{-1}$	
		N_{min}	N_{max}	C_{min}	C_{max}	Q_{min}	Q_{max}
	N_2, O_2, CH_4	10^{18} high temp.	10^{19} STP	10^{-28}	10^{-27}	10^{-10}	10^{-8}
	light oil vapour	10^{17}	10^{19}	10^{-27}	10^{-25}	10^{-10}	10^{-6}
	soot	10^{8}	10^{15}	10^{-25} diam. 1 nm	10^{-12} diam. 200 nm	10^{-10}	10^{-4}
	light oil droplets	1	10^{7}	10^{-10} diam. 1 μm	10^{-4} diam. 1000 μm	10^{-10}	10^{-3}
	PAH fluorescence $\lambda = 514.5$ nm $\lambda^° = 555$ nm	–	10^{15}	10^{-25} $\Delta\lambda = 0.3$ nm	10^{-19} $\Delta\lambda = 0.3$ nm	10^{-15} $\Delta\lambda = 0.3$ nm	10^{-8} $\Delta\lambda = 0.3$ nm
	Raman N_2, O_2, CH_4	–	10^{19}	10^{-31}	10^{-30}	–	10^{-11}

CROSS SECTION AND COEFFICIENTS FOR INELASTIC EFFECTS

Inelastic light scattering include Raman and fluorescence excitation of small molecules and radicals and of larger molecules present in the fuel or generated by pyrolysis. Quasi-elastic scattering effects, which are detectable with interferometric or optical-mixing techniques, are considered elsewhere.[5]

The diagnostic potentialities of spontaneous Raman to combustion have been described by different authors[6],[7]. More recently, Eckbreth et al.[8],[9] have exhaustively discussed them together with non-linear effects as Coherent Antistokes Raman Spectroscopy (CARS). Therefore, CARS spectrometry will not be considered here and only some comments on spontaneous Raman effect will be made.

Vibrational spontaneous Raman cross sections are in the order of 10^{-31} cm^2 sr^{-1} for small diatomic molecules, while larger linear and aromatic hydrocarbons have cross sections bigger than 10^{-30}. 10 Consequently, maximum scattering coefficient is below 10^{-11} cm^{-1} sr^{-1} when the detection of major species is considered, since their concentration is between 10^{17} and 10^{19} cm^{-3}; Raman coefficents as low as 10^{-15} cm^{-1} sr^{-1} have to be considered for minor species detection at high temperature.

The Raman emissions take place with a definite shift from the incident wavelength and, thus, are easily distinguished from elastic scattering effects; however, a rejection ratio of the detection optics higher than 10^8 is necessary in presence of relevant amount of particles in the scattering volume.

Polynuclear aromatic hydrocarbons (PAH) and conjugated double bonds linear molecules have partially or completely delocalized π electrons and, therefore, their absorption spectra range from the near u.v. to the visible region depending on their size and structure.[11],[12] Figure 2 reports, as an example, the absorption spectra for a series of linear polynuclear compounds obtained by Clar,[13] at room temperature and in liquid phase. The absorption cross sections vary between 10^{-15} cm^2 in the u.v. to 10^{-19} cm^2 in the visible, and their peak values shift toward the red for heavier compounds. Although absorption measurements of PAH in gas phase are not available, a shift toward the red of the absorption edges is to be expected due to other vibronic levels that become appreciably populated. Furthermore, a vibronic coupling with the upper excited state will eventually increase the radiative transition probability.[14] The maximum concentration of PAH, measured in premixed or diffusion flames [15],[16],[17] are in the range 10^{13}-10^{15} cm^{-3} and, consequently, their absorption coefficient should span between 10^{-6} and 1 cm^{-1}.

The extinction due to soot particles has values of 10^{-4}-10^{-3} cm^{-1} in the nucleation zone and it rises up to 10^{-1} cm^{-1} in the last coagulation zone.[18] Thus, the PAH absorption is often masked by soot extinction in the visible, whereas its contribution might be significant in the ultraviolet region. However, high concentration of PAH in soot-free regions are found in diffusion flames, as it will be discussed later, and it is to be expected that their contribution to the absorption should be relevant in this case.

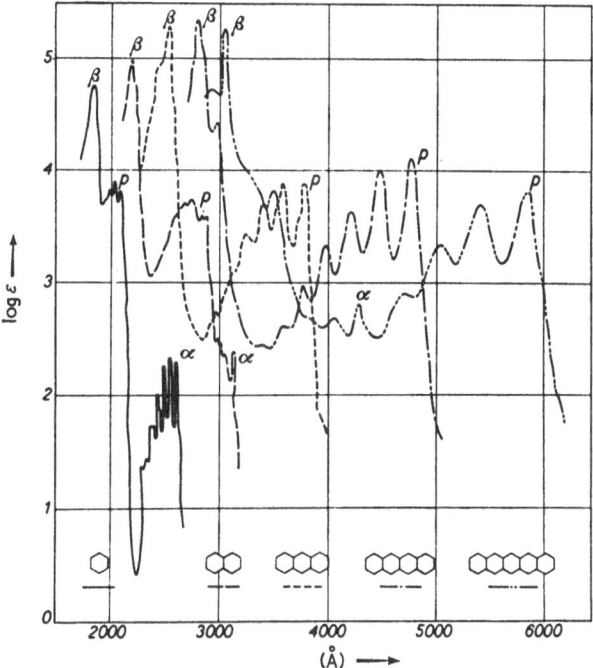

Fig. 2.   Molar extinction coefficient as function of wavelength
          for linear polycyclic aromatic hydrocarbons. [13]

As a consequence of a phonon absorption, aromatic molecules exhibit
a noticeable fluorescence emission, because their rigid ring
structures hamper the vibrational dissipation of the excitation
energy, thus decreasing the importance of the internal conversion
processes. When fluorescence is studied in gas phase, and at high
temperature, a broad range of vibronic levels are populated in
the ground, as in the excited states, thus favouring a coupling
between the electronic states. Internal conversion processes are
reversible: when inverse electronic relaxation and reverse
intersystem crossing take place, it is possible to observe fluorescence also form upper electronic states (dual fluorescence
effects).[19] Furthermore, the normal S_1-S_0 emission may have as
upper level a higher vibronic state, giving origin to hot bands
shifted toward higher energy (antiStokes shifts), as it has been
pointed out by Coe et al.[20] Fluorescence quantum yield is
defined as the ratio of the number of photons emitted to the
number of photons absorbed, and in a dilute system, is given by
the expression:

$$\phi_f = \frac{4\pi \int_0^\infty Q_\lambda^f \cdot \Delta\lambda \ d\lambda/\Delta\lambda}{k_{abs}} \qquad (7)$$

where Q_λ^f is the monochromatic fluorescence coefficient ($cm^{-2} sr^{-1}$) and $\Delta\lambda$ is the instrumental bandwidth; the integral is extended to the whole band and the 4π factor takes into account the isotropic nature of the emission.

An estimate of the maximum value of the monochromatic fluorescence cross section ($cm\ sr^{-1}$) is approximately given by the relation:

$$C_{\lambda\ max}^f \simeq \frac{C_{abs} \cdot \phi_f}{4\pi\ \Delta\lambda_{1/2}}$$

where $\Delta\lambda_{1/2}$ is the halfwidth of the fluorescence band. Numerical values of the quantum yield in liquid phase are given by Berlman[11], and generally they range between 5×10^{-2} and 1. The fluorescence band detected in rich flames are normally very broad with $\Delta\lambda_{1/2}$ of the order 10^{-5} cm, so that the monochromatic coefficient $C_{\lambda max}^f$ spans between 10^{-13} and 10^{-18} cm sr^{-1}. A typical detector spectral bandwidth is $\Delta\lambda = 10^{-8}$ and, therefore, $Q_\lambda^f \cdot \Delta\lambda$ may range from 10^{-6} to 10^{-12} $cm^{-1}\ sr^{-1}$ if the number concentration of the emitters are changed from 10^{13} to 10^{15} cm^{-3}. This order of magnitude calculation shows that fluorescence due to large aromatic molecules is certainly predominant in respect with spontaneous Raman scattering in fuel rich flames; also, it may be larger than the Rayleigh scattering due to molecules or very small soot particles.

The absorption cross sections are strongly dependent upon the wavelength and, therefore, the choice of the spectral region of excitation is critical since remarkable increase of this fluorescence signals may be obtained.

On the other side, in regions with high soot loading, the intense Mie scattering may generate high levels of stray light which interfere even with fluorescence signals, particularly when spectral zones near the elastic line are analyzed. High values of the monochromator and filter rejection factors are also necessary in this case, although the problem is not so severe as in the case of spontaneous Raman scattering.

In Table I, are also reported the Number Concentration, Cross Section and Scattering Coefficient for the Raman and fluorescence effects.

SIGNAL DETECTION AND PROCESSING PROCEDURES

The detectability of the different effects is related to their intensities, to the background interference, and to the required time and space revolution.

Rich flames have obviously high luminosity and this is often
the most relevant problem which imposes limits on the light
source characteristics and on the electronic processing procedure.
Continuous and pulsed laser with different powers and wavelengths
are available. The scattering effects, as discussed in the previous
paragraph, are supposed to be linearly dependent upon the energy
flux, although the scattering coefficients are very different;
the background emission due to soot particles follows roughly a
blackbody-type curve with its peak intensity in the near-infrared.

Fig. 3 accounts, in a very schematic way, for these combined
effects. In the lower sector on the right, the common fixed
frequency laser sources are located as function of their typical
irradiance and wavelength; the pulsed lasers have a much higher
power and are reported as black blocks. The upper right sector
illustrates how the scattered power is related to the incident
flux for elastic scattering, spontaneous Raman and broad-band
fluorescence, for a scattering volume of 10^{-3} cm^{-3}. It is worth
noting that the limit to the input power is given by the gas break-
down, as it is evidenced on the right side of the figure. The
left upper part of the figure reports the background luminous
emission for different fuels and different conditions as function
of the wavelength, using an emission volume of 10^{-2} cm^3 for the
computation.

Curves of emission for the combustion of coal and heavy fuel
oil were obtained in very large utility boiler[21], whereas the
results for light oil and propane were obtained on smaller scale
burner.[8,22] The blackbody curves for 1000 and 2000 K are also
reported for comparison. Very high signal to background luminosity
ratios are obtained for elastic scattering effects when high power
pulsed laser are employed; single pulse measurements are very
useful for time resolved analysis. However, it is to remark that
an excessive laser power might cause particle incandescence or
vaporization, as it has been discussed by Eckbreth et al.[8]

Single pulse measurements of broadband fluorescence in a
luminous environment are possible while they seem to be rather
difficult for spontaneous Raman, in addition to the interference
effects discussed in the previous paragraph.

Better signal to background ratios are obtained in the near
u.v. than in the visible, where the luminosity is much higher,
but it is not easy to find suitable lasers in this spectral
region. Frequency quadruplied Nd-YAG laser with $\lambda = 265$ nm
appears to be a promising source for the detection of fluorescence
from small aromatic molecules. On the other side, in the infrared
CO, lasers might be useful for the detection of soot and other
particles in the micronic range lowering the size to wavelength

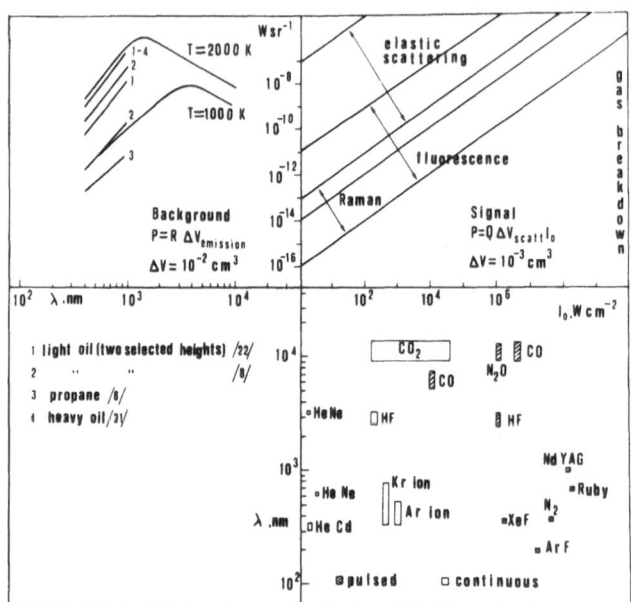

Fig. 3.   Upper left quarter: Background spectral emission from
          different authors, compared with blackbody emission at
          two temperatures.
          Upper right quarter: Elastic and inelastic scattering
          signal per unit solid angle as function of energy flux
          of incident light.
          Lower right part: Energy flux of the most widely
          utilized laser sources with their typical wavelength
          range.

ratio, although the noise of the infrared detectors might
determine the limit of sensitivity.[23] The signals might be
enhanced , in respect with the background, up to two orders of
magnitude by using an analogue averager or a transient recorder/
computer combination when a pulsed laser is used as source. In
this case also the signal waveform is followed, and it is very
useful for the determination of fluorescence lifetime.

     Obviously the increase in signal to background ratio
results in a loss of time resolution which is not acceptable
when the averaging procedure does not follow, in a meaningful
way the evolution of the system.

     For the detection of signals which are slowly varying in
time, a substantial increase of the signal/background ratio may be
obtained by decreasing the bandwidth of the detection electronics

and thus eliminating a great part of the flicker noise in the flame background. It is normally obtained with a lock-in amplifier, which is able to improve the signal-to-noise ratio more than four orders of magnitude.[23] Therefore, it is possible to extract signals of intensity much lower than the flame intensity background, with acceptable signal to noise ratio, when the measurement time is long enough.

It is worthwhile noting that in practical flames the noise arises more from the luminosity fluctuation than from the photomultiplier properties; photon counting techniques, which are indicated when the light intensity is particularly low, do not offer great advantages as compared to the analogue detection techniques.

SELECTED EXPERIMENTAL RESULTS

Premixed and Diffusion Gaseous Flames

In the last few years a great number of optical investigations on rich flames have been carried out. Most of the literature on the subject has been reviewed recently by Haynes and Wagner.[25],, therefore, only specific topics will be examined in the present discussion with reference to gaseous premixed and diffusion flames.

A typical evolution of elastic and inelastic scattering effects in a premixed flat flame at atmospheric pressure is reported in Fig. 4. The results were obtained on a rich CH_4/O_2 flame diluted with nitrogen in order to decrease the burning velocity and thus expand the first oxidation zone.

The profile of the elastic scattering coefficient at $\lambda = 514.5$ nm, in the initial part of the flame, gives a good example of interference between the scattering by gas phase compounds and by soot particles. The decreasing part and the plateau of the Q_{VV} curve is dominated by the Rayleigh scattering of gaseous reagent and/or products, the total density of which is decreasing as they pass through the preheated and reaction zones. This scattering background ranges between $2 \cdot 10^{-9}$ and $4 \cdot 10^{-9}$ cm^{-1} sr^{-1} and does not allow to detect soot particles in the early nucleation zone, where their concentration is lower than 10^{13} cm^{-3}. Extinction measurements have larger potentialities and the minimum detectable number concentration of soot nuclei is of the order of $4 \cdot 10^{-12}$ cm^{-3} for particles with 1.0 nm diameter. However, the soot particle extinction interferes with the absorption of high molecular mass aromatic compounds as will be discussed later.

Later on in the flame, the increase of the scattering coefficient is of several orders of magnitude and the ratio between scattering and extinction furnishes the mean size and number

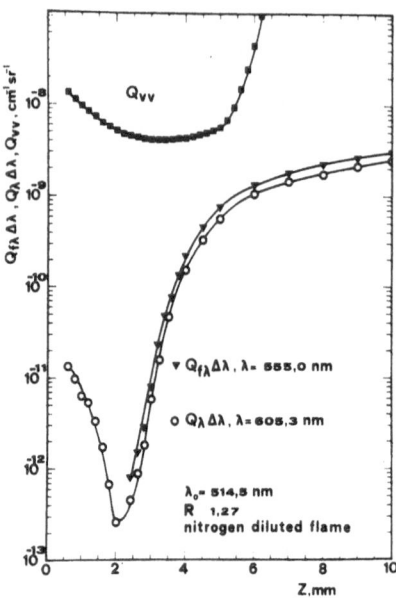

Fig. 4. Axial profiles of elastic scattering, fluorescence and
methane Raman Stokes scattering coefficients for a
nitrogen diluted premixed CH_4/O_2 flame. $\Delta\lambda = 0.32$ nm.

concentration of soot particles; this procedure has been extensively
used to follow the surface growth and coagulation processes in
different flames.[18,26,27,28,29] An analysis of the angular and
polarization properties of the light scattered by soot aggregates,
and their interest for diagnostics, have been presented elsewhere.[4]

      The lower part of Fig. 4 gives rise to a comparative discussion
of fluorescence and spontaneous Raman in the early regions of rich
premixed flames. The laser excitation wavelength was at $\lambda_o = 514.5$
nm, and the measurements at $\lambda = 605.3$ nm were taken with a 3.2 nm
bandwidth, in order to include both the Stokes vibrational Raman
emission of methane and broadband fluorescence, while the measure-
ments at $\lambda = 555$ nm carried out with a $\Delta\lambda = 0.32$ nm are in a region
where there is only a fluorescence emission. The light scattered
at $\lambda = 605.3$ nm has a value of 10^{-11} cm^{-1} sr^{-1} near the burner,
which is just that expected for CH_4 Raman at room temperature.[6]
It decreases down to z = 2 mm because of methane oxidation and
temperature increment; the subsequent strong enlargment of the
inelastic signal is due to PAH fluorescence as it is indicated
by the parallel evolution of the light scattered at $\lambda = 555$ nm.
Therefore, it is evident that fluorescence background in rich
flames overlaps the Raman scattering. In an atmospheric pressure
premixed flame the fuel is brought at the highest temperature in

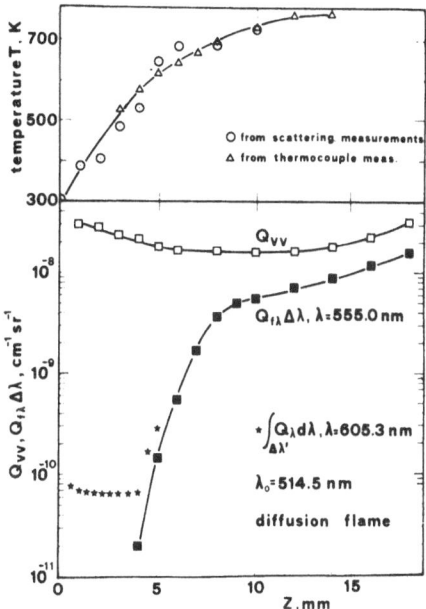

Fig. 5.   Lower part: Axial profiles of elastic scattering, fluo-
          rescence and methane Raman Stokes scattering coefficients
          for a diffusion CH_4/O_2 flame. $\Delta\lambda = 0.32$ nm.
          Upper part: Axial temperature profiles determined by
          Rayleigh scattering and thermocouple measurements.

a very short time and, therefore, it is not possible to evidence
a time lag between PAH and soot formation from light scattering
and fluorescence axial profiles; furthermore, the elastic scattered
signals are almost always much more intense than the fluorescence
excited in the visible.

        The formation of PAH on the fuel side of a diffusion flame
takes place at lower temperature than in a premixed flame, and
in this case fluorescence and Raman effects can be followed by
a less severe interference of the elastic scattering.   This is
shown on Fig. 5 for the case of a cylindrical methane diffusion
flame, where the fuel in the central region is kept at a moderate
temperature for a long residence time.

        Axial measurement of the $Q_{VV}(90°)$ scattering coefficient at
$\lambda = 514.5$ nm provides values which are always higher than 10^{-8}
cm^{-1} sr^{-1} and, up to z = 12 mm, are determined exclusively by
methane.   Temperature profile determined by means of the Rayleigh
scattering agree remarkably well with that measured with Cr/Al
thermocouple, and it increases from 300 to 800 K.   Methane Raman

scattering is of course higher than in the premixed case. Also in this condition fluorescence intensity overlaps Raman signal starting from 5 mm above the burner surface.

It is worthwhile noting that aromatic fluorescing species are present in the pyrolysis of methane even at such low temperatures. It is probably due to the catalysis effects of atoms and/ or radicals diffusing from the external reaction zones.

Additional information of the chemical and physical evolution of the system are obtained by following, theoretically and experimentally, the wavelength variation of both elastic and inelastic scattering effects. Fig. 6 reports the normalized to the maximum scattering intensity at $\theta = 20°$, in the wavelength range between 200 and 600 nm; the measurements were obtained along a premixed CH_4/O_2 flame using a high pressure 450 W xenon lamp as excitation source and a f = 100 mm monochromator on the scattered beam.[30] In the same figure is shown the theoretical Mie profile computed for a 20 nm particle diameter; the optical properties of the glassy carbon, obtained by Taft[31], are used. The scattering theory predicts reasonably well the position and shape of the experimentally observed maximum in the u.v., but it fails to model the measured intensity between 400 and 600 nm in the first part of the flame. Therefore, soot at high temperature seems to exhibit a resonance in the near u.v. which is similar to that exhibited by carbon and graphite.[1]

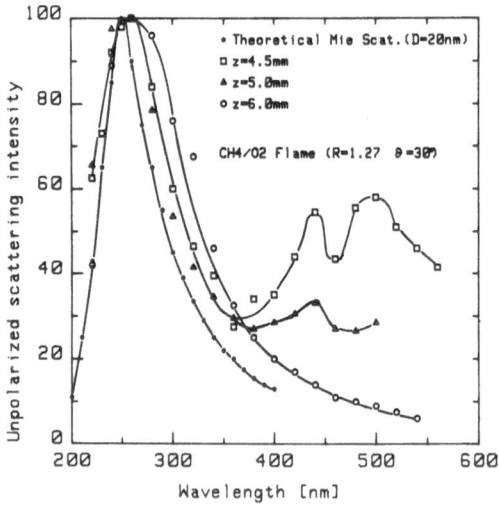

Fig. 6.   Unpolarized scattering intensity as function of wavelength at three heights above the burner; theoretical Mie scattering for a 20 nm dia. glassy-carbon particle is also reported.

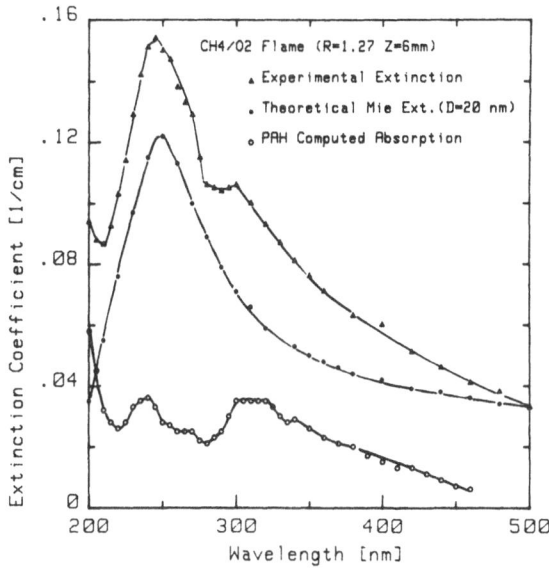

Fig. 7.   Spectral PAH absorption coefficient evaluated as difference
between experimental extinction coefficient and the predicted
one from Mie theory for a 20 nm dia. glassy-carbon particle.

The same interference between elastic and inelastic effects
in this region is evidenced by the spectral variation of the
extinction coefficient in the same wavelength range as is shown
on Fig. 7. The full line represents the computed Mie extinction
for a 20 nm diameter particle while the molecule absorption is
given by the difference with the total measured extinction. It
is to note that the computed soot extinction presents also a peak
in the u.v. region, when the proper set of optical constants is
considered. Therefore, the experimental peak near 250 nm is to
be attributed pre-eminently to soot particles rather than to
PAH species as was said previously.[15] A comparison between the
experimental quantum efficiency and that evaluated for PAH alone
shows that the molecular absorption accounts for less than 1% of
the total extinction at λ = 514.5 nm.[4] However, the importance
of PAH absorption is increasing when moving toward the u.v.
region; it peaks at 310 nm and 240 nm, and , near 200 nm, it seems
to become more relevant than the particle extinction.

It is reasonable to assume that different compounds are
responsible for this behaviour: the absorption in the 200-260 nm
range is to be attributed to benzene or other mono-ring aromatic
compounds; the absorption around 300 nm is related to compounds
like phenanthrene, pyrene, etc., whereas in the visible the

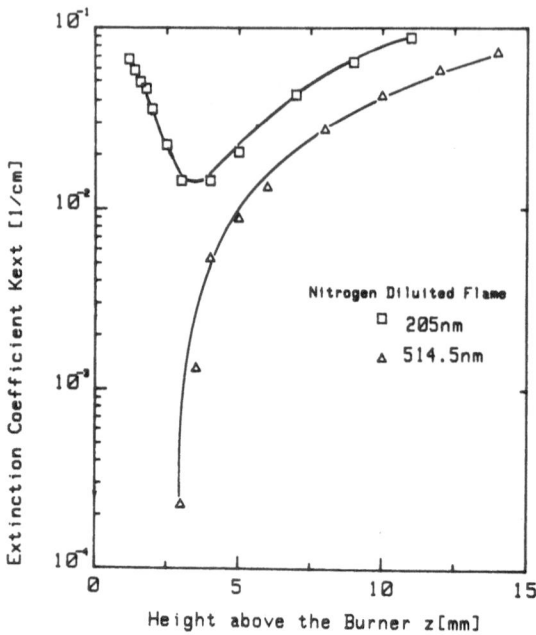

Fig. 8. Axial profiles of extinction coefficient measured at
two wavelengths for CH_4/O_2 premixed flame.

absorption is probably due to high temperature red shifted spectra
of acenaphtylene and fluoranthene, as it has been suggested by
Coe, Haynes and Steinfeld.[20] This point of view is supported by
the axial profiles of the extinction coefficient at λ = 205 nm and
λ = 514.5 nm obtained for the precedently-discussed nitrogen diluted
flame (Fig. 8). In fact, the visible extinction profile parallels
the visible fluorescence and scattering profiles while it has a
different trend in the u.v. region, where the initial decline of
this absorption is probably related to the formation and oxidation
of light aromatic species, which take place in the primary reaction
zone.

The normalized fluorescence spectra excited in the green by
an argon ion laser source are reported in Fig. 9 for high temperature condition ($\sim$ 1500 K) in a premixed flame, and for a low
temperature condition ($\sim$ 600 K) in a diffusion flame. The low
temperature gas phase spectrum shows a relatively good resolution;
peaks at 520 nm may be identified. At higher temperature the spectrum is much broader but it is still possible to identify some
of these peaks. The room temperature acenaphtylene spectrum
shows exactly the same peaks although the spectral resolution is
poorer in respect to the gas phase conditions. This comparison
strengthens the previous quoted attribution of fluorescence in
flames to acenaphtylene when an argon ion laser is used as exciting
source in the blue-green.

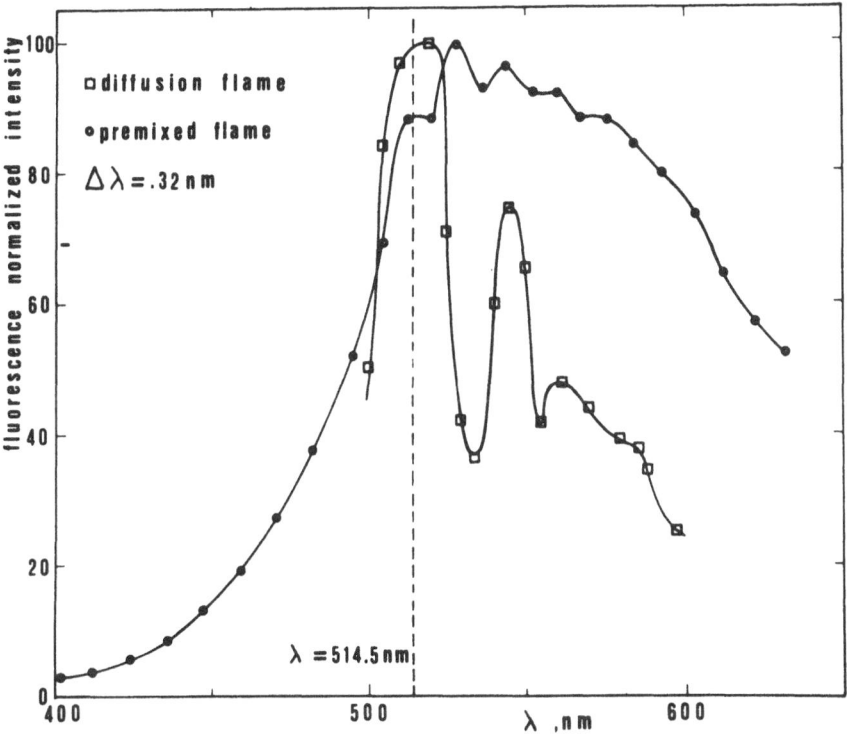

Fig. 9. Spectral distribution of the Ar-ion laser excited
        fluorescence (λ = 514.5 nm) emitted from diffusion and
        premixed flames.

        Acenaphtylene is almost always one of the most abundant PAH
in flame. Coe and Steinfeld [32] have evidenced that its quantum
efficiency in the visible was much higher in gas phase at moderate
temperature, than in liquid phase at room temperature. Consequently
there is no reason to attribute this effect to very heavy PAH with
more than 5-6 rings, which have generally a very low concentration
in flame. It is also reasonable to predict that other PAH exhibit
fluorescence spectra excited at higher frequency; more particularly
at the shorter wavelengths (in the u.v. region), biphenyl and
other two rings compounds should give their maximum contribution
whereas, in the region between 350 and 450 nm, it is likely that
alternative 3-4 rings PAH should contribute predominantly to the
fluorescence.

Oil Spray Flames

        A spray flame is by some aspects similar to a gaseous turbulent
diffusion flame but presents also peculiar effects due to droplets
formation, distribution and evaporation, in the first stages of
combustion. The problem of distinguishing, with optical techniques,

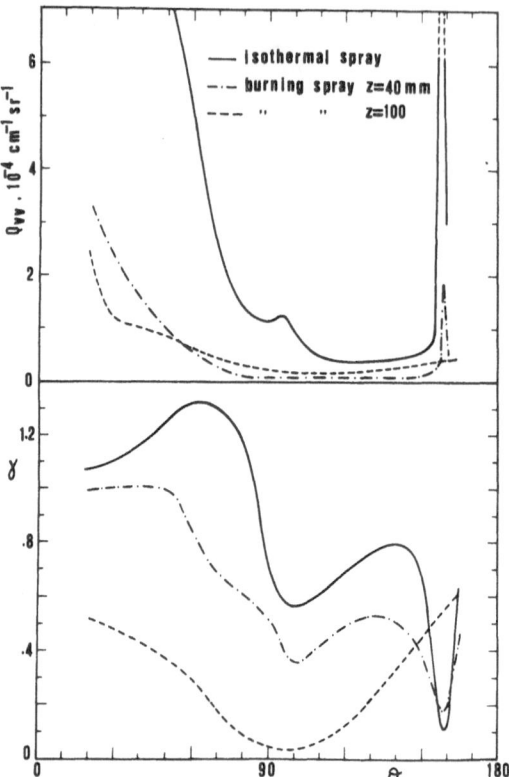

Fig. 10. Angular pattern of Q_{VV} vertically polarized scattering
          coefficient and polarization ratio measured on a iso-
          thermal and burning spray at two heights above the
          burner.

different classes of particles has been discussed in a previous
paper on the basis of theoretical predictions.[1] In this section,
some experimental results obtained on swirled flames of light
oil will be briefly discussed, as an example.

     The angular distribution of the scattering coefficients on
the vertical and horizontal polarization planes gives a good
starting point for discussing the different behaviour of the fuel
micronic droplets and the smaller soot particles.

     Fig. 10 reports the absolute values of the $Q_{VV}(\theta)$ scattering
coefficients and of the polarization ratio $\gamma = Q_{HH}(\theta)/Q_{VV}(\theta)$, in
the range of scattering angles between 20° and 160°. The results
obtained with isothermal conditions are in good agreement with
the prediction of the geometrical optics for non absorbing large
spherical particles. The forward lobe (20° < θ < 90°) intensities
are determined by the transmitted light contribution; the

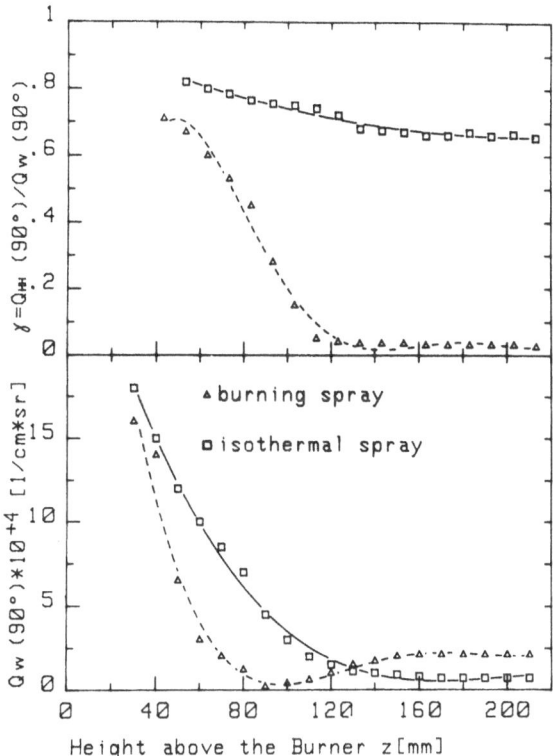

Fig. 11. Axial profiles of polarization ratio and Q_{VV} vertically
polarized scattering coefficient measured on isothermal
and burning spray.

polarization properties of the incident light are preserved so
that the γ ratio reaches values near one, or slightly higher.
The primary rainbow at θ = 157° is very sharp, the secundary one
is detectable at θ = 95°; their angular position is that expected
in a medium with a refractive index of 1.5, which is quite a
reasonable value for a light oil. The presence of the primary
rainbow is also evidenced in the diagrams by the presence of a
deep minimum in the backward region. The Q_{VV} coefficients are
relatively low in the dark side region between the two rainbows
where the light scattered is mostly due to external reflection.
The initial decrease of the γ ratio is determined by the combined
effects of the refraction and reflection, whereas the subsequent
increase in the side region is substantially governed by the
Fresnel reflection laws. A more detailed discussion about the
level of agreement between ray theory and experiments has been
presented elsewhere, also in connection with the other elements
of the scattering matrix.[3]

The angular patterns of Q_{VV} and γ, for flame conditions of nearly completed droplets evaporation, show the typical behaviour of the scattering of small soot particles in the Mie regime, while intermediate properties of scattering are evidenced in both polarization planes when measurements are taken in the presence of partially vaporized fuel.

Therefore, it appears that both angular and polarization measurements are able to give characteristic signatures for droplets and soot and, in the following examples, the γ ratio at $\theta = 90°$ has been chosen as a discrimination parameter.

The axial profiles of Q_{VV} and γ have been reported in Fig. 11 for a spray in isothermal and burning conditions, and with a moderate swirled air flow field ; the scattering decrease for the cold conditions depends upon the dispersion of the droplets cloud whereas the initial faster decline of the scattering in burning conditions should be attributed to the intense vaporization process. The subsequent increase of the scattering beyond the minimum is related to soot particles build-up; in fact, in correspondence to the minimum, the γ ratio falls from around 0.7, a typical value for droplets, to $2 \cdot 10^{-2}$, a value which is normally found when soot particles are the only scatterers.

Therefore, the ambiguity in the interpretation of scattering data may be greatly reduced by the additional measurements of the light scattered on the horizontal plane. The γ criterium on the polarization ratio has been used quantitatively for determining the scattering coefficients of droplets and soot for different flame configurations, including when the droplets scattering coefficient was lower or comparable to the soot coefficient.[33,34]

More recently, measurements on the fluorescence, excited by argon ion laser, have been extended to spray oil combustion too.[33] The presence of aromatic compounds, both in the inlet fuel and in the pyrolysis products, further complicates the interpretation of the effect. However, these preliminary measurements show that it is possible to distinguish among the different contributions on the basis of the ratio between the Stokes and anti-Stokes fluorescence.

An example of the results is shown in the diagrams of Fig. 12, where the scattering coefficients of droplets and soot, and the monochromatic fluorescence, have been measured along the radius in the inlet zone of a low swirled flame. It is easy to see that the fuel droplets are still concentrated in the central region, and that the high temperature pyrolysis develops in an external annulus where both soot and PAH have their maximum; the decay of both soot and PAH in the external region is due to their oxidation in the external combustion zone.

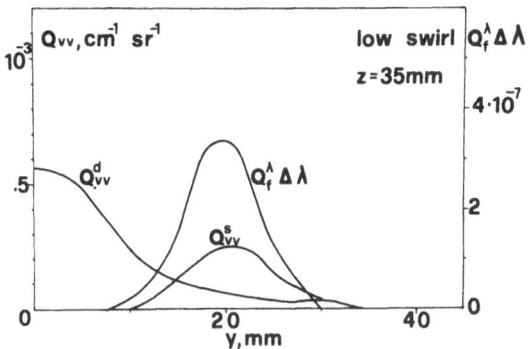

Fig. 12. Radial profiles of fluorescence coefficient and of vertically polarized scattering coefficient for soot (Q_{vv}^s) and droplets (Q_{vv}^d).

It is worth noting that this is a particular flame condition and that a large variety of distribution of soot, droplets and PAH may be obtained, depending on the atomization, fluid dynamic field and different combustion stages.

REFERENCES

1. A. D'Alessio, A. Cavaliere and P. Menna, Theoretical Models for Interpretation of Light Scattering by Particles Present in Combustion Systems , presented at the workshop "Soot in Combustion Systems and Its Toxic Properties" sponsored by N.A.T.O. and the University of Haute-Alsace, Le Bischenberg, France September 1981.

2. H. C. van de Hulst, "Light Scattering by Small Particles", J. Wiley, New York (1957)

3. F. Beretta, A. Cavalere, and A. D'Alessio, "Experimental and Theoretical Analysis of the Angular Pattern Distribution and Polarization State of the Light Scattered by Isothermal Sprays and Oil Flames", ASME Winter Annual Meeting, November 15-20, 1981, Washington, D.C., ASME paper 81-WA/HT-49

4. A. D'Alessio, Laser Light Scattering and Fluorescence Diagnostics of Rich Flames Produced by Gaseous and Liquid Fuels, in: "Particulate Carbon: Formation during Combustion", D. C. Siegla and G. W. Smith, Eds, Plenum Press, New York (1981)

5. G. Gouesbet and G. Grehan, The Quasi-Elastic Scattering of Light: A Lecture with Emphasis on Particulate Diagnosis, presented at the workshop "Soot in Combustion Systems and Its Toxic Properties" sponsored by N.A.T.O.

and the University of Haute Alsace, Le Bischenberg,
France, September 1981

6. M. Lapp and C. M. Penney, Eds., "Laser Raman Gas Diagnostics",
    Plenum Press, New York (1974)

7. R. Goulard, Ed., "Combustion Measurements", Part II, Session
    6, Academic Press, New York (1976)

8. A. C. Eckbreth, P. A. Boneziek and S. F. Verdieck, Combustion
    Diagnostics by Laser Raman and Fluorescence Techniques,
    Prog. Energy Combustion Sci. 5:253 (1979)

9. R. S. Hall and A. C. Eckbreth, Coherent Anti-Stokes Raman
    Spectroscopy (CARS): Application to Combustion Diagnos-
    tics, to be published in Laser Applications, Vol. 5

10. D. A. Stephenson, Raman Cross Sections of Selected Hydro-
    carbons and Freon, S. Quant. Spectrosc. Radiat. Transfer
    14:1291 (1974)

11. I. B. Berlman, "Fluorescence Spectra of Aromatic Molecules",
    Academic Press, New York (1971)

12. H. N. Jaffè and M. Orchin, "Theory and Application of
    Ultraviolet Spectroscopy, J. Wiley, New York (1962)

13. E. Clar, "Polycyclic Hydrocarbons", Vol. 1 and 2, Academic
    Press, London (1974)

14. J. D. Winefordner, S. G. Schulman and T. C. O'Haver,
    "Luminescence Spectrometry in Analytical Chemistry",
    J. Wiley, New York (1972)

15. A. Di Lorenzo, A. D'Alessio, V. Cincotti, S. Masi, P.
    Menna and C. Venitozzi, "U.V. Absorption, Laser Excited
    Fluorescence and Direct Sampling in Rich CH_4/O_2 Flames",
    Eighteenth Symposium (Intern'l) on Combustion, The
    Combustion Institute, Pittsburgh (1981)

16. R. Barbella, F. Beretta, A. Ciajolo and A. D'Alessio,
    "Laser Excited Fluorescence and Chromatographic
    Techniques for the Determination of PAH on a Spray
    Oil Flame", paper presented at the Sixth International
    Symposium on Polynuclear Aromatic Hydrocarbons,
    Battelle Laboratories, Columbus, Ohio, October 1981

17. G. P. Prado, M. L. Lee, R. A. Hites, D. P. Hoult and J. B.
    Howard, "Soot and Hydrocarbon Formation in a Turbulent
    Diffusion Flame", Sixteenth Symposium (Intern'l) on
    Combustion, The Combustion Institute, Pittsburgh (1981)

18. A. D'Alessio, A. Di Lorenzo, A. F. Sarofim, F. Beretta,
    S. Masi and C. Venitozzi, "Soot Formation in Methane-
    Oxygen Flames", Fifteenth Symposium (Intern'l) on
    Combustion, The Combustion Institute, Pittsburgh (1975)

19. T. Deinum, D. J. Werkhoven, J. Langelaer, R. P. H.
    Rettschuick and J. D. W. Van Voorst, Sequence Congestion
    and Temperature Effects on the Fluorescence of Isolated
    Pyrene Molecules, Chem. Phys. Lett. 27:206 (1974)

20. D. S. Coe, B. S. Haynes and J. I. Steinfeld, Identification

of a Source of Argon-Ion-Laser Excited Fluorescence
in Sooting Flames, Combustion and Flame 43:211 (1981)

21. F. Beretta, A. D'Alessio and C. Noviello, in the Research
    Report "Sugli studi di Fiamme con Metodi Chimico-
    Fisici" conducted by the Istituto di Ricerche sulla
    Combustione - C.N.R. , Naples under ENEL/CRTN contract
    (1975)

22. F. Beretta, A. Cavaliere, A. D'Alessio and C. Noviello,
    Visible and U.V. Spectral Emission and Extinction
    Measurements in Oil Spray Flames, Comb. Sci. Techn.
    22:1 (1980)

23. K. Sassen, Infrared (10.6 nm) Scattering and Extinction
    in Laboratory Water and Ice Clouds, Applied Optics
    1:185 (1981)

24. P.A.R. Tech. Publ. T - 198A

25. B. S. Haynes and H. Gg. Wagner, Soot Formation, Progr.
    Energy Comb. Sci. 7:229 (1981)

26. A. D'Alessio, A. Di Lorenzo, A. Borghese, F. Beretta
    and S. Masi, "Study of the Soot Nucleation Zone of
    Rich Methane-Oxygen Flames", Sixteenth Symposium
    (Intern'l) on Combustion, The Combustion Institute,
    Pittsburgh (1977)

27. B. S. Haynes, H. Jander and H. Gg. Wagner, "The Effects
    of Metal Additives on the Formation of Soot in Premixed
    Flames", Seventeenth Symposium (Intern'l) on Combustion,
    The Combustion Institute, Pittsburgh (1979)

28. G. Prado, J. Jagoda, K. Neoh and J. Lahaye, "A Study of
    Soot Formation in Premixed Propane/Oxygen Flames by
    In-Situ Optical Techniques and Sampling Probes",
    Eighteenth Symposium (Intern'l) on Combustion,
    The Combustion Institute, Pittsburgh (1981)

29. K. Müller-Dethlefs, ph. D. Thesis, Imperial College,
    London (1979)

30. P. Menna, unpublished results

31. E. A. Taft, private communication (1969)

32. D. S. Coe and J. I. Steinfeld, Fluorescence Excitation
    and Emission Spectra of Polycyclic Aromatic Hydrocarbons
    in an Atmospheric Flame, Chem. Phys. Lett. 76:485 (1980)

33. F. Beretta, A. Cavaliere and A. D'Alessio, Laser Excited
    Fluorescence Measurements in Spray Oil Flames for the
    Detection of Polycyclic Aromatic Hydrocarbons and
    Soot, Comb. Sci. Techn. 27:113 (1982)

34. F. Beretta, A. Cavaliere, A. Ciajolo, A. D'Alessio, C.
    Langella, A. Di Lorenzo and C. Noviello, "Laser Light
    Scattering, Emission/Extinction Spectroscopy and
    Thermogravimetric Analysis in the Study of Soot Behaviour
    in Oil Spray Flames", Eighteenth Symposium (Intern'l)
    on Combustion, The Combustion Institute, Pittsburgh (1981).

DISCUSSION

<u>F. Fetting</u> (Technische Hochschule, Darmstadt)

<u>Comment</u>: By evaluation of laser ligth scattering of sooting flames
we found that the refraction index m = n - ki is depending on the
C/O ratio and also on the position in the flame. With increasing
life time of the particles k will increase (Meyer, U. : Dissertation,
Techn. Hochschule Darmstadt, 1979)

THE STATUS OF THE ART IN SOOTS DIAGNOSIS BY MEANS OF DIFFUSION

BROADENING SPECTROSCOPY

M. E. Weill, P. Flament and G. Gouesbet

Laboratoire de Thermodynamiqye - L.A. C.N.R.S. n°230
Faculté des Sciences - BP 67 - 76130 MONT-SAINT-AIGNAN

INTRODUCTION

Soot sizing, with for instance relevance to soot growing study, is usually achieved by means of quasi-elastic (linear) scattering (QES) of light or of particulate sampling.

Using QES, sizes and concentrations are deducible from coupled turbidimetry and scattered intensities measurements.[1,2] Nevertheless, the knowledge of the complex refractive index of soot is needed to achieve data analysis, a well-known point also briefly discussed in another paper of the present workshop.[3] Furthermore, measurements are not local since turbidimetry data are integrated along a length of the incident laser beam.

Particulates sampling, on the other hand, is (as an example) achieved by means of cascade impactors put inside the flame under investigation. Soots are then transferred on a grid and examined using electronic microscopy.[4,5] Unfortunately, this technique is intrusive and delicate with many respects.

The main aim of the present paper is to discuss another non-intrusive (optical) method which is local and does not demand the knowledge of the complex refractive index to proceed to sizing analysis - in case of monodispersity - namely the Diffusion Broadening Spectroscopy (DBS), or self-beating spectro- scopy. Well-known for diagnosis of submicronic particles in suspensions at rest,[6] the DBS has been used for moving aerosols at low temperatures by Hinds and Reist,[7] then for soots diagnosis by Penner et al.,[8] Driscoll and Mann,[9], Driscoll et al.[10]

Preliminary investigations with discussion of the way to simultaneously measure concentrations have also been reported by Gouesbet
et al.[11]

THEORY OF THE DBS

Main features of the DBS-theory are now pointed out and
mathematical formulae are usually given without demonstration.
The reader should refer to earlier literature for details.

Particles under study are illuminated by a TEM_{00} laser
beam, as sketched on the self-explanatory Figure 1. The
incident light is scattered by the submicronic particles that we
suppose embedded into the flow and collected, then directed
on to the photocathode of the photomultiplier. The electronic
signal is then processed by an electronic device.

Each scattering particle is animated with a non-random
velocity (turbulent flows are not considered) over which is
superimposed a random motion due to the brownian process. The
electric field which is scattered by a given set of viewed
particles appears as a stochastic variable, the autocorrelation
function $R_E(\tau)$ being defined as:

$$R_E(\tau) = < E^*(o) E(\tau) > \tag{1}$$

where $E(\tau)$ is the scattered electric field at time τ, the star
meaning the "conjugate", and the symbol $< >$ referring to a set
average. Furthermore, only steady scatterings are discussed.

The anode photocurrent $i(\tau)$ delivered by the quadratic
photodetector at time τ is proportional to the square of $|E(\tau)|$
and thus appears also as a stochastic variable, the autocorrelation
function $R_i(\tau)$ of which being defined as:

$$R_i(\tau) = < i(o) i(\tau) > \tag{2}$$

According to Edwards et al.[12], the random process is a
gaussian one. Consequently, the τ-dependent part $\tilde{R}_i(\tau)$ of the
intensity autocorrelation function $R_i(\tau)$ is proportional to
$|R_E(\tau)|:$[2,13]

$$\tilde{R}_i(\tau) = \alpha |R_E(\tau)|^2 \tag{3}$$

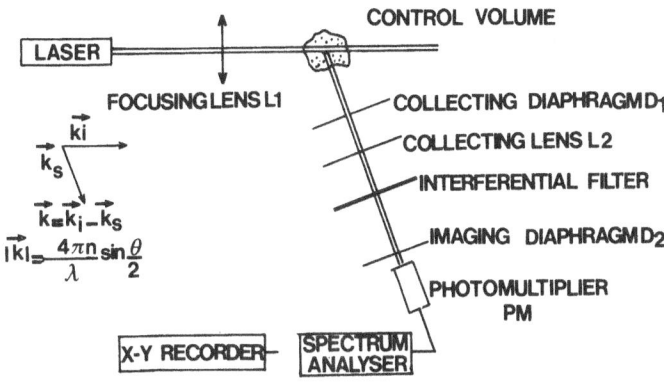

Fig. 1. The experimental set-up.

Two basic kinds of processing of the electronic signal are possible. The first one consists in directly recording and analyzing the intensity autocorrelation function by means of a photocorrelator. The second one (that has been used in the present work) consists in studying the power spectrum of the photocurrent. According to the Wiener-Khinchine theorem, this spectrum is the Fourier transform of $R_i(\tau)$. It contains three contributions due to (i) a non τ-dependent part of $R_i(\tau)$, (ii) the shot noise always associated with the signal and (iii) the variable spectrum $\tilde{S}(2\pi\nu)$, that we shall only consider here, defined as Fourier-transform of $\tilde{R}_i(\tau)$:

$$\tilde{S}(2\pi\nu) = 2\,\alpha \int_0^\infty \left| R_E(\tau) \right|^2 \cos 2\pi\nu\tau\cdot d\tau \qquad (4)$$

According to Van de Hulst,[14] the scattered field reads:

$$E = \sum_{j=1}^N E_j \exp i \left[k_i\, r_{j,i}(t) - \omega_o t \right] \qquad (5)$$

where the individual waves scattered by the N particles present in the control volume are summed up. E_j is the time-independent amplitude for the jth particle depending on its shape, its size, its nature and its location; k_i is the scattering vector (Fig. 1):

$$k_i = k_{s,i} - k_{i,i} \tag{6}$$

such as:

$$\left| k_i \right| = 4 \pi \sin\left(\frac{\theta}{2}\right) / \lambda \tag{7}$$

where λ is the wave-length of the light in the medium surrounding the particle. Finally, ω_o is the incident angular frequency, $r_{j,i}(t)$ the position vector of the jth scatter center at time t. Note that the Einstein rule of summation is used in relation (5).

From relations (1) and (5), and assuming that the motion of the jth scattering center is not correlated with the motion of the kth particle $(j \neq k)$, $R_E(\tau)$ reads* :

$$R_E(\tau) = \exp(-i\omega_o\tau) \sum_{j=1}^{N} <E_j(0)E_j(\tau)\exp\{ik_i(r_{j,i}(\tau)-r_{j,i}(0))\}> \tag{8}$$

Let us assume that all particles are identical, so that the time-independent amplitude only depends on the location of the scatter center. Thus:

$$E_j(r_{j,i}) = E \cdot A(r_{j,i}) \tag{9}$$

$A(r_{j,i}(t))$) is the incident time-independent amplitude at location $r_{j,i}(t)$ of the jth particle at time t, and E is the scattered amplitude for an incident amplitude equal to 1.

Furthermore:

$$r_{j,i}(\tau) - r_{j,i}(0) = \Delta r_{j,i}(\tau) + v_{j,i}\tau \tag{10}$$

where the two terms in relation (10) correspond to the random brownian motion and the non-random velocity respectively. The velocity $v_{j,i}$ is (generally) allowed to depend on the particle location.

With an obvious slight change of notation, relation (8) becomes:

*
$E_j(0) E_j(\tau)$ stands for $E_j(r_{j,i}(0)) E_j(r_{j,i}(\tau))$

$$R_E(\tau) = \exp(-i\omega_o\tau) \sum_{j=1}^{N} E^2 << A(r_{j,i}(0)) . A(r_{j,i}(0) + \Delta_{1j,i}(\tau) + v_{j,i}\tau)$$

$$\exp(ik_i\Delta r_{j,i}(\tau)) \exp(ik_i v_{j,i}\tau) >> \qquad (11)$$

where the double set average involves averaging on the location of the particles at time 0 and averaging on hte brownian displacements.

The second set averaging can be achieved knowing that, according to the general theory of brownian motion[15], the probability of a displacement $\Delta r_{j,i}(\tau)$ due to the brownian random process is:

$$P(\Delta r_{j,i}(\tau)) = (4\pi D\tau)^{-3/2} \exp(-\Delta r_{j,i}^2(\tau) / (4D\tau)) \qquad (12)$$

where D is the isothermal coefficient of diffusion of particles, expressed as :

$$D = k_B T B \qquad (13)$$

where k_B is the Boltzmann constant, T the temperature and B the mobility.

The mobility reads [16] :

$$B = (1 + A\frac{\ell}{r_p} + Q\frac{\ell}{r_p} e^{-br_p/\ell}) / (6\pi\mu r_p) \qquad (14)$$

where A, Q, b are fairly known constants (we shall use A = 0.86, Q = 0.29, b = 1.25), μ is the dynamic viscosity of the surrounding fluid, ℓ the mean-free path of molecules and r_p the radius of a particle. When the Knudsen number ℓ/r_p tends to 0, D tends to the Stokes-Einstein coefficient $k_B T / (6\pi\mu r_p)$. When the Knudsen number becomes very high, D tends to its expression for the molecular regime. For soots in flames (small particles in high temperature fluids), the expression (14) has been used in its complete form.

The first set averaging can be simply achieved when particles concentration is assumed uniform in space, the probability for a particle to be at time t = 0 in a small volume $dr_{j,i}(0)$ around the location $r_{j,i}(0)$ being $dr_{j,i}(0)/V$, where V is the volume of the control volume.

The sample volume is illuminated by a gaussian beam. For a convenient evaluation we shall choose to measure $r_{j,i}(o)$ with respect to the center of the gaussian volume. Accordingly, $A(r_{j,i})$ will be written as:

$$A(r_{j,i}) = A_o \exp(-r_{j,i}^2 / (4\sigma^2)) \qquad (15)$$

where 4σ is the diameter at $1/e^2$-intensity of the laser beam (at the waist).

We assume that the velocity is uniform, so $v_{j,i} \equiv v_i$ and after some computations it is found that :

$$R_E(\tau) = C_1 \exp(-i\,\omega_0\,\tau + i\,k_i\,v_i\,\tau)\exp(-v_i^2\,\tau^2/(8\sigma^2)-k_i^2 D\tau) \quad (16)$$

where C_1 contains terms which do not matter for our purpose and where we have assumed that $D\tau \ll 2\sigma^2$ (an inequality checked in realistic situations).

Thus, relation (4) becomes :

$$\tilde{S}(2\pi\nu)=C_2 \int_0^\infty \exp(-v_i^2\,\tau^2/(4\sigma^2))\exp(-2k_i^2 D\tau)\cos 2\pi\nu\tau d\tau \quad (17)$$

This is a Voigt profile, equal to the convolution of a gaussian profile (Fourier transform of gaussian profile) :

$$(\sigma\sqrt{2}/v)\exp(-4\pi^2\sigma^2\nu^2/v_i^2) \quad (18)$$

with a Lorentzian profile (Fourier transform of the decreasing exponential profile) :

$$\frac{2\,k_i^2\,D}{|\,2\,k_i^2\,D|^2 + (2\pi\nu)^2} \quad (19)$$

Equation (17) can be rewritten :

$$\tilde{S}(2\pi\nu) = C_2 \int_0^\infty \exp(-\tau/\tau_v)^2 \exp(-\tau/\tau_d)\cos 2\pi\nu\tau d\tau \quad (20)$$

where two characteristic times appear explicitly :

- a transit time $\tau_v = 2\sigma/\sqrt{v_i^2}$

- a diffusion time $\tau_d = 1/(2\,k_i^2\,D)$

Three cases must be considered :

1) $\tau_d \gg \tau_v$:   the transit contribution dominates, and the spectrum is gaussian. The Diffusion Broadening Spectrometer acts like a pure transit time velocimeter.

2) $\tau_d \ll \tau_v$:   which means that the velocity can be considered as equal to zero. The spectrum is Lorentzian. The Diffusion Broadening Spectrometer acts like a sizing system.

3) $\tau_d \sim \tau_v$: contributions from transit and diffusion times must be decoupled, usually by a numerical procedure.

Remark : In a given situation, it can be necessary to compromise since too large the width of the control volume means increasing the disturbing influence of possible velocity gradients while too small this width means making sizing impossible due to the dominating effect of transit time.

EXPERIMENTS

The experimental set-up

It is sketchedon Figure 1 and closely follows the chronology of previous theoretical analysis. The illuminating light source is an Argon laser, working in the TEM_{00} mode and delivering 1.2W on the 514.5 nm-line. The lens L_1 (f = 20 cm) focuses it in the medium under study. The light scattered by the submicronic particles present in the control volume is directed on to the quadratic photodetector PM through the collecting optics. An interferential filter enables us to get rid of most of the parasitic light. All optical components are located on an optical bench and may be pivoted around a vertical axis permitting to carefully achieve the required adjustments.

The photomultiplier (RTC, XP, 2008) anode current is then analyzed by means of a HP.3582 A. spectrum analyzer, using an efficient Fast Fourier Transform algorithm. Spectra displayed on the spectrum analyzer scope are also plotted on a X-Y recorder for convenient storage and further processing.

Measurements have been achieved on suspensions at rest, on moving suspensions and on soots in flames. In the latter case, a standard burner, as previously investigated by D'Alessio et al[1] or Lahaye and Prado[2], has been used in order to produce results with available comparisons.

This flat flame burner is described on Figure 2. The gaseous methane/oxygen mixture passes through a bronze porous plate (pores diameter $\sim$ 50 μm), cooled by flowing water, to produce an atmospheric pressure premixed flame. In order to prevent it to get a diffusion character, the fuel-oxygen reactive medium is protected by a nitrogen surrounding flow.

A 110 mm − diameter water - cooled plate is placed 3 cm above the burner in such a way that the flame is nearly cylindrical. Gas flow rates are measured using calibrated sonic orifice flowmeters. The flame is flat and steady · Operating conditions are well controlled and measurements can be reproduced with a very satisfactory accuracy.

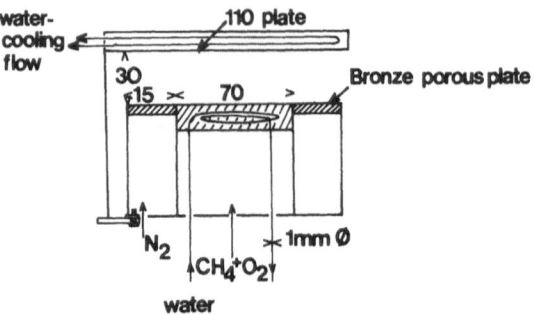

Fig. 2.   The burner.

Spectra processing

The spectrum analyzer permits a 256 spectra RMS averaging,
reducing considerably the statistical uncertainty on power values,
the analysis time for each individual spectrum being 10 ms for a
horizontal 25 k Hz -scale display. Due to the above averaging process,
experimental spectra generally appear rather smooth.

The Figure 3 shows a quite typical example of spectrum such as
obtained on the X-Y recorder. Note that the spectrum analyzer
actually displays the square root of the power spectrum.

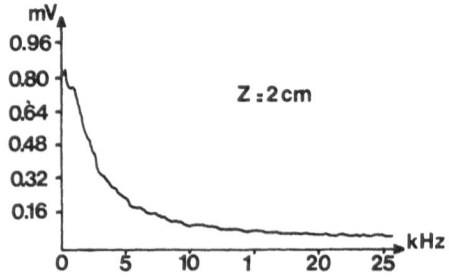

Fig. 3.   A typical spectrum.

Computer programs have been built to fit the experimental power
spectra with theoretical Lorentzian, Gaussian and Voigt profiles
respectively, using a least square procedure. When the Voigt profile
fitting is used, the transit time τ_v must be introduced, that requires knowing (or estimating) σ and $\sqrt{v_i^2}$. Note that it should be
possible to fit experimental spectra by Voigt profiles without
knowing τ_v, but the corresponding algorithmic procedure has not been
developed (or used) at the present stage of development. It will be
studied in a near future since the required knowledge of $\sqrt{v_i^2}$ is a
shortcoming of the present work. The Figure 4 gives an example of
the fitting process result. The experimental spectrum values taken
from a X-Y recording, squared to obtain the actual power spectrum,
are fitted by a Voigt profile. The width parameter σ was computed
from the focusing optics geometry and $\sqrt{v_i^2}$ was taken as the velocity
of the non-burning gas, an estimation which is probably not too crude.
The Figure 5 gives another example corresponding to measurements on
very small particles (d $\sim$ 200 Å) which produce large data spreading,
although it should be noticed that this spreading is not to be
interpreted straight away as an increase in the physical noise. For
soot studies, the inaccuracy in estimating the width of the Lorentzian contribution to the spectrum appears to be less than, say, 10%.

Results for suspensions in a liquid at rest

     The velocity $\sqrt{v_i^2}$ being zero, the spectrum is a Lorentzian
profile (relation 19) in which D is the Stokes-Einstein coefficient
$k_B T/6\pi\mu r_p$) since the Knudsen number ℓ/r_p is very small in liquids.
The (half) half-width (at half height) of the spectrum is readily
found to be:

$$\delta\nu_L = k_i^2 D / \pi \tag{21}$$

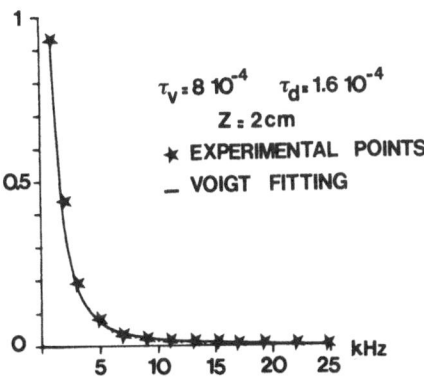

Fig. 4. A fitting process result.

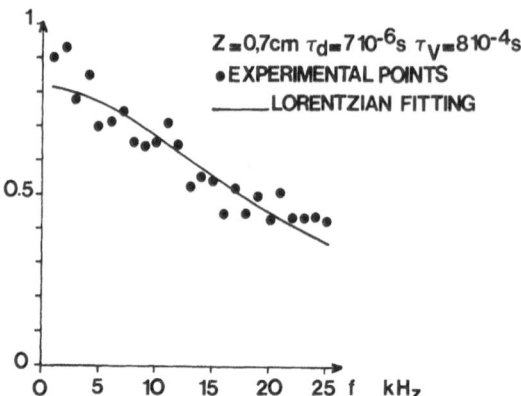

Fig. 5.   A fitting process result.

that means :

$$\delta \nu_L = \frac{16\, k_B\, T}{3\, \mu\, \lambda^2\, d}\; \sin^2\left(\frac{\theta}{2}\right)$$ (22)

Measurements have been carried out on calibrated monosized
latex particles suspended in water. The temperature T is measured
by means of a mercury thermometer from which the viscosity is also
known using Handbook tables. $\delta \nu_L$ is measured using the above described spectra processing (specified for Lorentzian profiles). For
θ = 90°, d has been then measured using relation (22) for three
different calibrated diameters giving 107, 120 and 183 nm respectively for particles announced by the manufactory as having diameters
equal to 91, 109 and 176 nm giving disagreements equal to 9%, 9%
and 4% respectively, corresponding mainly to the inaccuracy in the
fitting analysis procedure. For the 91 nm -particles, $\delta \nu_L$ has been
measured as a function of $\sin^2(\theta/2)$. Results are shown on the Figure
6. In agreement with relation (22), a straight line fits satisfactorily the experimental points. The diameter d can be measured
from the slope of the line, giving d = 92 nm , corresponding to only
a 1% disagreement.

Results for suspensions in a moving liquid flow

Measurements have been carried out on ludox particles embedded
in a Poiseuille flow in a water pipe. The diameters of the particles were in the range 12-13 nm . The scattering angle θ has been
chosen small enough so that the Lorentzian contribution to the
spectrum can be neglected for a velocity $(v_i^2)^{1/2}$ higher than a
certain critical value (see relation 22). In these conditions, the
half-width at (1/e) of the gaussian profile $\delta \nu_g$ is measured from

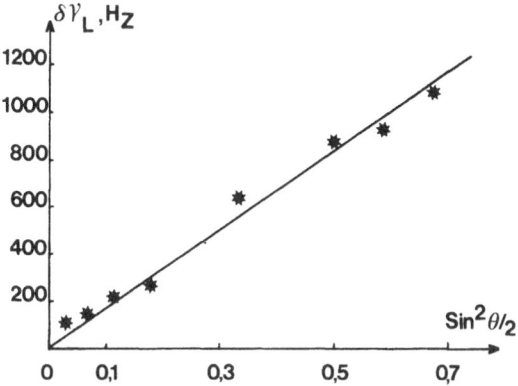

Fig. 6. The (half) half-width $\delta \nu_L$ versus $\sin^2(\theta/2)$ for
         monosized latex particles.

the experimental spectra using the above described spectra processing (specified for Gaussian profiles). According to relation (18),
$\delta \nu_g$ reads :

$$\delta \nu_g = \sqrt{v_i^2} / (2 \pi \sigma) \qquad\qquad (23)$$

     Relation (23) shows a linear relation between $\delta \nu_g$ and $\sqrt{v_i^2}$ as
confirmed by the experimental results plotted on Figure 7. Note that
the slope of the curve corresponds to $\sigma = 24$ µm while, according to
the theory of propagation of gaussian beams [17], σ is found to be
23µm.

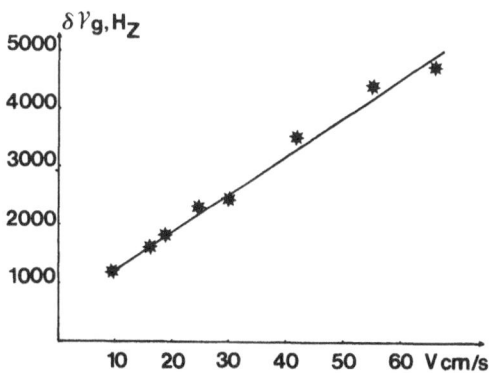

Fig. 7. $\delta \nu_g$ versus transit velocity.

Soots results

The (half) half-width $\delta\nu_L$ of the Lorentzian contribution is
deduced from the experimental spectra using the above described
spectra processing, from which D is known using relation (21). The
scattering angle has been choosen very small ($\theta = 5°$) so that all
the obtained spectra could fit into the maximal display range of
the spectrum analyzer (25 k Hz). This fact is not convenient for
accuracy but is not demanded by the method itself, being only a
consequence of spectrum analyzer limitations. The relative error
$\delta D/D$ is estimated equal to roughly 20%. This could be tremendously
decreased by increasing θ but requires to use a spectrum analyzer
not limited to the audiorange. The isothermal diffusion coefficients
D are shown versus z on the Figure 8.

Diameters d are then known from the coefficients of diffusion D
using relations (13) and (14). V is the velocity of the non burning
gases and R the feed ratio here defined as the concentrations ratio
$[CH_4]/[O_2]$ over the stoichiometric ratio. For V = 6.20 cm/s and
R = 2.65, the temperature T, measured using the Kurlbaum's method,
has been found to evolve from 1750 ± 50K at z = 0,6cm down to
1400 ± 50K at z = 2cm. The mean free path ℓ has been computed using
the elementary theory of the transport phenomena giving $\ell = 1/(\pi n \sigma^2 \sqrt{2}$
where σ is the diameter of molecules considered as rigid spheres,
and n is the number-density of molecules computed from the perfect
gases law. Here, σ has been typically taken as 4 Å. The viscosity
has been estimated from Wilke's expression [18] giving the viscosity
of a mixture as a function of the viscosities of simple gases and
components concentrations. The species concentrations have been
taken from D'Alessio measurements [19] and the viscosities of simple
gases (depending on the temperature) have been taken from Svehla

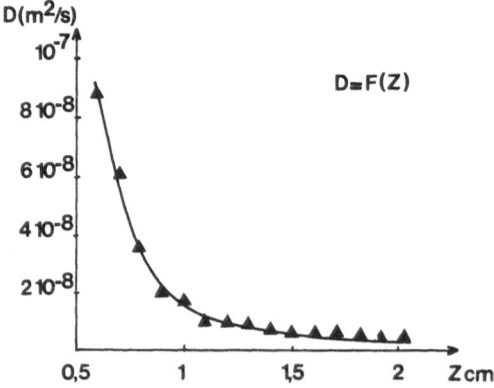

Fig. 8. The diffusion coefficient D versus z.

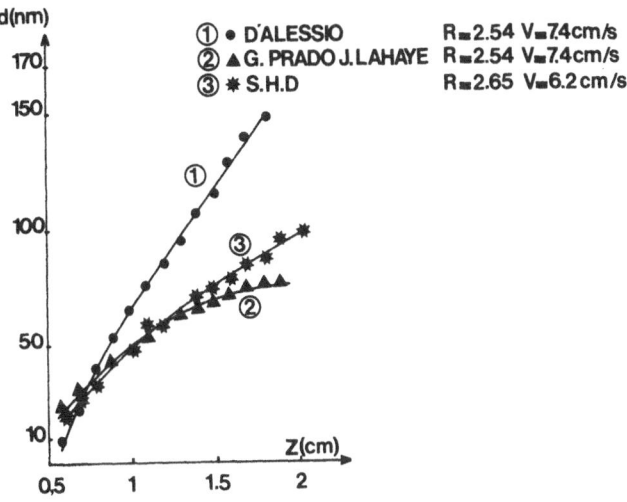

Fig. 9. Soot diameters versus z.

table.[20] The obtained results are shown on Figure 9 and compared
against Prado and Lahaye[2] and D'Alessio.[21]

In the present stage of development, one of the main errors
in measuring d appears to come from the estimation of the mean-free-
path ℓ. To illustrate this point, taking σ equal to 3 Å instead of
4 Å , the diameter d on profile 3, figure 9, changes from 1310 to
1000 Å at z=2cm, and from 310 to 210 Å at z=0.6cm. It is worthwhile
to point out that preliminary measurements of TBr_p should avoid
the above difficulty (and similar connected ones), considerably
increasing the accuracy of the technique.

According to Gouesbet et al,[11] the number-density N of particles over the number-density N_0 at a given point can be estimated by
a simple formula when some assumptions are made, one of them being
that the coefficient of dispersion reads like the Stokes-Einstein
coefficient. In the present work where a more general expression of
D must be used, the above mentioned simple theory should be gene-
ralized. Approximate computations have shown that the evolution of
number-density versus z closely follows the D'Alessio et al's one.[21]

CONCLUSION

The present status of the art in Diffusion Broadening Spectroscopy applied to soot diagnosis in flames has been described. Further
developments must be achieved, for instance it should be attempted
to obtain polydispersity informations and, although it is not clear
whether it could be possible to succeed, non-laminar turbulent
flames situations should be investigated. As far as accuracy is
concerned, two main problems have to be pointed out. The first one

is knowing the velocity in order to take into account for the transit time contribution in interpreting measurements, a way to proceed being possibly to use DBS itself as a transit velocimeter (see previous section). The second one is in estimating TBr_p since we have shown that the main error at the present time is involved in that term. Again, one of the way to proceed could be to use DBS itself. These further developments are currently under progress in Rouen. Just now, the obtained results are comparable to the ones obtained from the well-known scatter-absorption technique; they may provide us with valuable data.

REFERENCES

1. A. D'Alessio, A. DiLorenzo, A. Borghese, F. Beretta and S. Masi, 16th Symp. (Intern'l) on Combustion, The Combustion Institute, Pittsburgh, p.695 (1977)

2. G. Prado and J. Lahaye, A.T.P. Combustion et Turbulence, Rapport de fin d'étude, Mars (1979)

3. G. Gouesbet and G. Gréhan, The quasi-electric scattering of light; a lecture with emphasis on particulate diagnosis. Present workshop.

4. B. L. Wersborg, J. B. Howard and G. C. Williams, 14th Symp. (Intern'l) on Combustion, The Combustion Institute Pittsburgh, p.929 (1973)

5. J. Lahaye and G. Prado, Particulate carbon: Formation during Combustion, D.C. Siegla and G.W. Smith, Eds., Plenum Press, New-York - London, p.143, (1981)

6. G. B. Benedek, in:"Polarisation, matière et radiation", P.U.F., Paris (1969)

7. W. Hinds and P.C. Reist, J. of Aerosol Sc. 3:501 (1972)

8. S. S. Penner, J. M. Bernard and T. Jerskey, Acta Astronautica 3:69 (1976)

9. J. F. Driscoll and D. M. Mann, Submicron particle size measurements in an acetylene/oxygen flame, Proceedings of the 3rd International Workshop on Laser Velocimetry (LV-III), Purdue University, p.438, July 11-13 (1978)

10. J. F. Driscoll, D. M. Mann and W. K. Mc Gregor, Comb. Sci. and Technology 20:41 (1979)

11. G. Gouesbet, P. Flament, M. E. Weill and G. Gréhan, La Rivista dei Combustibili 25(1):50 (1981)

12. V. E. Edwards, J. C. Angus, J. F. French and J. W. Dunning, J. of Appl. Phys. 42(2):837 (1971)

13. B. J. Berne and R. Pecora, Dynamic light scattering, John Wiley and Sons, New York (1976)

14. H. C. Van de Hulst, Light scattering by small particles, John Wiley and Sons, New York (1957)

15. S. Chandrasekhar, Rev. Mod. Phys. 15:1 (1943)

16. M. A. Fuchs, The Mechanics of Aerosols, Pergamon Press
     (1964)
17. H. Kogelnik, The Bell System Tech. J., p.455 (1965)
18. C. R. Wilke, J. Chem. Phys. 18:517 (1950)
19. A. D'Alessio, A. DiLorenzo, F. Beretta and C. Venitozzi,
     14th Symp. (Intern'l) on Combustion, The Combustion
     Institute, Pittsburgh, p.941 (1973)
20. R. A. Svehla, Technical Report R.132, N.A.S.A. (1962)
21. A. D'Alessio, A. DiLorenzo, A. F. Sarofim, F. Beretta,
     S. Masi and C. Venitozzi, 15th Symp. (Intern'l) on
     Combustion, The Combustion Institute, p.1427 (1975)

DISCUSSION

S. Galant (Société Bertin)

Since soot particles can be used as tracers, can you get
measurements of both average and fluctuating values of velocity?

Gouesbet

The Diffusion Broadening Spectroscope can be made to act
as a transit velocimeter and thus average velocities measurements
are quite possible. Some examples of such measurements have been
discussed during the presentation. Nevertheless, a more precise
answer would need to consider a specific situation, with figures.
Concerning fluctuating values, the problem is more complex since
you consider turbulent flows. The point is that, at the present
stage of development, we do not know exactly what is to be done
if the flame were turbulent. Thus, this is an open question,
although I guess that the answer might be positive. Furthermore,
there always remains the possibility of coupling a Diffusion
Broadening Spectroscope with a more classical LDV-technique,
probably not in the real fringe optical set-up but rather in a
referential set-up.

THE QUASI-ELASTIC SCATTERING OF LIGHT: A LECTURE WITH EMPHASIS

ON PARTICULATE DIAGNOSIS

G. Gouesbet and G. Grehan

Laboratoire de Thermodynamique - L.A. C.N.R.S. n°230 -
Faculté des Sciences et Techniques de Rouen
B.P. 67 - 76130 MONT-SAINT-AIGNAN (France)

INTRODUCTION

The Quasi-Elastic (linear) Scattering (QES) of light has
been used for more than a hundred years for particulate diagnosis.
It is particularly relevant to achieve particulate measurements
(sizes, concentrations, shapes, velocities, refractive indices...)
in combustion systems, and has been the object of a very large
renewal of interest after the advent of lasers.

A lot of very valuable books are available to get precise and
extensive informations on the QES and underlying electromagnetism,
such as those by Kastler[1], Van de Hulst[2], Stratton[3], Kerker[4],
Born and Wolf[5]. There exists also a very large literature of which
we shall only take out the recent review by Jones[6]. Let us also
mention books devoted to radiative transfer and multiple scattering
by Chandrasekhar[7] and Van de Hulst.[8]

The aim of the authors is not to attempt to summarize the
above writings but to focus the attention of the reader on recent
new developments, with emphasis on particulate diagnosis.
Mathematical relations will be avoided as far as possible, the
main spirit of the present work being to present a general framework useful for discussion and to dispatch the reader on relevant
references. The first part of this paper will be devoted to
the theory itself and the second one to measurements techniques.

(QES)-THEORY

Generalities

    The QES of the light by a scatter center refers to the case
where no frequency change is involved in the light/matter interaction but that one due to the Doppler effect and the other
(singular) one from ν-frequency to 0-frequency due to a photon
absorption. The light will be here basically considered as an
electromagnetic wave (although the scattering phenomena is a
random quantum process), with a rectilinear polarization (a case
from which more general states of polarization can be readily
investigated).

    We shall assume that each particle in a cloud scatters
independently (no coherent scattering); this demands a separation
between particles greater than about three radii.[4] No multiple
scattering phenomena are discussed here, although an elementary
discussion by Jones[6] shows that this assumption is not always true
for soot in flames. We shall consider clouds containing large
numbers of randomly located particles so that the scattered energies
can be summed up directly. Although, the present workshop is
mainly focussed on soot in combustion systems, this paper will
also discuss larger particles up to sprays because of their
interest in, say, fuel pulverized combustors. Particles will be
assumed to be spherical in most of these pages, although some
insights on non-spherical scatter centers will be incidentally
provided.

    The basic scattering problem first addressed is the scattering
of a polarized TEM-wave by a spherical, homogeneous, isotropic,
non-magnetic particle (a Mie scatter center) having a diameter
d and a complex refractive index m = n(1 - ik) located at the
point 0 of a Cartesian coordinate system (Fig. 1). The incident
wave propagates from the negative z to the positive z, and the
electric field E vibrates in the plane (xOz). Only axisymmetric
incident light profiles, defined by $|\vec{E}| = E(\rho, z)$, are considered.
The scattered light is observed at the point P(r, θ, ϕ).

    We may be interested by the following scattering properties:

(i) Scattered intensities
    The scattered intensity at P in the far field (far enough
with respect to the light wave-length λ), per unit of incident
intensity at 0, reads:

$$I_{diff} = I_{\theta} + I_{\phi} = \frac{\lambda^2}{4\pi^2 r^2} [S_2^*(\theta) S_2(\theta) \cos^2\phi + S_1^*(\theta) S_1(\theta) \sin^2\phi] \quad (1)$$

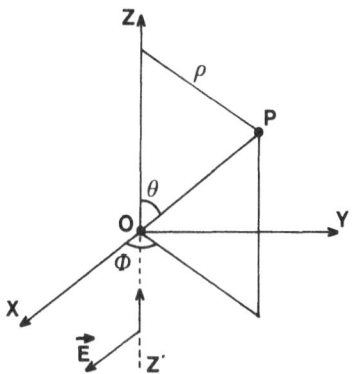

Fig. 1.   The basic scattering problem.

where I_θ refers to an electric field E_θ vibrating in the plane
(Oz, OP) and I_ϕ to an electric field E_ϕ vibrating in the perpendicular plane.   Thus, the scattered intensity at a given point P is
known from the quantities $|s_1(\theta)|^2$ and $|s_2(\theta)|^2$, the $s_i(\theta)$ being
the so-called amplitude functions.   It is consequently sufficient
to know what are the intensities in the parallel plane ($\phi = 0$)
and the perpendicular plane ($\phi = \pi/2$) corresponding to $|s_2|^2$ and
$|s_1|^2$ respectively. $|s_2|^2$ and $|s_1|^2$ plotted versus the scattering
angle θ are the parallel and perpendicular diagrams respectively.

ii) Phase angle
      Between the electric field E_θ and E_ϕ of the elliptic scattered
light.   It will not be discussed here.

iii) Cross-section and efficiency factors
      The cross sections for scattering, absorption and extinction
(C_{sca}, C_{abs}, and C_{ext} respectively) are defined as the total
energy which is scattered, absorbed or scattered and absorbed,
respectively, per unit of time and per unit of incident intensity.
The respective efficiency factors Q_i are cross-sections C_i (where
i stands for 'sca', 'abs', or 'ext') over $(\pi d^2/4)$.   When
$\alpha = (\pi d/\lambda) \to \infty$, $Q_{ext} \to 2$, meaning that the particle scatters or
absorbs 200% of the light it receives geometrically.   This is
known as the Mie-paradox (although it is not paradoxal and can
be explained) or as the extinction anomaly.

iv) Turbidity
      When the light propagates through a length L of a medium

containing homogeneously identical particles, the energy is
attenuated as $e^{-\tau L}$ where the turbidity τ is equal to NC_{ext}, N
being the number-density of the particles (Note that the turbidity
is sometimes defined as τL). The turbidity τ over the total
volume V of particles per unit volume reads $\tau'_s = 3\, Q_{ext}/(2d)$,
depending only on d and m.

v) Pressure of radiation
   It will not be discussed here.

The Generalized Lorenz-Mie Theory (GLMT)

In the present work, the GLMT refers to the scattering
problem defined in the section II-1 (although more generalizations
are possible). In this section, we shall limit ourselves to
intensities computations. These intensities can be obtained by
solving the Maxwell equations, with adequate boundary conditions.[9]
A more concise discussion, mainly limited to the far field (that
we only consider) is available from [10]. The obtained formalism
is very similar to the one given by Kerker[4] for Lorenz-Mie Theory
(LMT). The amplitude functions read:

$$S_1 = \sum_{n=1}^{\infty} \frac{2n+1}{n(n+1)}\, g_n\, \{a_n \pi_n(\cos\theta) + b_n \tau_n(\cos\theta)\} \qquad (2)$$

$$S_2 = \sum_{n=1}^{\infty} \frac{2n+1}{n(n+1)}\, g_n\, \{a_n \tau_n(\cos\theta) + b_n \pi_n(\cos\theta)\} \qquad (3)$$

where π_n and τ_n are the Legendre functions, and a_n and b_n are
the so-called scattering coefficients (see Kerker for details
of notations). The correcting scattering coefficients g_n
contain the new features provided by the GLMT. The relevant
arguments contained in the scattering coefficients are the size
parameter $\beta = \pi d/\lambda$, and the 'second' size parameter $\beta = m\alpha$.
Formulae for g_n are not given here.

One aim of the above way of research is to be able in the
future to provide a general theory of visibility for particle
sizing (see a later section). Another aim is the precise interpretation of optical levitation experiments carried out (with
thermonuclear fusion purposes) by Ashkin,[11] and Roosen,[12] and
(for pure scattering) by Gréhan and Gouesbet.[13]

Nevertheless, no numerical computations of the coefficients
g_n are available at the present time, although the work is under
progress.

The Lorenz-Mie Theory (LMT)

Basic papers are from Lorenz [14,15], Mie [16] and Debye. [17] The plane wave being a special axisymmetric light profile, the LMT will be here deduced from the GLMT. For an incident plane wave, the coefficient g_n become all equal to 1, [9,10] so that Relations (2) and (3) are readily specified for the LMT.

The efficiency factors read [4] :

$$Q_{sca} = \frac{2}{\alpha^2} \sum_{n=1}^{\infty} (2n+1) \{ |a_n|^2 + |b_n|^2 \} \tag{4}$$

$$Q_{ext} = \frac{2}{\alpha^2} \sum_{n=1}^{\infty} (2n+1) \{ Re \{ a_n + b_n \} \} \tag{5}$$

$$Q_{abs} = Q_{ext} - Q_{sca} \tag{6}$$

The advent of powerful computers has been required to obtain extensive numerical calculations in the framework of the LMT, a lot of them being aimed at astrophysical problems. Let us mention works from Wickramasinghe [18], Dave [19], Diermendjian et al [20], Cherdron et al.[21] Yet, a new algorithm to compute the ratios of Bessel functions (involved in scattering coefficients) without recurring formulae, has been recently published by Lentz.[22] A new computer program, the so-called SUPERMIDI, has been built up in Rouen with the aid of the Lentz algorithm [23,24], possibly constituting the first one of a new generation of computer codes for LMT. In its more recent version, SUPERMIDI computes scattered intensities, phase angle, efficiency factors, turbidity and pressure of radiation for particles having a size parameter as high as 2000 (and probably larger) and imaginary parts of m as high as 10^5, 10^6 (an academic case, but proving the stability of the Lentz algorithm). Two other programs derive from SUPERMIDI, the first one (POLYMIDI) to handle clouds of polysized particles (unpublished) and the second one (HOLOMIDI) being devoted to aerosol diagnosis by micro-holography.[25] The Lentz algorithm has also been recently discussed by Wiscombe. [26]

As an example of numerical results, the Figure 2 shows a polar perpendicular scattering diagram $|S_1|^2$ versus θ corresponding to, say, a fuel particle (m = 1.4 - 0.1 i). Length scale is logarithmic. Other results will be discussed later.

The limiting theories

Although numerous limiting scattering theories are available, discussion will be here limited to four of them, particularly relevant according to the authors' opinion.

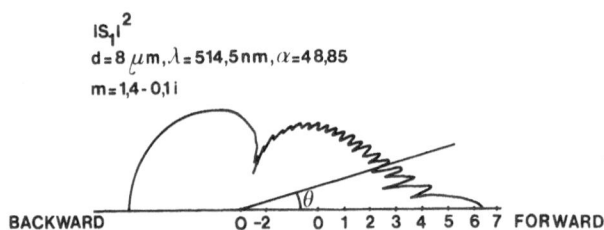

Fig. 2.   A scattering diagram for a fuel Mie scatter center

(i) The first one concerns the case of small particles (α and $|\beta|$ small) considered without change of concepts, that means we just deduce new formulae from the LMT with the aid of mathematical approximations. The basic idea is then to compute the series S_1 and S_2 as sums of a few terms and write the involved functions as limited expansions. The scattering coefficient a_1 is then found to be [4]:

$$a_1 = \frac{2}{3} i \left(\frac{m^2 - 2}{m^2 + 2} \right) \alpha^3 \tag{7}$$

while a_2 and b_1 involve α^5, and so on, leading to $S_1 = -\frac{3}{2} a_1$ and to $S_2 = -\frac{3}{2} a_1 \cos \theta$ and thus giving :

$$I_{diff} = \frac{16 \pi^4 r_p^6}{r^2 \lambda^4} \left| \frac{m^2 - 1}{m^2 + 2} \right|^2 \left[\cos^2 \theta \cos^2 \phi + \sin^2 \phi \right] \tag{8}$$

where $r_p = (d/2)$. The case of transparent particles is obtained when $m = n$, where n is a real number. For natural light, it can be readily shown than $[\cos^2\theta \cos^2\phi + \sin^2\phi]$ must be changed to $(1 + \cos^2\theta)/2$. For N particles, the scattered intensity thus contains $N d^6$. The scattering diagrams are perfectly symmetric with respect to forward and backward directions : $I(\theta) = I(\pi - \theta)$.

Based on the turbidity concept, it is possible to set a criterion of validity of this theory. Such a commonly admitted criterion is $r_p < 0.05\lambda$, that means $r_p < 250$ Å when $\lambda = 0.5\mu$m. When $r_p > 0.05 \lambda$, the departure from the exact theory can become fastly dramatically important, as we shall discuss it later.

From (4) and (5) the efficiency factors Q_{sca} and Q_{ext} read respectively $6a_1^*a_1/\alpha^2$ and $6\,Re\,(a_1)/\alpha^2$ that means, for transparent particles $(m = n)$:

$$Q_{sca} = \frac{8}{3} \left(\frac{n^2 - 1}{n^2 + 2} \right)^2 \alpha^4 \neq 0 \tag{9}$$

$$Q_{ext} = 0 \tag{10}$$

While the relation (9) is perfectly correct (see Rayleigh theory), the relation (10) is a completely stupid one! This strange behaviour of mathematical approximations has been recently extensively discussed and explained by Chylek and Pinnick.[27]

In limited ranges of n and nk, Q_{ext} can be computed with the aid of the Penndorf approximations $(Q_{ext})_n$.[28] The first approximation, valid for $n \sim \sqrt{2}$ and $nk \sim n$ reads :

$$(Q_{ext})_1 = \frac{24\,\alpha\,n^2\,k}{(n^2 + n^2 k^2)^2 + 4\,(n^2 - n^2 k^2 + 1)} \tag{11}$$

and is very extensively used by soots workers.

(ii) The Rayleigh theory concerns (strictly speaking) the case of small transparent particles and involves the concept of a Hertz dipole. It leads to the same formulae as given above (for m = n). Let us note that we obviously find $Q_{ext} = Q_{sca}$ (since $Q_{abs} = 0$ in this case).

(iii) When $m \sim 1$, it is possible to consider a (large) scatter center as a juxtaposition of elementary Hertz dipoles, the waves emitted by each of them being summed up taking into account for the path-length differences. This limiting theory is very powerful to understand in terms of plain physics some main features of the LMT, such as the lobes pattern associated with constructive and destructive interferences between the waves emitted by the elementary dipoles (see discussion of the case of the sphere in [1]).

(iv) Geometrical optics. When $\alpha \to \infty$, the series appearing in the LMT exhibit a separation between terms which do depend on the nature of the particle, and terms which do not depend on it, corresponding to reflections and refractions, and diffraction, respectively.[2] At a given point of observation P, geometrical optics combines the reflected ray (order p=0), refracted rays leaving the sphere after 1, 2, ... p, ... internal reflections, and diffracted waves, taking into account for phase differences. In the general case, such a theory would prove to become more tedious than the LMT.[29] Nevertheless, for transparent particles and in near-forward scattering ($\theta < 20°$), a simplified theory where only rays of orders p = 0 and 1, and diffraction, are considered, can be used. This simplified theory has been extensively used at Sheffield University, by Chigier's team, for particle sizing in combustion systems. Systematic comparisons against the LMT have been recently achieved.[30,31] The Figure 3 shows an example for d = 25 μm : the comparison is very satisfactory.

For 'random'-spherical particles (particles roughly spherical but with random imperfections at the surface), A. Ungut suggested to replace the scattered intensities I by averaged intensities I_{av} defined by :

$$I_{av} = \frac{1}{\lambda} \int_{d-\lambda/2}^{d+\lambda/2} I(d) \cdot dd .$$ (12)

The Figure 4 shows what happens when the results of Figure 3 are averaged using the above relation. Again, agreement is satisfactory. More results and discussions are available from [30] and [31].

The geometrical optics also enables us to understand the Mie paradox : when $\alpha \to \infty$, the contribution $(Q_{ext})_r$ due to reflections and refractions tends to 1, while the diffraction contribution $(Q_{ext})_d$ also tends to 1, that means as a whole $Q_{ext} \to (Q_{ext})_r + (Q_{ext})_d = 2$.[32]

DIAGNOSIS

Generalities

The reference [33] contains a rather exhaustive overview of particulate diagnosis techniques, but the present discussion is limited on optical techniques based on QES. Yet, even so, the scope

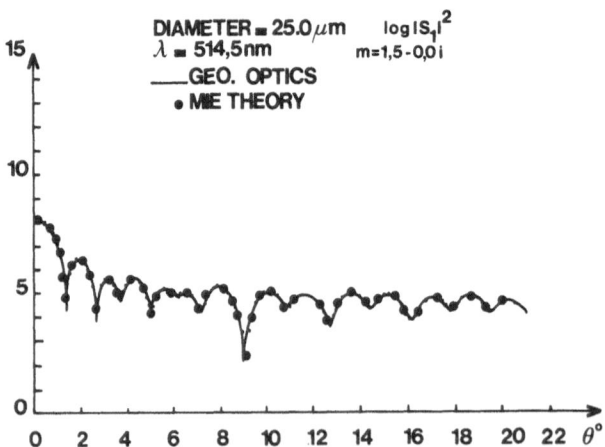

Fig. 3.   Comparison geometric optics/LMT.

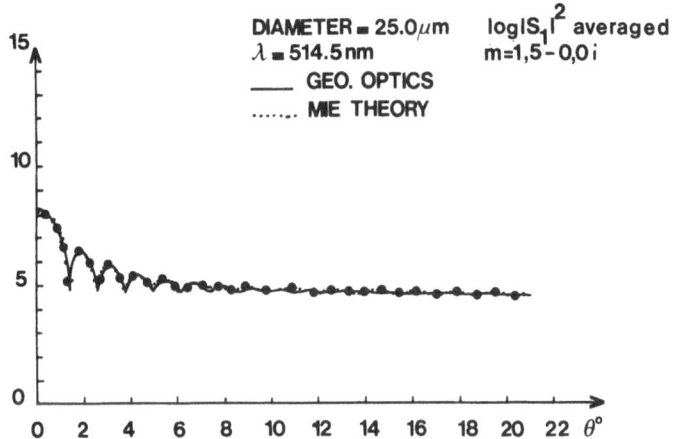

Fig. 4. Comparison geometrical optics/LMT for a 'random' –
spherical particle.

remains too large. Conventional non-laser techniques such as optical
microscopy, ultraviolet or spark photographs, cinematography, etc...
will be dismissed, and interest will be focused on laser techniques
(although some of these could in principle be used with conventional
sources).

Some laser techniques use a frequency analysis of the scattered
light to achieve particle sizing. Let us mention the Doppler Shift
Spectroscopy [34] or the Diffusion Broadening Spectroscopy, the latter
being extensively discussed elsewhere in the present workshop.[35]
Nevertheless, attention will be focused on techniques achieving
sizing by processing amplitudes informations (scattering or
absorption).

A partition between the techniques can be achieved depending
whether they permit or not simultaneous velocimetry. Such a partition is not completely non-ambiguous but it will nevertheless be
preserved as convenient. By 'methods permitting simultaneous velocimetry', we shall here refer explicitly to the visibility concept,
the pedestal calibration technique, and to a system using a corrected laser beam currently under study in Rouen.

Even so, it remains here impossible to attempt an exhaustive
listing of the relevant measurements techniques, leading us to
achieve more or less arbitrary choices. Complementary informations
and discussions are available from [36] and [37].

Without simultaneous velocimetry

Let us begin that section by mentioning the microholography
[38,39], the numerical inversion of diffraction patterns[40] and its

tomographic variant,[41] or the speckle photography.[42] The 'complete'
scattering diagrams of clouds are recorded by Laug and Delfour[43]
with the aid of a nice system using the ellipse properties and a
mathematical inversion procedure. As a matter of fact, some measurements have been carried out in rocket exhausts. Papers from
Bartholdi et al.[44,45] discuss variants, working on a single particle
recording possibly a small angular part of the scattering diagrams.
The above sizing systems are aimed to rather large particles
measurements. Let us go now to the discussion of the diagnosis
of rather small particles.

     Turbidimetry (extinction measurements) is commonly used for
soots and smokes diagnosis, often simultaneously with scattered
intensities measurements, since a single kind of techniques is not
sufficient to measure separately sizes (say diameters d) and
number-densities N (or concentrations c), even if m is known,
as can be seen from 'limiting theories' sections. It is particularly true in the Rayleigh domain where both τ and I_{diff} depend on
Nd^6 (or cd^3). For soots diagnosis, simultaneous measurements of
light absorption and scattered intensities permit to obtain
simultaneously d and N (or c) and when m is known, as achieved
and/or discussed in the references [46, 47, 48, 49, 50] (a list
which does not pretend to be exhaustive).

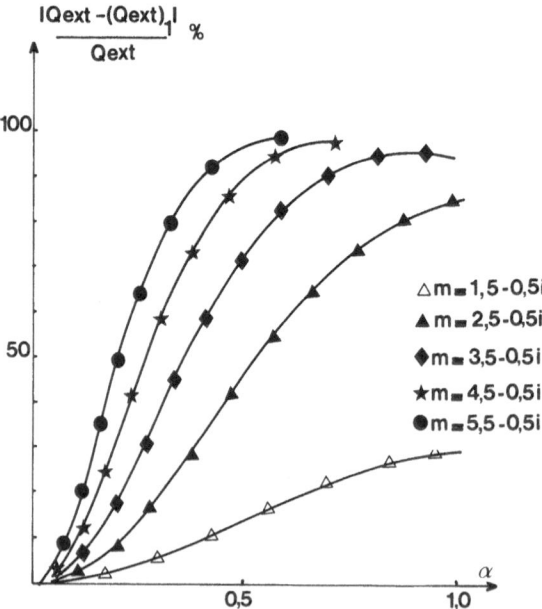

Fig. 5.   Examples of errors produced by using the first Penndorf
          approximation.

Q_{ext} is expressed by some authors as $(Q_{ext})_1$. Nevertheless, care should be taken when using the relation (11) since its range of validity is limited.

As a matter of fact, Figure 5 shows the error $[Q_{ext} - (Q_{ext})_1]$ /Q_{ext} produced by using the first approximation from Penndorf, Q_{ext} having been computed with the aid of the SUPERMIDI computer program, for different values of m. The error becomes sensible for, say, $\alpha \sim 0.1$, corresponding to $d \sim 100$ Å when visible light is considered. More results, including other values of m and discussion of higher Penndorf approximations, are available from [52].

Thus, except for very small soots, the question must be asked whether the LM-Theory should not be used. Computer programs as efficient as SUPERMIDI are not required to the purpose since (i) we are here concerned with small particles and (ii) the knowledge of Q_{ext} is basically sufficient to interpret extinction measurements. Lorenz-Mie theory has been used (for instance) by Bonczyk[52], who Chippett and Gray [53] who calculated the extinction cross-sections of soot agglomerates, or by Roessler and Faxvog [54] then [55], who concluded that 'a simple model of acetylene smoke in which the particles are simulated by small unagglomerated carbon spheres is inadequate to explain the observed optical properties in both the visible and infrared regions', or by Lowes and Newall [56] to discuss emissivities of soot clouds, a choice which again does not pretend to exhaustivity.

The above conclusion from Roessler and Faxvog reminds us that soots are not generally Mie scatter centers (spherical, homogeneous, isotropic particles). The anisotropy in the material (polarizability) or in the shape produces depolarization effects which can be used as a diagnostic tool. See for instance Jones and Wong who detected the presence of elongated agglomerates in sooty flames [57].

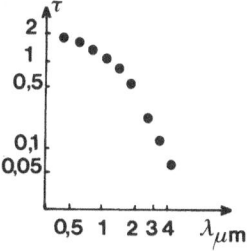

Fig. 6. Agglomerated wood pyrolysis: log τ versus log λ . (experiments)

Another admitted major feature of soots turbidimetry is that τ depends on λ roughly as λ^{-1}. This well-known fact, observed by many authors, can be checked with the aid of the relation (11) in which values for n and nk (versus λ) from Dalzell and Sarofim [58], for instance, are introduced. Yet it is worthwhile to point out that this statement only holds for restricted ranges of diameters and λ, as discussed by Portscht. [59] The λ^{-1} behaviour has again been recently observed in Rouen for polypropylene flame soots and during poplar pyrolysis, but has been found to fail in agglomerated wood pyrolysis for which a slope-breaking has been observed in the range $\lambda = 0.5 - 4$ μm, as shown on Figure 6. [60] This slope-breaking has been 'qualitatively' explained by SUPERMIDI computations, using values of n and (nk) versus λ taken from Dalzell and Sarofim. [58] Results are shown on Figure 7, and illustrate very well the slope-breaking behaviour.

Furthermore, the location λ_c of the slope-breaking depends on d. An elementary analysis of the phenomenon, based on the functions $Q_{ext} = f(\alpha)$, leads to the conclusion that d is roughly equal to (λ_c/π), permitting a fast estimate of d. [60]

Coupled turbidimetry and light scattering techniques present two main disadvantages, one being that the absorption technique is not local but integrates informations on a line-of-sight, the second one being that the complex refractive index must be known in order to correctly interpret experimental data. It is fair to say that these problems do not exist in Diffusion Broadening Spectroscopy (that does not mean that DBS is free of problems). Measurements of m have been carried out by some soot workers on sampled particles [58,61,62,63] among others. Let us also mention the nice experiment by Pluchino et al. [64] in which 'a single spherical particle is electrostatically suspended by an automatic version of

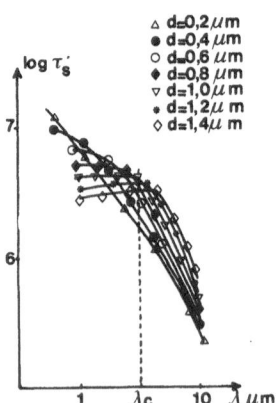

Fig. 7.   Logarithm of specific turbidity τ_s' versus λ for different diameters: the slope-breaking behaviour.

the Millikan oil drop experiment' in order to measure the complex refractive index of particulates. These measurements are carried out at ambient temperature. Extrapolations at high (in-situ) temperatures have been carried out on theoretical basis.[62,65] Errors on m could produce errors on measurements by a factor of 2.[66] Furthermore, values of n and (nk) are closely connected with the ratio H/C of soots[67], meaning that optical properties fastly evolve for young soot particles as does the ratio H/C.[68] It is thus worthwhile to point out the work from Ishizu and Okada who discussed the determination of particle size distribution of small aerosol particles, such as cigarette or incence smoke, by a light-scattering method, when the refractive index is not known.[69]

With simultaneous velocimetry

We are here concerned with systems using the laser-Doppler anemometry for velocimetry, the aim being to be able to measure simultaneously the size of the scatter centers when they cross the control volume. A lot of solutions have been imagined and tested, and the reader can go to the references [36,37] to find a rather exhaustive review of the matter, although it should be sometimes up-to-dated. Again, we shall focus our interest on restricted points.

When a particle crosses the optical LDA-probe of a real fringe system, it produces, behind a quadratic detector, a Doppler burst, such as exhibited on Figure 8, here corresponding to a rather small particle (with respect to the fringe spacing i) following a median trajectory in the control volume.

Such a signal (considered as perfect, that means without noise, and classical, that means no photon resolved) contains a high frequency modulation permitting the velocity measurement and a 'low frequency' part basically corresponding to the gaussian character of the laser beam used as a source. This 'low frequency' is expected to contain a scatter center size information. The first basic idea which

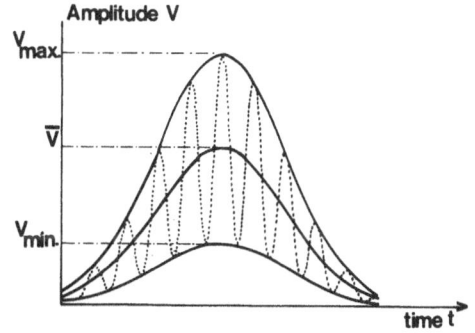

Fig. 8. A typical Doppler signal.

has been mainly developed was to attempt to simultaneously get the velocity and the size of the scatter center by analysis of this single burst (number-densities measurements can always be achieved by bursts counting).

Two main ways of research have been developed in that respect. The first one used the visibility concept and was investigated first by Farmer whose pioneer's paper, in 1972,[70] is very well-known, then by many other authors (see [36] and [37]), up to the last papers by Ghezzi et al [71], Farmer [72] and Farmer et al [73]. According to Farmer (among other) the visibility V defined, say, as $(V_{max}-V_{min})/(V_{max} + V_{min})$ depends on the size of the scatter center. When certain (numerous) assumptions are complied with, V simply reads as $2 \left| J_1 (\pi d/i) \right|/(\pi d/i)$. Nevertheless, more refined analysis produce much more complicated expressions. The second way (pedestal calibration techniques) of research is to link $\overline{V}$ and d and has been mainly developed by the Chigier's team [74]. From an experimental point of view, calibration is needed. From a theoretical point of view, a simplified geometrical optics is used to achieve the analysis. These two kinds of methods (visibility or 'pedestal' calibrations) can be now considered as classical, and are extensively discussed in many places. Let us just mention a main disadvantage for each of them.

The visibility techniques should be used with care due to the complexity of a rather full analysis, as clearly appearing from the literature. As a matter of fact, there is no general theory of visibility available since nobody is able at present (as far as we are aware of) to (numerically) compute the scattering of a gaussian beam by a Mie scatter center thus *a fortiori* to know what must theoretically be a Doppler signal produced by a given Mie scatter center having a given trajectory in a given control volume. Thus we share the conclusion given by Ghezzi et al [71] who observed that the visibility technique cannot be considered as a standard one and that 'many other reliability tests, improvements and modifications are necessary'.

Now, the main disadvantage of the present pedestal calibrations methods is as following : a large particle passing near the edge of the control volume can produce the same size measurement as a small one passing through the center of the control volume. To overcome this difficulty (known as 'trajectory ambiguity'), a mathematical inversion (as used also by Don Holve and Self[75]) is needed with the result that simultaneous measurements of velocities and sizes distributions are obtained, not really the simultaneous measurement of the velocity and the size of a given particle.

To reach the more precise aim just above pointed out, another way of research is under investigation in Rouen, based on the following ideas :

i) Try to find monotonic relationships between scattered powers and diameters with the aid of the SUPERMIDI computer program coupled with an INTIMI program[76] which integrates the scattered intensities over a collecting surface defined around the point P by $\Delta\theta = \Delta\phi$. Such monotonic relationships have been found and published.[77,78]

ii) Use (for instance) a 'green' laser optical probe for sizing and a 'blue' optical probe for LDA, that means do not try to achieve the simultaneous measurements on a given Doppler burst, but use two different spatially superimposed optical probes to separately achieve sizing and velocimetry when the particle crosses simultaneously the two probes.

iii) Get rid of the gaussian character of the sizing beam by correcting it to obtain a top-hat intensity profile, in order to avoid trajectory ambiguities. When a particle crosses such a top-hat intensity profile, it will produce a top-hat amplitude signal, the height of which being linked to size through the above mentioned relationships, the simultaneous velocity measurements being independently obtained from the superimposed LDA-probe.

Such a system produced measurements, recently reported[79] while more extensive experiments are currently running. It is expected that a commercial device based on these ideas could be possibly put on to the market, say, at the beginning of 1984.

REFERENCES

1. A. Kastler, La diffusion de la lumiere par les milieux troubles, influence de la grosseur des particules, Herman et Cie, Paris (1953)

2. H. C. Van de Hulst, Light scattering by small particles, John Wiley & Sons, London (1957)

3. J. A. Stratton, Electromagnetic theory, Mc Graw Hill, New-York (1941)

4. M. Kerker, The scattering of light and other electro-magnetic radiation, Academic Press, New-York (1969)

5. M. Born and E. Wolf, Principles of optics, electro-magnetic thoery of propagation, interference and diffraction of light, Pergamon Press (1959)

6. A. R. Jones, Prog. Energ. Combust. Sci. 5:73 (1979)

7. S. Chandrasekhar, Radiative transfer, Oxford at the Clarendon Press, London (1950)

8. H. C. Van de Hulst, Multiple light scattering, Tomes 1 and 2, Academic Press, New-York (1980)

9. G. Gouesbet and G. Gréhan, Internal Report TTI/GG/80/06/IV

10. G. Gouesbet and G. Gréhan, Sur la généralisation de la

théorie de Lorenz-Mie. To be published in Journal of Optics.

11. A. Ashkin, Phys. Rev. Letters 24(4):156 (1970)
12. G. Roosen, Canad. J. of Physics 57(9):1260 (1979)
13. G. Gréhan and G. Gouesbet, Applied Optics 19(15):2485 (1980)
14. L. Lorenz, Vidensk. Selk. Skr. 6:1 (1980)
15. L. Lorenz, Oeuvres scientifiques de L. Lorenz revues et annotées par H. Valentiner, Librairie Lekman et Stage, Copenhague, p.405 (1898)
16. G. Mie, Ann. der Phys. 25:377 (1908)
17. P. Debye, Ann. der Phys. 30:57 (1909)
18. N. C. Wickramasinghe, Light scattering functions for small particles, Adam Hilger, London
19. J. V. Dave, Report 320-3237, IBM Scientific Center, Palo-Alto (1968)
20. D. Diermendjian, R. Clasen and W. Viezee, J. Opt. Soc. of America 51:620 (1961)
21. W. Cherdron, F. Durst and R. Richter, Sonderforchungsbereich 80, Ausbreitungs und Transportvorgänge in Strömungen. Universität Karlsruhe SFB 80/TM/121, Januar (1978)
22. W. J. Lentz, Appl. Opt. 15:668 (1976)
23. G. Gréhan and G. Gouesbet, Internal Report TTI/GG/79/03/20
24. G. Gréhan and G. Gouesbet, Appl. Opt. 18:3489 (1979)
25. G. Gréhan, F. Slimani and G. Gouesbet, Internal Report TTI/GSG/81/3/III
26. W. J. Wiscombe, Appl. Opt. 19:1505 (1980)
27. P. Chylek and R. G. Pinnick, Appl. Opt. 18;1123 (1979)
28. R. B. Penndorf, J.O.S.A. 52(8):896 (1962)
29. H. M. Nussenzveig, J. Of Math. Phys. 10(1):82 (1969) and 10(1):125 (1969)
30. A. Ungut, G. Gréhan and G. Gouesbet, Joint Sheffield/Rouen Internal Report TTI/UGG/80/06/III
31. A. Ungut, G. Gréhan and G. Gouesbet, Comparisons between geometrical optics and Lorenz-Mie theory for transparent particles in forward directions, Appl. Opt., to be published
32. L. Brillouin, J. of Appl. Phys. 20:1110 (1949)
33. M. J. Groves, Edit., Proceedings of a conference organized by the Analytical Division of the Chemical Society, University of Salford, Sept. 12-15 (1977)
34. I. Chabay and D. S. Bright, J. of Colloïd and Interface Science 63:304 (1978)
35. M. Weill, P. Flament and G. Gouesbet, The status of the art in soots diagnosis by means of Diffusion Broadening Spectroscopy, at the present workshop
36. G. Gouesbet and G. Gréhan, Proceedings of the 4th Intern. Symposium on Plasma Chemistry, Zürich, 27 août-1er septembre (1979)

37. A. Boulos, F. Cabannes, P. Fauchais and G. Gouesbet, Thermal treatment of particles under plasma conditions. International Report IUPAC.,to be published
38. H. Royer, Nouv. Rev. Optique 5:87 (1974)
39. M. Gohar, D. Allano, M. Trinite and M. Ledoux, Intern. Symposium on "Applications of fluid mechanics and heat transfer to energy and environmental problems", Patras, June 29-July 3 (1981)
40. J. Swithenbank, J. M. Beér, D. S. Taylor, D. Abbot and G. C. Mc Creath, "Progress in Astronautics and Aeronautics" , B. T. Zinn, Edt., AIAA, New-York, 53:421 (1977)
41. A. J. Yule, C. A. Seng, P. Felton, A. Ungut and N. A. Chigier, A laser tomographic investigation of liquid fuel sprays, Combustion Aerodynamics Research Laboratory, University of Sheffield, England, Report (1980)
42. O. F. Genceli, J. B. Schemm and C. M. Vest, J. Opt. Soc. Am. 70:1212 (1980)
43. M. Laug and A. Delfour, Proceedings of the Symposium on Long range and short range optical velocity measurements, ISL, France, Sept. 15-18 (1980)
44. M. Bartholdi, G. C. Salzman, R. D. Hiebert and G. Seger, Optics Letters 1:223 (1977)
45. M. Bartholdi, G. C. Salzman, R. D. Hiebert and M. Kerker, Appl. Opt. 10:1573 (1980)
46. A. D'Alessio, A. DiLorenzo, A. Borgese, F. Beretta and S. Masi, 15th Symp. (Intern'l) on Combustion, The Combustion Institute, Pittsburgh, p.1427 (1975)
47. A. D'Alessio, A. DiLorenzo, F. Beretta, and S. Masi, 16th Symp. (Intern'l) on Combustion, The Combustion Institute, Pittsburgh, p.695 (1977)
48. G. Prado and J. Lahaye, "ATP Combustion et Turbulence 1976", Rapport de fin d'étude, Mars (1979)
49. B. S. Haynes, H. Jander and G. Gg. Wagner, Ber. Bunsenges. Phys. Chem. 84:585 (1980)
50. B. S. Haynes, and G. Gg. Wagner, Ber. Bunsenges. Phys. Chem. 84:499 (1980)
51. P. A. Bonczyk, Combustion and Flame 35:191 (1979)
52. G. Gréhan, Thèse de 3ème cycle, Rouen, 26 février (1980)
53. S. Chippet and W. A. Gray, Combustion and Flame 31:149 (1978)
54. D. M. Roessler and F. R. Faxvog, Appl. Opt 18:1399 (1979)
55. D. M. Roessler and F. R. Faxvog, J. Opt. Soc. Am. 70:230 (1980)
56. T. M. Lowes and A. J. Newall, Combustion and Flame 16:191 (1971)
57. A. R. Jones and W. Wong, Combustion and Flame 24:139 (1975)
58. W. H. Dalzell and A. F. Sarofim, Transactions of the ASME, p.100 (1969)

59. R. Portscht, Straub Reinhalt Inft 32(7):277 (1972)
60. G. Gréhan, P. Vervisch, G. Gouesbet and D. Puechberty,
     "Experimental and theoretical investigation of fire
     smoke absorption", available upon request
61. J. J. Mc Cartney and S. Ergun, Fuel 37:272 (1958-59)
62. V. R. Stull and G. N. Plass, J. of the Optical Society
     of America 50:121 (1960)
63. P. J. Foster and C. R. Howarth, Carbon 6:719 (1968)
64. A. B. Pluchino, S. S. Goldberg, J. M. Dowling and C. M.
     Randall, Appl. Opt. 19:3370 (1980)
65. C. R. Howarth, P. J. Foster and M. W. Thring, Proceedings
     of the Third International Heat Transfer Conference
     5:122 (1966)
66. F. G. Roper and C. Smith, Combustion and Flame 36:125
     (1979)
67. T. R. Johnson, Ph.D. Department of Chemical Engineering
     and Fuel Technology, University of Sheffield, (1971)
68. B. L. Wersborg, L. K. Fox and J. B. Howard, Combustion
     and Flame 24:1 (1975)
69. Y. Ishizu and T. Okada, J. of Colloid and Interface
     Science 66:234 (1978)
70. W. M. Farmer, Appl. Opt. 11:2603 (1972)
71. U. Ghezzi, A. Coghe and F. Miot, Paper presented at the
     4th International Symposium on Air Breathing Engines,
     Florida, U.S.A., April 1-6 (1979)
72. W. M. Farmer, Appl. Opt. 19:3660 (1980)
73. W. M. Farmer, F. A. Schwartz, E. S. Stallings and R. Belz,
     A.I.A.A. 16th Thermophysics Conference, Palo-Alto,
     California, June 23-25 (1981)
74. A. J. Yule, N. A. Chigier, S. Atakan and A. Ungut,
     J. Energy 1:220 (1977)
75. D. Holve and S. A. Self, Appl. Opt. 18:1632 (1979)
76. G. Gréhan and G. Gouesbet, Internal Report TTI/GG/78/12/5
77. G. Gréhan, G. Gouesbet and C. Rabasse, Proceedings of the
     Symposium on Long range and short range optical velocity
     measurements, ISL, France, Sept. 15-18 (1980)
78. G. Gréhan, G. Gouesbet and C. Rabasse, Appl. Opt. 20:796
     (1981)
79. G. Gouesbet and G. Gréhan, A.I.A.A. 16th Thermoplastics
     Conference , Palo Alto, California, June 23-25 (1981)

SYNTHESIS AND RECOMMENDATIONS FOR FUTURE WORK

INTRODUCTION

Organic particulates are produced in a large variety of combustion
equipments from gaseous, liquid or solid fuels. These particulates
consist of solid carbonaceous material, referred to as soot, and
extractible organics containing polycyclic aromatic hydrocarbons
(PAH) and other molecules which are usually poorly identified.

From a practical point of view soot is beneficial in equipment designed to extract heat from combustion, as it enhances
radiative heat transfer from the flame to the load. On the other
hand, in equipment producing mechanical energy (internal combustion
engines, gas turbines) soot is detrimental since it might lead
to serious material damage (excessive heat radiation or fouling).

Independently of the equipment, it is important to minimize
organic particulate emissions since unburnt materials reduce global
yields, and, more importantly, cause environmental concerns. In
addition to impact on visibility and general air quality problems,
several of the extractible organics are known mutagens and suspected
carcinogens. Even though the soot particles do not appear toxic
by themselves, they might have a role of carrier of toxic substances
through the natural defences of the organism.

Emerging trends accentuate the problem. For example, the development of low NO_x technologies for conventional oil and coal-fired
boilers may ultimately be limited by carbonaceous emissions.
Increased wood and coal burning in small residential furnaces,
increased use of stoker coal-fired systems, development of diesel
engines for small cars, are expected to add to the total organic
particulate emissions.

413

A better understanding of the mechanisms of production and destruction of these particulates is needed, in order to assess the impact of combustion parameters and operating conditions, and to design equipments minimizing these emissions. Furthermore, the health effect of organic particulates does not appear clearly defined, and should be further documented.

In order to address these questions, three groups were formed at the end of the session. Their aim was the synthesis of the various aspects of the problem and recommendations for directions of research:
- toxicity of organic particulates (discussion leaders Prof. W. Thilly and Dr. L. Appelman)
- practical systems and aerodynamics (discussion leaders Prof. J. M. Beér and Dr. S. Galant)
- mechanisms and optical diagnostics (discussion leaders Prof. A. Sarofim and Dr. A. D'Alessio).
A general synthesis was subsequently made by all the participants (discusion leaders Profs. H. Gg. Wagner and G. Prado).

I.   TOXICITY OF ORGANIC PARTICULATES

W. G. Thilly   Department of Food and Nutrition Science
               Massachusetts Institute of Technology
               Cambridge, Mass. USA
L. Appelman   Institute CIV)-Toxicology and Nutrition TNO
               P.O. Box 360
               3700 AJ ZEIST   The Netherlands

Organic particulates are not the only, or even the major, repository of toxic agents in combustion system emissions. Present concern focuses on their demonstrated mutagenicity and suspected carcinogenicity. The toxicity appears to be essentially related to the uncondensed material, the solid soot particles having no demonstrated adverse effect on health. However, these particles carry toxic substances, adsorbed on their surface, and for that reason, deserve special attention.

Mutagenic assays provide a simple and convenient means of evaluating possible health problems related to chemicals. It is important, however, to realize that there is no demonstrated relation between these tests and human cancer. Consequently, they are no more than "early warning signals".

Lectures and discussions during the workshop demonstrated the following points :

1) Uncondensed material in the emission gases accounts for the majority of genotoxic materials as determined by

bacterial mutation assay in the exhaust of common home
oil burners.

2)   Therefore, it is strongly recommended that collection of
       all effluent material by use of filters, resins and cold
       solvent traps or similar devices be used in evaluating
       the emissions of any combustion system.

3)   Benzo(α)pyrene is apparently a very minor contributor
       to the genotoxic properties of combustion system emis-
       sions. Therefore, air quality standards based on benzo
       (α)pyrene levels would not be expected to reflect the
       genotoxicity of combustion emissions.

4)   Bacterial mutation studies confirmed by human cell muta-
       tion assays have thus far shown fluoranthene, cyclo
       penteno[c,d]pyrene and certain monomethyl phenanthrenes
       to be sufficiently mutagenic and to be found in sufficient
       quantities to account for a large portion of the genotoxic
       activity of a number of different combustion systems,
       including home oil burners and light-duty diesel engines.

5)   It is expected that other important contributors to
       genotoxic action, such as alkyl fluorenes and other
       alkyl phenanthrenes, will be found to account for the
       as yet unaccounted genotoxicity of combustion system
       emissions. The discovery of the actual genotoxic agents
       in the complex mixtures of combustion emissions seems
       to be possible by direct interaction of genetic toxicolo-
       gists and analytical chemists.

6)   The use of any single solvent in the extraction of particu-
       late material cannot be expected to mimic the physiologic
       conditions found in human tissues. Care should be taken
       to examine a series of solvents, such as methylene chloride,
       cyclo-hexane-toluene mixtures and others, in order to
       gauge the possible variations to be expected biologically.

7)   Experiments to directly test the ability of mammalian
       cells to desorb and be effected by soot-bound chemicals
       should be designed and executed. Cell culture medium
       desorption experiments fail to account for the process
       of endocytosis which would be expected to effect inhaled
       or ingested small particles.

8)   It is still too early to generalize about which combustion
       conditions produce which genotoxic agents. However, the
       means exist to evaluate chemically and biologically the
       effect of design modifications or modifications in
       combustion parameters. Such efforts should lead to

decreasing potential health hazards while maintaining
fuel efficiency in the combustion systems in most common
use today.

II. PRACTICAL SYSTEMS AND AERODYNAMICS

J. M. Beer   Department of Chemical Engineering
             Massachusetts Institute of Technology
             Cambridge, Mass. USA
S. Galant   Société BERTIN
             Zone Industrielle
             Plaisir, France

The discusions of this session are divided in two sections :
- Emissions from practical equipments
- Aerodynamics and modelling.

A)   Emissions from practical equipments

The problem.

     Most of the reliable information on particulates emission comes
from the utility industry; data which relate combustion operating
variables such as fuel type (e.g., asphaltene content), atomization
quality, excess air and mixing of fuel and air with the amount of
cenospheres and soot formed and emitted. More recently, information
has become available also on condensible (extractable) polycyclic
aromatic hydrocarbon species both for petroleum and coal derived
heavy fuel combustion.

     Utility applications are of great interest because of the
large amount of fuel used, but the emissions are relatively low
due to the long residence time of the fuel in the combustor at
well controlled conditions of oxidizing atmospheres at high temperatures. The emission levels of organic particulates are significantly
higher, however, in diesel engines and in small residential boilers
burning liquid and solid fuels. Studies on the health effects of
these pollutants - which are emitted close to the ground in densely
populated areas have just begun; a concentrated effort is needed
to determine the potential health hazards and develop technologies -
combustion process modifications - for their reduction.

Recommandations

     Sampling equipment and procedures need to be developed for
field tests to ensure that data on particulates and condensible
hydrocarbon species reported from different field tests can be
compared, and related also to data obtained from pilot plant and
laboratory scale studies. Special attention should be paid to the

better understanding of these details of the combustion - heat transfer processes which may be responsible for the quenching of the oxidation reaction of these organic compounds.

B)   Aerodynamics and practical systems

The problem

Most practical flames are turbulent, and with a few exceptions diffusion flames. The general problem of calculating the distribution of chemical species concentration in turbulent diffusion flames is due to strong coupling between the mean and fluctuating concentration and temperature fields.

The difficulty is increased by the usually complex geometry of the enclosure (combustion chamber) which additionally introduces secondary flows and heat extraction at the walls into the analysis.

In order to simplify the description of such systems the axisymmetrical enclosed turbulent diffusion flame can be considered as a representative flame type relevant to a wide range of practical combustion systems.

A major problem of industrial combustion is the increasing use of highly aromatic fuels (coal, coal derived liquid and gaseous fuels, shale oil, etc.) which also have high concentration of heterocyclic nitrogen compounds. To prevent quantitative oxidation of these compounds to NO_x in lean flames the admixing of the combustion air to the fuel has to be delayed and the fuel pyrolysed in an oxygen deficient atmosphere followed by completing the combustion in a fuel lean atmosphere. Staged air introduction or delayed oxidant entrainment into highly aromatic fuel flames, however, is conductive to the formation of increased amounts of soot and PAH.

Significant amount of information on soot concentration distribution in turbulent diffusion flames of gaseous, liquid and solid hydrocarbon fuels is available from different sources, such as the International Flame Research Foundation. The data were gathered in connection with research on the emission of thermal radiation from the flame. The data are presented in terms of integral properties of turbulent jets (e.g., linear and angular momentum flux), fuel type, rate of heat extraction at the wall and burner geometry.

There are no equivalent data available for PAH and other organics ; the techniques for their sampling and analysis has been developed only recently.

In recent years, computational fluid dynamics has significantly progressed. For the simple axisymmetrical enclosed diffusion

flame, it is now possible to give a priori predictions on major
species concentration and temperature distribution in turbulent
diffusion flames of hydrocarbon fuels including preliminary
results on pulverized coal. While this represents a significant
advance, it does not allow for i) calculations of PAH and soot
concentration distributions in the flame with the accuracy and
detail required for heat transfer calculations, ii) prediction
of their emission as pollutants from the combustion system.
Furthermore, extension to fully tridimensionnal complex reacting
flows has yet to be made.

The objectives

Two main objectives of the turbulent diffusion flame studies
can be identified :

a) Develop prediction methods for determining soot concentration distribution in the flame as a function of fluid dynamics
and thermal input variables for the purpose of predicting the
distribution of radiative heat flux from the flame to the bounding
surfaces of the enclosure.

b) Predict PAH and soot emissions from the flame for the
effects of fluid dynamic and thermal input parameters (including
fuel type).

It is thought that the chemical kinetic information required
for both objectives is different. For the specific calculation
of soot concentration distribution in the flame a chemical kinetic
scheme of soot formation preferably not exceeding four steps would
be desirable. Information would also be required to give soot
oxidation rates as a function of the local mass concentration of
soot, temperature and oxidant. For PAH concentrations, further
basic kinetic investigations are still needed before proposing
any reliable kinetic scheme.

For the calculations of the unburned PAH and soot concentrations emitted from the flame turbulent combustion processes which
lead to the quenching of oxidation reactions in small flamelets,
both by aerodynamic means e.g., excessive stretch, or by cooling,
would have to be better understood and quantitatively described.

Research needs

Experimental studies which will help the better understanding
of soot formation and burn-out in turbulent diffusion flames
include:

a) Critical experiments to shed light on the destruction
mechanism of PAH and soot combustion in "fine structures" formed

in the shear layer of turbulent reacting flows.

b) Detailed mapping of turbulent diffusion flames for time
average and time resolved velocity, temperature and (when possible)
species concentration distributions. Such studies can be used to
obtain local rates of formation of PAH and soot, relationships
between soot precursors, PAH and soot in forming and soot burning
regions of the flame and to test mathematical models on a rigorous
basis.
Both probe measurements and non intrusive diagnostics (e.g.,
LDV, laser scattering techniques, laser diffraction methods,
laser holography, etc...) should be used in detailed flame mapping
studies.

In parallel, continuous improvement in mathematical models
should be pursued to include more complex kinetic schemes which
are coherent with the chosen turbulent flow models.

III. MECHANISMS AND OPTICAL DIAGNOSTICS: STATUS AND RESEARCH NEEDS

A.F. Sarofim   Department of Chemical Engineering
                Massachusetts Institute of Technology
                Cambridge, Mass U.S.A.

D. D'Alessio   Laboratorio de Ricerche sulla Combustione
                C.N.R. , Piazza V. Tecchio
                Napoli, Italy

The discussion followed the sequence of the technical
presentations during the preceding sessions and briefly reviewed
the status of current understanding, major gaps, and research
needs in the following areas: chemical kinetics, nucleation
and coagulation of particles, mass growth of soot, charged
species in flames, soot burn-out, and optical diagnostics.

A) Chemical kinetics

Detailed kinetic modeling of the oxidation of aliphatic
hydrocarbons has progressed significantly and can now provide
reasonable estimates of the concentration profiles of soot
precursors, up to C_4 species, in premixed flames. It is desirable
to extend the models to bridge the gap in our ability to predict
soot formation from the lower molecular weight species. It would
also be desirable to consider aromatic fuels. It was recognized
that the kinetics of many of the important reactions are unknown
and others are known only at temperatures and pressures far
removed from those of interest in practical combustion systems.
A systematic evaluation of the kinetics of all the reactions of
interest is not feasible, and the studies should focus on classes

of reactions that are critical. The selection of these reactions
should be guided by either detailed measurements of intermadiates
in flames or by screening of reaction mechanisms using kinetic
models.

It was recommended that the following classes of reactions
be studied:

a) The oxidation of reaction intermediates and products.
These reactions are important in defining the limits of soot
formation because of the competition between the formation and
destruction of soot precursors.

b) Reactions of biradicals, particularly with aromatic
compounds. These reactions appear to provide the major path
to the formation of multiring compounds.

c) Pyrolysis reactions. These reactions provide a route
to soot formation different from those encountered in premixed
flames and are important in the pyrolytic zone of diffusion flames.

Recommendations were also made on the development of experimental methodologies. These included:

a) Extensions of the pressure range and the mass discrimination capabilities of molecular-beam sampling systems. Mechanistic
studies, most of which have been performed on low pressure flames,
need to be extended to higher pressure. By limiting studies to
the presooting regime smaller nozzles may be used and it should
be possible to develop molecular-beam sampling systems for atmospheric pressure flames. Improvements of the sensitivity of the
mass spectrometer for the measurement of higher molecular species
in low concentrations, possibly by use of photoionization techniques,
should be explored.

b) Consideration should be given to selecting a standard
burner configuration and standard combustion conditions to permit
resolution of apparent conflicts in the experimental results
reported by different groups.

B)   Nucleation and coagulation

The formation of soot particles is a logical consequence of
the growth of molecules through collisions. Definition of a soot
nucleus is somewhat arbitrary and the size of a nucleus to a large
extent depends upon the resolution of the method used for characterizing particle size. The term nucleation is misleading as it implies
a reversible phase change which is not applicable to the soot
formation process with the possible exception of condensation
processes encountered in low-temperature pyrolytic systems. The

selection of a less precise term such a <u>particle inception</u> might
avoid the problems with the connotations associated with the
term <u>nucleation</u>.

The particle inception process is currently poorly understood.
Kinetic models have been developed primarily for the formation
of multiring molecules and more attention needs to be given to
the mechanisms leading to three dimensional networks. However,
knowledge of the details of the "nucleation" process is not
critical to the determination of soot size. It has been demonstrated that, if rates of particle inception are high enough, as
is usually the case in combustion systems, coagulation theory
predicts that the particle number density becomes, after a short
time, independent of the initial number of particles produced.

C) <u>Mass growth of soot</u>

Rates of surface growth show a decay with increasing height
above the burner plate for premixed flames, even in the presence
of high concentrations of acetylene and other hydrocarbons. One
rationalization of the data is provided by assuming that soot
precursors equal in amount to the final soot yield are produced
early in the flame and are depleted by conversion to soot in a
first order reaction. For atmospheric pressure flames and
conventional combustion temperatures the rate constant is of the
order of 100 sec^{-1}. The identity of the soot precursors, and the
mechanism of their deposition on soot, are, however, uncertain.
We need to know more upon the molecular identity of the soot
adducts and on the effect of the nature of the surface crystalline
organization and chemical state on their net capture efficiency,
recognizing that surface growth is the balance of surface deposition and removal.

D) <u>Charged species in flame</u>

Measurements of ions in premixed flames show a bimodal
distribution with height of both positively and negatively
charged particles with a major peak occurring above the oxidation
zone and a secondary peak downstream. Charged particles are also
produced in diffusion flames. The charges are determined by
chemi-ionization reactions, more significantly by surface ionization, and may be influenced by addition of salts of alkali and
alkaline earth metals. Charges on particles provide a means of
controlling particle motion, and hence soot growth, by the application of external electrical potentials to flames.

Fast ion-radical reactions may be important in the formation
of soot precursors but their role is still being discussed. There
is need for quantitative estimates of the rates of the competing
ionic and electrically neutral pathways to resolve the question.

E) Soot burn-out

Reliable measurements of soot oxidation by molecular oxygen
or hydroxyl radicals are available. A better mechanistic understanding of the oxidation process that takes into account variations
in the crystalline structure of soot surface is needed. This is
particularly important for less reactive oxidants such as O_2, which
will diffuse into soot particles to an extent dependent upon the
soot structure and subsequently increase the internal surface area
and rate of soot oxidation. An important area for future research
is the investigation of the use of catalysts to promote the oxidation
of soot at the lower temperature of importance in automotive exhaust
systems.

F) Optical diagnostics

Soot

Determination of mean soot size and concentrations in flames
using absorption-scatter techniques have become routine. Uncertainty in the measurements results from (a) polydispersity of the
soot particle size, (b) non-sphericity of the particles, and
(c) variations in the optical constants of soot. Diffusion
broadening spectroscopy might improve the accuracy of the determination. Additional experiments are needed, however, to make clear
whether this technology can be applied to concentrated aerosols
of small soot particles.

Polycyclic aromatic hydrocarbons (PAH)

Fluorescence provides a useful composite measure of the
concentration of polycyclic aromatic hydrocarbons in flames. At
present, insufficient information is available on the spectroscopy
of these compounds to enable the inference of detailed chemical
composition from the fluorescence signals. In view of the
complexity of the mixtures of polycyclic aromatics in flame it
is unlikely that fluorescence signals can be used in the foreseeable
future for the determination of detailed composition but this
should not detract from their utility in locating the dominant
regions of PAH formation and burnout in flames.

Complex flames

Interest in coal and synthetic fuels provides motivation for
developing techniques for measuring soot and PAH in the presence
of ash, char, oil droplets, and coked oil cenospheres. Properties
which discriminate between the different classes of particles can
be utilized to measure soot in the presence of other particles.
For example, the polarization ratio has been successfully applied

to separate the scatter by soot and oil droplets in oil spray
flames. Other characteristics of particles radiation, such as
the spectral dependence of emission, may provide the additional
information needed to differentiate the various constituents of
more complex flames.

Advanced diagnostics

The use of advanced diagnostics including spontaneous Raman
and CARS should be investigated whenever they can provide information on high molecular compounds or surface-adsorbed species which
can provide insight on mechanism and cannot be readily obtained
by more conventional techniques.

Whatever optical diagnostic is used in a given experiment,
it is important that the results be critically evaluated by comparison with those obtained by other techniques.

GENERAL RECOMMENDATIONS

H.Gg. Wagner   Institut für Physikalische Chemie
                Universität Göttingen
                Tammanstrasse 6
                Göttingen, West Germany
G. Prado   C.R.P.C.S.S.
                24, avenue du Président Kennedy
                Mulhouse, France

After the specialized meetings have been finished, a general
discussion took place with all the participants. The discussion
leaders of the specialized meetings reported about the results
of their meetings and they gave their recommendations. Their
reports were discussed again and, as a result of these discussions,
the following recommendations are being proposed:

1) More mapping of turbulent flame is needed in order to
    obtain a reliable mathematical model of a given burner-
    combustion chamber arrangement. It seems necessary to
    obtain spatial time-averaged mappings of turbulent
    diffusion flames as a function of time under well defined
    conditions.

2) In order to become able to mathematically model burners,
    it would be very useful to extract from kinetic experiments
    on soot formation a few characteristic properties of
    the process which can be used for model calculations
    (induction period for soot formation, average rates of

soot formation as a function of temperature, etc...)

3) Reliable data about oxidation of soot for conditions of
temperature, oxygen and OH radical concentrations, are
needed in order to be able to calculate the burnout of
soot under flame conditions, including the exhaust pipe
of the system. The catalytic aspects in the oxidation
of soot should be seriously considered. These considerations should also include the burnout of soot by intercalation of the oxidizing agent into the pseudo graphitic
part of soot.

4) The analysis of turbulent diffusion flames indicate that
extinction of flamelets, due to unmixedness and heat
extraction in turbulent shears are, to a large extent,
responsible for amount of soot and PAH emitted from
practical burners. It is necessary to perform detailed
studies about these extinction of flames and to find out
conditions where these effects can be minimized.

5) There are a series of investigation available about the
detailed structure of carbon black. It seems necessary
to do similar experiments about the surface cristalline
structure and chemical state of soot and to study the
intercation of these particles with PAH on their way
through the combustion system. This information is
particularly necessary for an evaluation of the mapping
of the turbulent diffusion flame (cf. 1)) and it would
give an access to the optical properties of these particles
which are required for the optical diagnostics of the
flames.

6) To understand and model mathematically soot formation in
combustion systems, further investigations of the kinetics
of the reaction of groups of molecules, especially at
high temperatures, are necessary.

   - oxidation reactions of PAH and of other intermediates
   - investigation of the formation and degradation of large
   molecules at high temperatures over a sufficiently large
   range of pressure.

7) A careful comparison of the experimental results obtained
in different laboratories clearly indicate a strong
dependence of the resulting curves on the experimental
conditions. Slight changes of pressure, temperature,
mixture ratio and flow velocity can be the reason for
large changes in the amount of soot formed. It is
strongly recommended to agree on certain standardized
experimental conditions. A group of participants agreed

to work out these standard conditions.

8) The presentation of the applications of modern measuring techniques (CARS, laser Raman, holography, photoionization mass spectrometry and photoelectron spectroscopy,...) did show that these techniques can give valuable additional information about the process of soot formation. Some could also be used for the control of technical burners. A complete agreement has been reached that further investigations of the processus related to soot formation are necessary in order to reach high thermal conversion efficiency and, at the same time, fulfill requirements necessary for the protection of the environment.

INDEX

427